4D Printing Technology

Scrivener Publishing
100 Cummings Center, Suite 541J
Beverly, MA 01915-6106

Advances in Additive Manufacturing Technologies

Series Editor: Bijaya Bikram Samal (bijaybikram@iitkgp.ac.in)

The main focus of this book series is to provide a scientific platform for researchers, practitioners, professionals, and academicians to discuss the most recent technological developments in Additive Manufacturing. It will cover various aspects of AM through different books which may include but not limited to AM processes, available standards, materials, recent developments in AM technologies, 4D printing, AM in Industry 4.0, AM applications, future possibilities in AM, etc.

Publishers at Scrivener
Martin Scrivener (martin@scrivenerpublishing.com)
Phillip Carmical (pcarmical@scrivenerpublishing.com)

4D Printing Technology

Principles, Materials and Applications

Edited by

Bijaya Bikram Samal
Advanced Technology Development Centre, Indian Institute of Technology, Kharagpur, West Bengal, India

Cheruvu Siva Kumar
Dept. of Mechanical Engineering, Indian Institute of Technology, Kharagpur, West Bengal, India

and

Shailendra Kumar Varshney
Dept. of Electronics and Electrical Communication, Indian Institute of Technology, Kharagpur, West Bengal, India

Scrivener Publishing

WILEY

This edition first published 2025 by John Wiley & Sons, Inc., 111 River Street, Hoboken, NJ 07030, USA and Scrivener Publishing LLC, 100 Cummings Center, Suite 541J, Beverly, MA 01915, USA

For more information about Scrivener publications please visit www.scrivenerpublishing.com.

Wiley Global Headquarters
111 River Street, Hoboken, NJ 07030, USA

For details of our global editorial offices, customer services, and more information about Wiley products visit us at www.wiley.com.

Library of Congress Cataloging-in-Publication Data

ISBN 978-1-394-21258-3

Front cover image generated using Generative AI (Claude AI)
Cover design by Russell Richardson

Set in size of 11pt and Minion Pro by Manila Typesetting Company, Makati, Philippines

Printed in the USA

10 9 8 7 6 5 4 3 2 1

Contents

Preface

The convergence of advanced manufacturing techniques and smart materials has ushered in a transformative era in engineering and design. Among the most groundbreaking developments is Four-Dimensional (4D) printing, a technological evolution that builds upon the established principles of 3D printing by integrating the element of time. This dynamic process enables printed structures to adapt, respond, and morph under external stimuli, paving the way for a new generation of functional, intelligent, and responsive products. This book comprehensively explores the principles, materials, and applications driving this revolution, providing a robust foundation for those seeking to innovate in this rapidly evolving field. Contributions from leading experts in additive manufacturing, materials science, photonics, and nanotechnology offer both breadth and depth in its coverage.

The initial chapters provide a thorough overview of foundational techniques in additive manufacturing before progressing into sophisticated methodologies that enable time-dependent behaviors, which define 4D printing. These include detailed discussions on programming methodologies and cutting-edge processes driving the field forward. Subsequent chapters delve into the science behind the smart materials essential for 4D printing, such as shape memory polymers, liquid crystal elastomers, and reinforced polymer composites. These materials form the core building blocks for dynamic structures, enabling the creation of adaptive and responsive systems. By examining the unique properties and applications of these materials, the book highlights how 4D printing can revolutionize the functionality of printed objects.

The later chapters transition from theory to real-world applications, emphasizing essential characterization techniques for assessing the performance and reliability of 4D printed structures. The book also explores advancements in micro- and nano-scale 4D printing, showcasing its impact in fields like photonics and soft robotics. Applications such as adaptive photonic devices and flexible soft robotic systems illustrate the transformative potential of 4D printing in creating dynamic, responsive technologies across various sectors.

This book is intended for researchers, academics, graduates, postgraduates, and industry professionals engaged in cutting-edge technologies, material sciences, engineering innovations, and the evolving landscape of manufacturing. As industries increasingly adopt smart manufacturing by integrating automation, advanced materials, and data-driven technologies, 4D printing emerges as a key enabler for creating programmable materials and intelligent products that can adapt to changing conditions. We hope this book will serve as both a comprehensive reference and an inspiration for future innovations in this rapidly advancing field.

Bijaya Bikram Samal
Cheruvu Siva Kumar
Shailendra Kumar Varshney

Acknowledgements

We extend our deepest gratitude to all those who have contributed to the development of this book, ***4D Printing Technology: Principles, Materials and Applications***.

First and foremost, we would like to thank the esteemed chapter contributors for their invaluable insights and expertise. Their dedication to advancing the field of 4D printing has greatly enriched this work, providing readers with a comprehensive understanding of this transformative technology.

We are grateful to the Wiley-Scrivener imprint for providing us an opportunity to edit this book. Our special thanks to Mr. Martin Scrivener, for his unwavering professionalism and guidance throughout this project. His commitment to excellence and support in navigating the publishing process have been instrumental in bringing this book to fruition.

We also appreciate our institutions, the Indian Institute of Technology Kharagpur, India and various other research organizations and labs that foster a culture of innovation and collaboration. The supportive environment created by our colleagues and students has been pivotal in our research endeavors and the writing process.

Finally, we wish to thank our families and friends for their patience and encouragement throughout this process. Their steadfast support has inspired us to dedicate the necessary time and effort to this endeavor. We hope this book serves as a valuable resource for researchers, industry professionals, and students alike, inspiring new ideas and innovations in the ever-evolving field of 4D printing technology.

Bijaya Bikram Samal
Cheruvu Siva Kumar
Shailendra Kumar Varshney

1

Importance of Additive Manufacturing in the Era of Industry 4.0

Bijaya Bikram Samal[1]*, Abhishek Kumar[2], Anita Jena[1], Debadutta Mishra[3], Shailendra Kumar Varshney[4], Ashish Kumar Nath[2] and Cheruvu Siva Kumar[2]

[1]*Advanced Technology Development Centre, Indian Institute of Technology, Kharagpur, West Bengal, India*

[2]*Department of Mechanical Engineering, Indian Institute of Technology, Kharagpur, West Bengal, India*

[3]*Department of Production Engineering, Veer Surendra Sai University of Technology, Burla, Sambalpur, Odisha, India*

[4]*Department of Electronics and Electrical Communication, Indian Institute of Technology, Kharagpur, West Bengal, India*

Abstract

The Industrial Revolution has undoubtedly shaped human history, with its continual development since the eighteenth century leading to four major transformations, and potentially a fifth yet to come. At the forefront of this revolution is Industry 4.0, embodying extensive digitization, interconnections, and data analytics. Within this modern industrial movement, Additive Manufacturing (AM) plays a crucial role, facilitating decentralized production, customization, and sustainability. This integration is made possible by a variety of technological enablers, including the Internet of Things (IoT), big data analytics, cloud computing, and artificial intelligence. With its ability to create intricate structures and designs layer by layer, AM perfectly aligns with the core principles of Industry 4.0. This chapter delves into the dynamic synergy between AM and digital innovations, exploring how it is changing the face of modern manufacturing. By incorporating AM alongside digital twins and simulation technologies, manufacturing efficiency is significantly improved with the ability for live monitoring and proactive maintenance.

**Corresponding author*: bijaybikram3991@gmail.com

Bijaya Bikram Samal, Cheruvu Siva Kumar and Shailendra Kumar Varshney (eds.) 4D Printing Technology: Principles, Materials and Applications, (1–28) © 2025 Scrivener Publishing LLC

***Keywords*:** Additive manufacturing, 3D printing, Industry 4.0, digital twin, machine learning, big data, Industry 5.0, 4D printing

1.1 Introduction

The Industrial Revolution refers to the transformative shift from a handicraft-based economy to a mechanized manufacturing industry driven by advancements in technology. This monumental shift began in the 18th century, spanning the years 1760 to 1840, and ushered in significant changes in economies around the world [1]. Prior to this revolution, economies primarily relied on simple handicrafts and agriculture. However, with the advent of the industrial revolution, economies were propelled forward by the emergence of factory systems, large-scale industries, and mechanized production [2]. The development of new industries showcased advanced power sources, cutting-edge machinery, and innovative methods for organizing departments within these industries. The Industrial Revolution has gone through four stages, each enhancing into a better, modern, and more innovative stage [3]. The industrial revolution began in the 18th century and was started by the introduction of mechanical production facilities powered by steam engines. The second industrial revolution began in the late 19th century and was characterized by the introduction of mass production facilities powered by electricity. The third industrial revolution began in the late 20th century and was characterized by the introduction of computerized production facilities. The industrial revolution also known as Industry 4.0, began in the early 21st century and is characterized by extensive interconnections, digitization, advanced analytics, and data collection [4]. Apart from this there is also a speculation of the fifth industrial revolution called Industry 5.0 [5], which is still in its infancy, but it is expected to be characterized by the integration of humans and machines [6], with a focus on human-centered production systems and also sustainability [7]. Evolution of industrial revolution from Industry 1.0 to Industry 5.0 has been depicted in Figure 1.1.

Additive manufacturing (AM), three-dimensional (3D) printing, rapid prototyping (RP), all these are the technological terms that involve manufacturing of a product through the addition of various layers of materials according to the geometrical data provided by the computer aided designs (CAD) [8]. AM is a technology that has revolutionized the manufacturing industry. It has brought about a paradigm shift in the way products are designed, developed, and manufactured. The technology has enabled industries and manufacturers to create complex geometries and structures

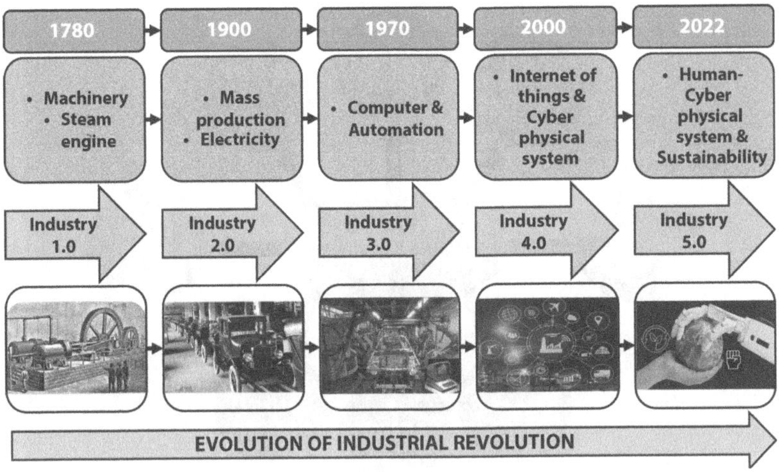

Figure 1.1 Evolution of industrial revolution from Industry 1.0 to Industry 5.0.

that were previously impossible to produce using traditional, conventional manufacturing processes [9]. AM has also enabled manufacturers to produce customized products that meet the specific needs of customers [9]. This has helped manufacturers to differentiate their products from those of their competitors and gain a competitive advantage in the market. Therefore, AM is a key technology in the era of industry 4.0, as it allows for mass customization, reduces waste generation, on demand manufacturing, low manufacturing lead time, promotes faster business digitalization, etc. [10]. The present AM technology requires continuous and effective links between the machine, software, generated data, monitoring, and analytics, etc., that is only possible with high level of digitalization of the manufacturing industries, enabled by industry 4.0. The realization of Industry 4.0 relies on several crucial factors, which drive the transformation of traditional manufacturing into a more intelligent, interlinked, and efficient system [11]. These enablers as shown in Figure 1.2, are essential in achieving this evolution:

- **Internet of Things (IoT), Industrial IoT (IIoT)**: IoT is a web of connected devices that are equipped with sensors and communication features. These devices have the ability to gather and share data, as well as initiate actuations, ultimately leading to a smarter and more adaptive environment. The complex network of smart devices, equipped with sensors and communication modules, facilitates the seamless

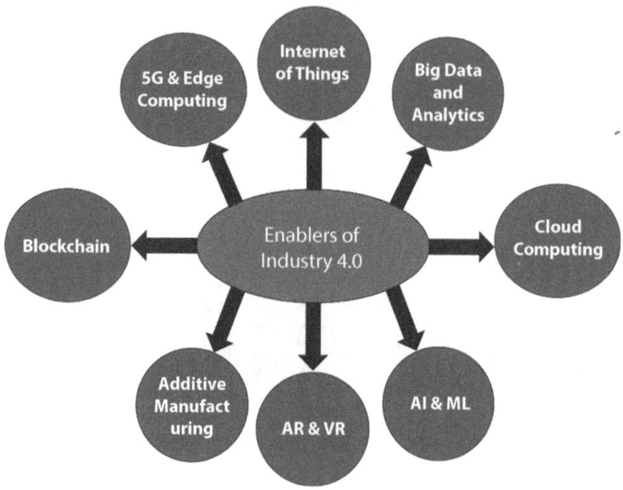

Figure 1.2 Major enablers of Industry 4.0.

exchange of real-time data. This interconnectedness extends from the factory floor to machinery and even the products themselves, creating a constantly evolving industrial ecosystem [11]. At the center of industry 4.0, the IoT plays a crucial role in driving data-driven decision-making. The abundance of data generated by these interconnected devices becomes a valuable resource for decision-makers. Through advanced analytics and machine learning algorithms, this data is carefully analyzed, providing valuable insights into production processes, equipment performance, and supply chain dynamics. This data-centric approach not only improves overall efficiency but also enables predictive maintenance [12].

- **Big Data and Analytics**: The vast amount of data produced by IoT devices in Industry 4.0 is effectively utilized through big data analytics. With the help of advanced analytics tools and algorithms, this data is processed and analyzed, providing valuable insights and enabling informed decision-making [13].
- **Cloud Fog Computing**: Cloud computing plays a vital role in providing adaptable access to computing resources. Its capabilities extend to facilitating the storage and processing of large amounts of data from IoT devices, promoting

real-time collaboration, and facilitating the distribution of software and applications in a variety of industrial settings [14].

- **Artificial Intelligence (AI) and Machine Learning (ML)**: They are revolutionizing industrial processes. These cutting-edge technologies empower machines to analyze data, make intelligent decisions, and adapt to dynamic situations. Industry 4.0 harnesses the power of AI and ML for predictive analytics, autonomous decision-making, and optimizing processes to improve industrial efficiency. By constantly learning from data and deriving valuable insights, these technologies elevate performance and drive progress in industrial systems [15].
- **Cyber-Physical Systems**: These are innovative systems that combine computing, communication, and physical processes [16]. By leveraging sensors and actuators, these systems seamlessly bridge the gap between the digital and physical worlds [17]. This integration is crucial for effectively controlling and monitoring industrial processes in real-time [18].
- **Augmented Reality (AR) and Virtual Reality (VR)**: These are game-changing technologies that elevate human-machine interaction [19]. With AR, digital information is superimposed onto the physical world, giving workers real-time insights. On the other hand, VR immerses users in virtual environments, providing a powerful tool for training, maintenance, and collaborative decision-making. These technologies greatly improve efficiency and decrease errors in industrial settings [20].
- **Additive Manufacturing**: Revolutionary manufacturing process known as AM has emerged as a game-changing technology [21]. Its ability to produce intricate and tailor-made components in a layered manner is in line with the fundamental principles of Industry 4.0 [22]. This breakthrough flexibility enables swift prototyping, decentralized production, and personalization of various goods [23].
- **Blockchain technology**: Its reputation for providing a secure and transparent platform has led to its integration into multiple industries [24]. By ensuring the integrity and traceability of data, blockchain enhances supply chain operations and enables seamless and visible transactions. In complex

industrial ecosystems, blockchain further enhances the credibility of data and promotes its reliability [25].

- **5G Connectivity**: With the emergence of 5G networks, there is now an opportunity for lightning fast and low-latency connectivity, meeting the communication demands of Industry 4.0 [26]. This breakthrough allows for instantaneous exchange of data between devices and systems, powering essential applications like remote monitoring, control, and collaborative robotics [27].
- **Edge Computing**: It plays a crucial role in Industry 4.0 by processing data at its point of origin, minimizing delays and maximizing the use of bandwidth. This is especially vital in scenarios where quick decision-making is crucial, such as with autonomous systems and decentralized industrial processes [28]. Together, 5G connectivity and edge computing are revolutionizing the potential of Industry 4.0 [29].

Together, these enablers serve as the building blocks of Industry 4.0, ushering in a new era of manufacturing that is characterized by intelligent, interconnected, and highly efficient industrial systems. By combining these cutting-edge technologies, companies are equipped to quickly respond to changing market needs, boost their productivity, and explore groundbreaking opportunities for advancement and development.

1.2 Additive Manufacturing as an Enabler of Industry 4.0

AM is a driving force behind the decentralization of production, which is a key aspect of Industry 4.0. In contrast to traditional manufacturing methods, which rely on large centralized factories and mass production, additive manufacturing enables distributed and localized production [30]. As the availability and capabilities of 3D printers have advanced, companies now have the opportunity to establish smaller, more agile manufacturing facilities that are in close proximity to their end-users. This not only significantly reduces lead times in transportation and distribution, but also has a positive impact on the environment by minimizing the need for extensive transportation networks [31]. In the era of Industry 4.0, where attention is placed on tailoring and customization, AM stands out as a vital instrument. By harnessing the power of 3D printing, small batches and even one-of-a-kind items can now be produced at a cost-effective rate [32]. This aligns perfectly with the evolving trend away from traditional mass production towards a more personalized and customer-oriented approach to manufacturing. Whether it's designing tailor-made medical implants that perfectly fit a patient's unique anatomy or creating custom consumer goods, additive manufacturing empowers the production of immaculate solutions that cater to each individual's distinct demands and desires [33]. The seamless incorporation of additive manufacturing into the industry 4.0 landscape is highlighted by its compatibility with advanced digital technologies and data-driven processes. With its inherent connection to digital design files, 3D printing easily integrates into the interconnected data-driven systems that define Industry 4.0 [34]. Various AM processes are depicted in Figure 1.3. The concept of digital twins, utilizing virtual models of physical objects or systems, perfectly complements the capabilities of additive manufacturing. Through the creation of a digital twin, manufacturers can simulate and enhance the design, production, and performance of a product or part, enabling a more streamlined and informed approach to manufacturing [35]. Additive manufacturing has become an essential component in addressing the pressing need for sustainability in today's industrial world. Unlike traditional methods that involve removing excess material, additive manufacturing ensures minimal waste by carefully depositing materials layer by layer. This technology also offers the use of recycled or eco-friendly materials, in line with the sustainability objectives of Industry 4.0 [36]. The power of additive manufacturing to contribute to the sustainability imperative cannot be overstated in the contemporary

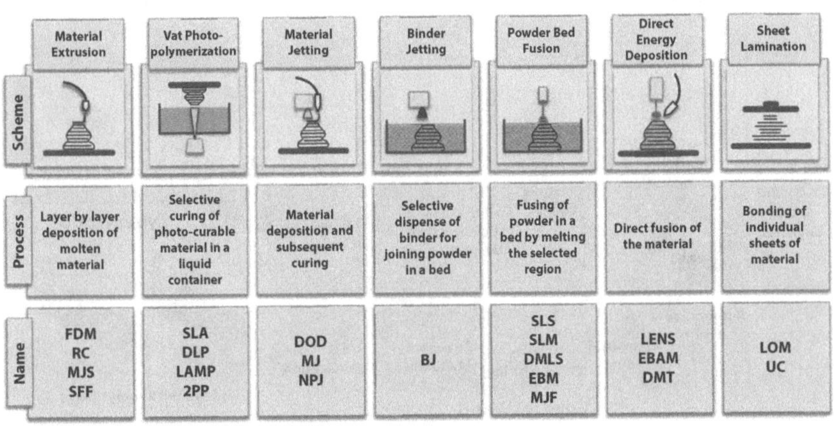

Figure 1.3 Various types of additive manufacturing processes. Reprinted with permission [10].

industrial landscape. Moreover, the continuous evolution of additive manufacturing techniques is constantly diversifying the types of materials that can be utilized, allowing for the incorporation of advanced composites and metals. As a result, the application spectrum of 3D printing is expanding, making it a viable option for the creation of functional and end-user components in various industries including aerospace, automotive, and healthcare [37]. The unparalleled capability to produce lightweight, yet durable parts with intricate inner workings solidifies additive manufacturing as a groundbreaking solution in the pursuit of streamlined and effective manufacturing processes, which is a fundamental principle of Industry 4.0 [38]. Smart factories require some specific properties required in Industry 4.0; those are depicted in Figure 1.4. Other advances in AM technologies like four-dimensional (4D) printing has led to the development of new smart materials and composites for dynamic behavior of the printed parts structures [39]. 4D printing is an AM process that utilizes smart materials to create a structure that can change its shape and size when an external fillip is applied to it [40]. Smart materials can be shape memory alloy, shape memory polymer, hydrogels, electroactive materials, magnetorheological fluids, etc. [41]. Recently, researchers have developed a very low cost DIY 4D printing technology, thereby expanding this area fast [42]. The new

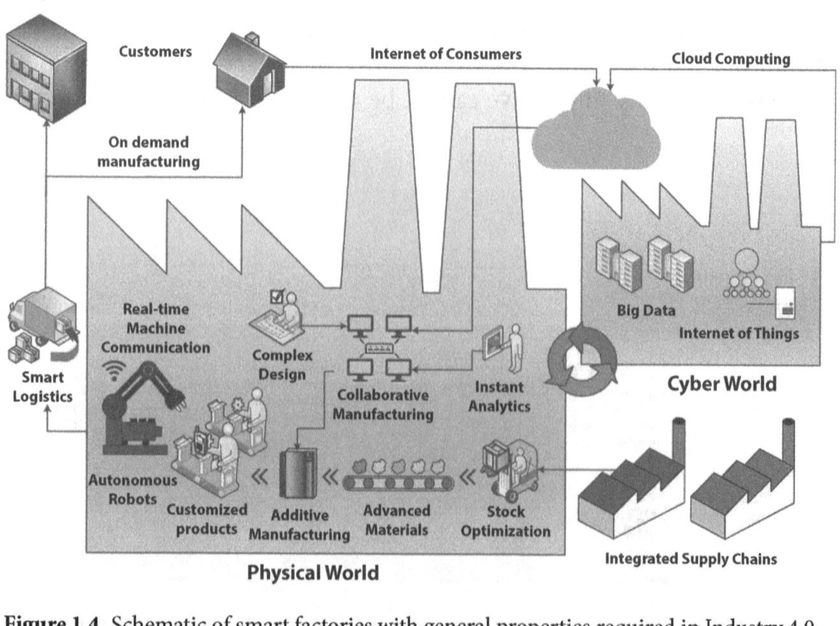

Figure 1.4 Schematic of smart factories with general properties required in Industry 4.0. Reprinted with permission [10].

kind of actuators and sensors getting developed from this technology is very promising [43]. Their developments can further improve the adaptability of AM as an important aspect of industry 4.0.

1.3 Synergies Between Additive Manufacturing and Digital Technologies

The intersection of AM and digital technologies has revolutionized conventional manufacturing methods, leading to unparalleled efficiencies [38]. In this section, we closely examine the dynamic interaction between AM and digital technologies, uncovering how their fusion enhances their individual capabilities to stimulate innovation and streamline manufacturing processes [44].

1.3.1 Flexibility and Customization

Thanks to layer-by-layer fabrication, AM offers inherent flexibility that perfectly complements the capabilities of digital technologies. The result is a seamless synergy, allowing manufacturers to quickly and efficiently transform their digital designs into physical prototypes through rapid prototyping [22]. This iterative process empowers designers and engineers to freely explore multiple iterations in a cost-effective way, leading to an accelerated innovation cycle. Additionally, the integration between AM and digital technologies facilitates customization, making it possible to create highly detailed and personalized components that would be difficult or impractical to produce using conventional manufacturing methods.

1.3.2 Decentralized Manufacturing and Localized Production

By implementing smart and connected systems, digital technologies give power to the decentralization of manufacturing. This trend is further enhanced by the integration of AM, allowing for localized production [45]. In the past, traditional manufacturing has heavily relied on centralized factories and lengthy supply chains, resulting in longer lead times and logistical difficulties. But with AM and digital technologies, smaller and more nimble manufacturing facilities can be strategically located near the end-users, reducing the environmental impact of long transportation networks and improving the ability to respond swiftly to local market demands.

1.3.3 Sustainability and Environmental Impact

The integration of additive manufacturing (AM) and digital technologies has a profound impact on promoting sustainable practices within the manufacturing industry. The seamless CAD design, simulation, and production processes allow for efficient material usage, reducing waste. The layer-by-layer approach of AM also allows for precise material deposition, ultimately reducing the environmental footprint. In addition, digital technologies combined with AM provide manufacturers with the necessary tools to analyze and optimize energy consumption, identify areas for sustainability improvements, and implement eco-friendly practices throughout the entire product lifecycle [45]. As the importance of sustainability continues to grow, this synergy between AM and digital technologies stands as a fundamental aspect of responsible and resource-efficient manufacturing. The seamless integration of AM and digital technologies results in adaptability, dispersed production, and eco-friendliness. By looking at practical implementations of these complementary forces in different sectors, we can witness firsthand how this merger is revolutionizing the manufacturing industry.

1.3.4 Continuous Innovation through Iterative Design

The combination of AM and digital technologies has revolutionized the design process, allowing for continuous innovation. In traditional manufacturing, making changes to a product design can be cumbersome and expensive due to required tooling changes. However, with the integration of AM and digital technologies, design modifications can be easily incorporated into the digital model, and the AM system can efficiently produce physical prototypes. This iterative approach enables a speedy and adaptable design process, promoting innovation and empowering manufacturers to quickly adapt to changing market demands and customer feedback [45].

1.4 Integration of AM with Digital Twins and Simulation

The fusion of AM and cutting-edge digital twins and simulation technologies marks a monumental leap forward in the world of manufacturing. In this section, we will dive deep into the dynamic contributions of digital twins and simulation in enabling seamless integration of AM. Our exploration will cover their critical role in manufacturing processes, the utilization of top-notch simulation technologies, and the invaluable benefits of implementing real-time monitoring and predictive maintenance [46]. The digital twin serves as a virtual counterpart, mirroring the behavior and performance of the physical printing process in real-time. It facilitates monitoring, analysis, and optimization, contributing to improved efficiency and quality control. On the other hand, the mechanistic model operates on established mathematical and physical principles, simulating interactions between printing parameters, material properties, and environmental conditions. By providing a detailed understanding of the underlying mechanisms, these representations play a crucial role in advancing additive manufacturing technologies and optimizing production processes. Combining both approaches allows for a synergistic utilization of real-time data from the digital twin with the predictive capabilities of the mechanistic model, leading to enhanced process control, reliability, and performance optimization in 3D printing operations as shown in Figure 1.5.

1.4.1 Role of Digital Twins in Manufacturing

Imagine having a digital clone of a machine or production line, constantly mirroring its behavior and adapting to any changes in real time. This is the power of digital twins, virtual replicas of physical objects or systems [47]. In the manufacturing industry, digital twins go beyond mere replication and offer a dynamic representation of the entire production environment. This integration with additive manufacturing is game-changing, as it creates a seamless connection between the physical and digital worlds. With digital twins, manufacturers gain a holistic understanding of the manufacturing process, making it easier to identify areas for improvement and optimize operations [48]. Digital twins are instrumental in the early stages of production, empowering designers and engineers to construct virtual prototypes. In the dynamic world of additive manufacturing, this translates to intricate 3D models being meticulously simulated and perfected before the physical production phase kicks off. This preemptive simulation is a vital tool in identifying potential obstacles, refining designs, and optimizing the overall efficiency of the additive manufacturing process. By streamlining the need for expensive trial-and-error techniques, this approach accelerates the development cycle and guarantees that the end result perfectly mirrors the intended design. The influence of digital twins spans beyond the initial stages of a product's development. By allowing for continual

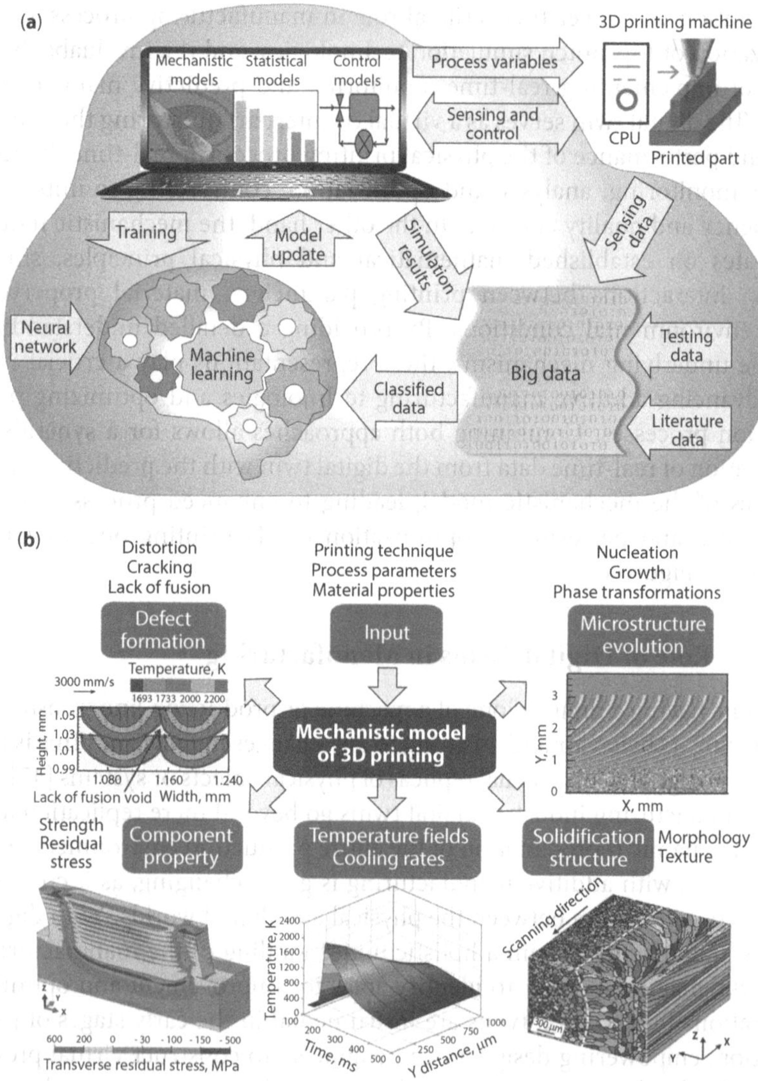

Figure 1.5 Schematic representation of (a) the digital twin and (b) mechanistic model of 3D printing. Reprinted with permission [54].

observation and data interpretation, they create a constant feedback loop, where the physical world enhances the digital model, and vice versa [49]. This harmonious bond between the physical and digital domains significantly improves manufacturing efficiency and overall effectiveness.

1.4.2 Simulation Technologies Enhancing AM Integration

Simulation technologies are essential for improving the integration of additive manufacturing. By creating a virtual representation of the AM process, these technologies offer valuable insights into the effects of various variables, such as material properties, printing parameters, and environmental conditions, on the end product [50]. They provide a digital environment where engineers and manufacturers can experiment and fine-tune their designs before moving on to actual production, which helps to minimize risks and minimize mistakes. In the world of AM, the use of simulation technologies plays a crucial role in ensuring the structural soundness of the end product. These innovative tools allow for a thorough examination of the behavior of selected materials throughout the step-by-step deposition process. By pinpointing concerns such as warping, thermal stresses, and structural vulnerabilities, manufacturers are empowered to fine-tune their designs for optimal performance and efficiency [51]. Simulation technologies play a crucial role in maximizing printing parameters. Through simulating various printing scenarios, such as temperature changes, layer thickness, and speed variations, manufacturers can pinpoint the most effective and consistent settings for a specific material and product. This not only enhances the excellence of printed components but also enhances the overall cost-efficiency of the additive manufacturing process.

1.4.3 Real-Time Monitoring and Predictive Maintenance

The seamless integration of AM technology with digital twins and simulation tools relies heavily on real-time monitoring and predictive maintenance. This process entails the constant collection and examination of data from the physical manufacturing process. By equipping AM machines with sensors that gather data on factors such as temperature, humidity, and printing speed, companies can gain real-time insights into the progress of production [52]. Once the real-time data is collected, it is integrated into the digital twin to form a dynamic and continuously updated virtual depiction of the manufacturing setting. With this digital twin, manufacturers can closely monitor the advancement of the additive manufacturing process, identify any irregularities, and make informed decisions to optimize production in the moment. Additionally, the integration of AM with digital twins allows for predictive maintenance, taking this concept to the next level [53]. Through analyzing information gathered from the physical production process and comparing it with the virtual model, manufacturers can anticipate when maintenance may be needed for machinery or equipment. This proactive approach aids in avoiding unexpected downtime, decreases the chances of equipment malfunction, and prolongs the lifespan of AM machines [44].

The digital twin serves as a virtual counterpart, mirroring the behavior and performance of the physical printing process in real-time. It facilitates monitoring, analysis, and optimization, contributing to improved efficiency and quality control. On the other hand, the mechanistic model operates on established mathematical and physical principles, simulating interactions between printing parameters, material properties, and environmental conditions. By providing a detailed understanding of the underlying mechanisms, these representations play a crucial role in advancing additive manufacturing technologies and optimizing production processes. Combining both approaches allows for a synergistic utilization of real-time data from the digital twin with the predictive capabilities of the mechanistic model, leading to enhanced process control, reliability, and performance optimization in 3D printing operations.

1.5 Applications Across Industries

Additive manufacturing (AM) is a cornerstone of Industry 4.0, revolutionizing traditional manufacturing with its flexibility and customization. Across industries like aerospace, automotive, healthcare, and consumer goods, AM enables rapid prototyping, on-demand production, and intricate designs unattainable through conventional methods. In aerospace, lightweight AM components enhance efficiency, while in automotive, it allows for rapid tooling and lightweight part production. Healthcare benefits from personalized medical devices, and consumer goods see streamlined development and customization. Some detailed industrial applications are provided as shown in Figure 1.6.

1.5.1 Automotive Lightweighting for Fuel Efficiency

The automotive industry is currently experiencing a groundbreaking shift in lightweighting; thanks to the integration of additive manufacturing (AM) and digital technologies. This powerful combination is driving the development of advanced lightweight structures that were previously unattainable. With the help of cutting-edge digital design tools, engineers are able to create incredibly intricate and lightweight designs with ease. The clever use of AM methods, such as selective laser sintering (SLS) and stereolithography (SLA), brings these designs to life by laying down precise

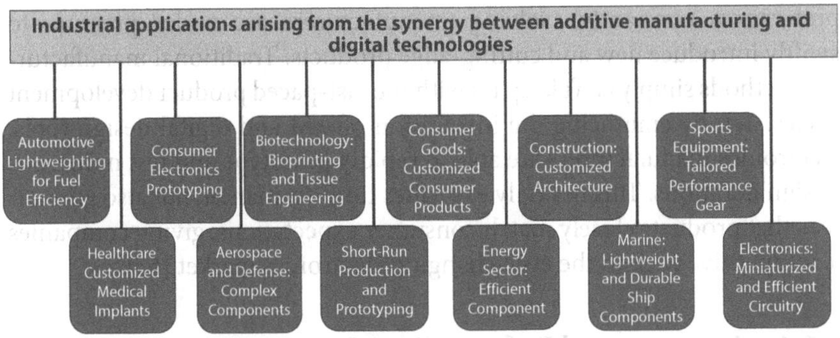

Figure 1.6 Application of additive manufacturing with digital technologies in Industry 4.0.

layers of materials. The end result is the highly optimized components that use less material and greatly improve fuel efficiency [45]. This real-life case study is a prime example of the innovative and industry-specific solutions that can be achieved through the synergistic use of AM and digital technologies[55].

1.5.2 Healthcare Customized Medical Implants

The healthcare industry is experiencing a revolutionary shift towards personalized medical components like dental and bone implants. With the integration of AM, materials like Co-Cr, Ti6Al4V with advanced post processing methods like laser polishing, it enables good surface finish required for the medical industries [56, 57]. Through advanced digital imaging, medical professionals are able to capture highly detailed anatomical data unique to each patient. This data seamlessly integrates with AM processes, enabling the production of custom medical implants tailored to the individual's need [58]. Whether it be orthopedic implants or dental prosthetics, this integration guarantees a perfect fit, shortens recovery periods, and improves overall patient outcomes [59]. This synergy between AM and digital technologies is a prime example of the move towards patient-focused and personalized approaches in healthcare [60].

1.5.3 Consumer Electronics Prototyping

The consumer electronics industry serves as a prime example of the remarkable impact of integrating AM processes like micro-nano printing and digital technologies into the design process [61, 62]. With consumer preferences constantly evolving, manufacturers are under pressure to swiftly introduce new and cutting-edge products. Traditional manufacturing methods simply can't keep up with the fast-paced product development cycles. Yet, by embracing the integration of AM and digital design tools, electronics manufacturers are able to rapidly prototype and test numerous design iterations. This not only expedites time-to-market, but also guarantees that products closely match consumer expectations, giving companies a competitive edge in the ever-changing electronics market [63].

1.5.4 Aerospace and Defense: Complex Component Manufacturing

In the ever-evolving world of aerospace and defense, the fusion of AM and digital technologies is transforming the realm of intricate component production. The demand for lightweight and durable parts is met by AM's proficiency in fabricating complex geometries and lattice structures. Through advanced digital simulations, these designs are fine-tuned for maximum performance, while the digital thread streamlines the transition from virtual models to tangible components. This dynamic collaboration yields components that not only meet the rigorous standards of the industry, but also enhance the efficiency and sustainability of aerospace and defense systems as a whole [63].

1.5.5 Biotechnology: Bioprinting and Tissue Engineering

The biotechnology field is undergoing a revolutionary transformation with the fusion of AM and digital technologies, specifically in the areas of bioprinting and tissue engineering. Through advanced digital imaging techniques, precise anatomical data is captured and used to create accurate 3D models of organs and tissues [63]. These models are then brought to life through the layer-by-layer construction process enabled by AM technologies. This remarkable combination has vast potential in improving organ transplantation, drug development, and individualized medical treatments. By incorporating AM and digital technologies in healthcare, a major shift towards personalized and patient-centric approaches is reshaping the boundaries of what is possible in the medical realm [63].

1.5.6 Automotive and Transportation: Short-Run Production and Prototyping

In the fast-paced world of automotive and transportation, AM and digital technologies are coming together to revolutionize short-run production and prototyping. By utilizing sophisticated design tools, manufacturers are now able to craft intricately detailed and lightweight automotive components that prioritize fuel efficiency and performance [64]. And with the help of AM, these digital designs can quickly become physical prototypes, giving companies the ability to test and refine their products at a rapid pace. Thanks to this agile approach to prototyping, automotive companies can now bring their newest designs to market faster and more efficiently than ever before [64].

1.5.7 Consumer Goods: Customized Consumer Products

The incorporation of AM and digital technologies has sparked a major transformation in the consumer goods industry, specifically regarding customizability. With the help of innovative digital design tools and input from consumers, it has become simpler to create personalized products [65]. Thanks to AM technologies, these digital designs can now be translated into tangible goods at a lower cost, making the production of customized consumer goods more accessible. Whether it's personalized clothing or unique home decor, this synergy between AM and digital technologies empowers consumers to become active participants in the design process, resulting in a more engaging and personalized consumer experience [66].

1.5.8 Energy Sector: Efficient Component Manufacturing

When AM technology is combined with the power of digital advancements in the energy sector, it leads to the highly efficient production of components. This fusion of technology enables digital design and simulation capabilities to optimize the performance of energy-related parts, ensuring they adhere to stringent efficiency standards [67]. Moreover, AM's ability to fabricate intricate geometries unlocks a world of design possibilities for components such as turbine blades and heat exchangers. This integration of AM and digital technologies not only increases the efficiency of energy systems but also drives the creation of sustainable and groundbreaking solutions within the sector [68].

1.5.9 Construction: Customized Architectural Components

AM and digital technologies have revolutionized the construction industry, bringing about a profound change. By leveraging this powerful combination, architects and engineers can now craft customized architectural components with intricate designs [69]. Whether it's a complex facade or a structurally optimized part, additive manufacturing offers a cost-effective approach for producing one-of-a-kind building elements. This integration simplifies construction processes, empowers greater design flexibility, and ultimately contributes to the creation of both sustainable and aesthetically striking structures [70].

1.5.10 Marine: Lightweight and Durable Ship Components

The maritime industry is experiencing a revolutionary collaboration between AM and digital technologies, completely transforming the manufacturing of ship components. Employing cutting-edge digital design tools, highly skilled experts produce lightweight and robust parts that are essential for maximizing fuel efficiency and overall vessel performance. Utilizing techniques such as powder bed fusion (PBF), intricate elements like propellers and brackets are produced with impeccable precision. This innovative approach not only elevates efficiency in the maritime sector, but also showcases the versatility of AM in various industries [71].

1.5.11 Sports Equipment: Tailored Performance Gear

The sports equipment manufacturing industry is experiencing a major transformation as AM and digital technologies begin to merge. This integration allows for a whole new level of customization in sports gear, from tailor-made running shoe soles to intricately designed bicycle components [72]. Thereby, athletes can now enjoy equipment that is specifically catered to their individual biomechanics and preferences, enhancing their performance and reducing the likelihood of injuries. It's truly a remarkable collaboration between AM and digital design, showcasing the perfect harmony between technology and solutions that prioritize the human experience in the sports industry [73].

1.5.12 Electronics: Miniaturized and Efficient Circuitry

The electronics industry is undergoing a dramatic transformation as AM and digital technologies join forces to revolutionize the production of electronic components. Thanks to precise AM methods, compact and highly efficient circuitry is now a reality, supporting the development of portable gadgets and IoT devices. With its intricate layering technique, AM opens up possibilities for intricate circuit designs, boosting the performance and dependability of electronic goods. Not only does this allow for rapid prototyping, but it also enables quick adaptation to the ever-changing demands of the vibrant electronics market [73].

1.6 Challenges and Opportunities

The fusion of AM and digital technologies has the potential to bring about transformative changes, but it is not without its difficulties. In this section, we delve into the complex and varied terrain of challenges and opportunities that come with this integration.

1.6.1 Technical Challenges in Integration

Integrating AM with digital technologies presents complex technical hurdles, namely material limitations and precision requirements [74]. While AM offers a wide range of material options, industries like aerospace and healthcare have specific needs for their materials. This creates a challenging gap that must be bridged between material capabilities and industry demands. Additionally, industries with strict standards also require continuous advancements in AM technologies to maintain superior precision and quality [75].

1.6.2 Adoption Barriers and Industry Transition

Integrating AM and digital technologies presents several challenges, including cost, education, and workforce training. The initial financial commitment for implementing AM technologies and digital infrastructure can be substantial, especially for small scale enterprises. Moreover, it is crucial to provide extensive education and training opportunities to prepare the workforce for effectively designing, operating, and maintaining these complex systems. To overcome these hurdles, a strategic approach and cooperation within the industry are necessary, along with initiatives that promote a smooth transition to integrated manufacturing methods.

1.6.3 Opportunities for Research and Innovation

While there may be obstacles to overcome, the infusion of AM into digital technologies presents a myriad of possibilities for breakthrough research and advancement. The emergence of cutting-edge technologies such as multi-material printing, continuous printing, and advanced materials holds the promise of addressing current constraints. Through collaborative research efforts involving academia, industry, and government bodies, we can chart an exciting course towards uncharted territories in manufacturing. The merging of AM with disruptive technologies like artificial intelligence (AI) and blockchain opens up new avenues for the development of intelligent, secure, and highly efficient manufacturing ecosystems. It is the researchers and innovators at the forefront of this transformative movement, propelling the evolution of integrated manufacturing technologies.

1.7 Conclusions

In the convergence of additive manufacturing (AM) and digital technologies within the Industry 4.0 landscape, a paradigm shift unfolds, redefining the essence of modern manufacturing. This symbiotic alliance not only addresses prevailing challenges, such as decentralized production and sustainability but also unveils a realm of transformative opportunities. The real-world impact of this integration reverberates across industries, showcasing the adaptability of AM. From crafting personalized medical implants to engineering lightweight components for the aerospace sector, the versatility of this technology becomes evident. Its ability to seamlessly align with the principles of Industry 4.0, fostering customization and sustainability, propels manufacturing into an era marked by unprecedented possibilities. At the core of this transformation lies the seamless interplay between digital twins, simulation technologies, and AM. Their collaboration creates a dynamic interplay between the physical and digital realms, offering manufacturers a holistic understanding of their processes. This not only optimizes efficiency and production but also propels the industry towards intelligent and sustainable practices. While challenges persist, including technical limitations and adoption barriers, the vast opportunities for research and innovation far outweigh them. The integration of AM with disruptive technologies like artificial intelligence and blockchain opens new frontiers, promising intelligent, secure, and highly efficient manufacturing ecosystems. As Industry 4.0 continues its evolution, the collaborative efforts of academia, industry pioneers, and innovative minds will shape the trajectory of additive manufacturing. Together, they propel the industry towards a future where boundaries are not constraints but gateways to unexplored possibilities. In this narrative, additive manufacturing stands not only as a technological force but as a catalyst for redefining the very essence of modern manufacturing in the digital age.

Acknowledgments

BBS, AK, and AJ acknowledge the Ministry of Human Resource Development, Government of India, for the graduate fellowship. All authors are grateful to the research scholars and staff of Direct Digital Manufacturing Lab, Department of Mechanical Engineering, & Fiber Optics Systems Lab, Department of Electronics and Electrical Communication Engineering, Indian Institute of Technology, Kharagpur, for their immense help throughout.

References

1. Yang, F. and Gu, S., Industry 4.0, a revolution that requires technology and national strategies. *Complex Intell. Syst.*, 7, 3, 1311–1325, 2021. doi: 10.1007/s40747-020-00267-9.
2. Alenizi, F.A., Abbasi, S., Hussein Mohammed, A., Masoud Rahmani, A., The artificial intelligence technologies in Industry 4.0: A taxonomy, approaches, and future directions. *Comput. Ind. Eng.*, 185, 109662, 2023. doi: https://doi.org/10.1016/j.cie.2023.109662.
3. Haddara, M. and Elragal, A., The Readiness of ERP Systems for the Factory of the Future. *Procedia Comput. Sci.*, 64, 721–728, 2015. doi: https://doi.org/10.1016/j.procs.2015.08.598.
4. Vaidya, S., Ambad, P., Bhosle, S., Industry 4.0 – A Glimpse. *Procedia Manuf.*, 20, 233–238, 2018. doi: https://doi.org/10.1016/j.promfg.2018.02.034.
5. Selvam, A., Aggarwal, T., Mukherjee, M., Verma, Y.K., Humans and robots: Friends of the future? A bird's eye view of biomanufacturing industry 5.0. *Biotechnol. Adv.*, 68, 108237, 2023. doi: https://doi.org/10.1016/j.biotechadv.2023.108237.
6. Zhang, C., *et al.*, Towards new-generation human-centric smart manufacturing in Industry 5.0: A systematic review. *Adv. Eng. Inf.*, 57, 102121, 2023. doi: https://doi.org/10.1016/j.aei.2023.102121.
7. Maddikunta, P.K.R., *et al.*, Industry 5.0: A survey on enabling technologies and potential applications. *J. Ind. Inf. Integr.*, 26, 100257, 2022. doi: https://doi.org/10.1016/j.jii.2021.100257.

8. Wong, K.V. and Hernandez, A., A Review of Additive Manufacturing. *ISRN Mech. Eng.*, 2012, 208760, 2012. doi: 10.5402/2012/208760.
9. Gao, W., *et al.*, The status, challenges, and future of additive manufacturing in engineering. *Comput.-Aided Des.*, 69, 65–89, 2015. doi: https://doi.org/10.1016/j.cad.2015.04.001.
10. Dilberoglu, U.M., Gharehpapagh, B., Yaman, U., Dolen, M., The Role of Additive Manufacturing in the Era of Industry 4.0. *Procedia Manuf.*, 11, 545–554, 2017. doi: https://doi.org/10.1016/j.promfg.2017.07.148.
11. Čater, T., Čater, B., Černe, M., Koman, M., Redek, T., Industry 4.0 technologies usage: motives and enablers. *J. Manuf. Technol. Manage.*, 32, 9, 323–345, Jan. 2021. doi: 10.1108/JMTM-01-2021-0026.
12. Manavalan, E. and Jayakrishna, K., A review of Internet of Things (IoT) embedded sustainable supply chain for industry 4.0 requirements. *Comput. Ind. Eng.*, 127, 925–953, 2019. doi: https://doi.org/10.1016/j.cie.2018.11.030.
13. Sharma, A. and Pandey, H., *Big Data and Analytics in Industry 4.0 BT - A Roadmap to Industry 4.0: Smart Production, Sharp Business and Sustainable Development*, A. Nayyar and A. Kumar (Eds.), pp. 57–72, Springer International Publishing, Cham, 2020, doi: 10.1007/978-3-030-14544-6_4.
14. O'Donovan, P., Gallagher, C., Leahy, K., O'Sullivan, D.T.J., A comparison of fog and cloud computing cyber-physical interfaces for Industry 4.0 real-time embedded machine learning engineering applications. *Comput. Ind.*, 110, 12–35, 2019. doi: https://doi.org/10.1016/j.compind.2019.04.016.
15. Rao, T.V.N., Gaddam, A., Kurni, M., Saritha, K., Reliance on Artificial Intelligence, Machine Learning and Deep Learning in the Era of Industry 4.0, in: *Smart Healthcare System Design*, pp. 281–299, John Wiley & Sons, Ltd, River Street, Hoboken, USA & Beverly, MA, USA, 2022, doi: https://doi.org/10.1002/9781119792253.ch12.
16. Zhou, K., Liu, T., Liang, L., From cyber-physical systems to Industry 4.0: make future manufacturing become possible. *Int. J. Manuf. Res.*, 11, 2, 167–188, 2016. doi: 10.1504/IJMR.2016.078251.
17. Lee, J., Bagheri, B., Kao, H.-A., A Cyber-Physical Systems architecture for Industry 4.0-based manufacturing systems. *Manuf. Lett.*, 3, 18–23, 2015. doi: https://doi.org/10.1016/j.mfglet.2014.12.001.
18. Krugh, M. and Mears, L., A complementary Cyber-Human Systems framework for Industry 4.0 Cyber-Physical Systems. *Manuf. Lett.*, 15, 89–92, 2018. doi: https://doi.org/10.1016/j.mfglet.2018.01.003.
19. Santi, G.M., Ceruti, A., Liverani, A., Osti, F., Augmented Reality in Industry 4.0 and Future Innovation Programs. *Technologies*, 9, 2, 33, 2021. doi: 10.3390/technologies9020033.
20. Damiani, L., Demartini, M., Guizzi, G., Revetria, R., Tonelli, F., Augmented and virtual reality applications in industrial systems: A qualitative review towards the industry 4.0 era. *IFAC-PapersOnLine*, 51, 11, 624–630, 2018. doi: https://doi.org/10.1016/j.ifacol.2018.08.388.

21. Prashar, G., Vasudev, H., Bhuddhi, D., Additive manufacturing: expanding 3D printing horizon in industry 4.0. *Int. J. Interact. Des. Manuf. (IJIDeM)*, 17, 5, 2221–2235, 2023. doi: 10.1007/s12008-022-00956-4.
22. Haleem, A. and Javaid, M., Additive Manufacturing Applications in Industry 4.0: A Review. *J. Ind. Integr. Manage.*, 04, 04, 1930001, Aug. 2019. doi: 10.1142/S2424862219300011.
23. Elhazmiri, B., Naveed, N., Anwar, M.N., Haq, M., II, The role of additive manufacturing in industry 4.0: An exploration of different business models. *Sustain. Oper. Comput.*, 3, 317–329, 2022. doi: https://doi.org/10.1016/j.susoc.2022.07.001.
24. Javaid, M., Haleem, A., Pratap Singh, R., Khan, S., Suman, R., Blockchain technology applications for Industry 4.0: A literature-based review. *Blockchain: Res. Appl.*, 2, 4, 100027, 2021. doi: https://doi.org/10.1016/j.bcra.2021.100027.
25. Bale, A.S., Purohit, T.P., Hashim, M.F., Navale, S., Blockchain and Its Applications in Industry 4.0, in: *A Roadmap for Enabling Industry 4.0 by Artificial Intelligence*, pp. 295–313, John Wiley & Sons, Ltd, River Street, Hoboken, USA & Beverly, MA, USA, 2022, doi: https://doi.org/10.1002/9781119905141.ch16.
26. Alhayani, B., *et al.*, 5G standards for the Industry 4.0 enabled communication systems using artificial intelligence: perspective of smart healthcare system. *Appl. Nanosci.*, 13, 3, 1807–1817, 2023. doi: 10.1007/s13204-021-02152-4.
27. Javaid, M., Haleem, A., Rab, S., Singh, R.P., Suman, R., Mohan, S., Progressive schema of 5G for Industry 4.0: features, enablers, and services. *Ind. Robot Int. J. Robotics Res. Appl.*, 49, 3, 527–543, Jan. 2022. doi: 10.1108/IR-10-2021-0226.
28. Sittón-Candanedo, I., A New Approach: Edge Computing and Blockchain for Industry 4.0 BT, in: *Distributed Computing and Artificial Intelligence, 16th International Conference, Special Sessions*, pp. 201–204, 2020.
29. Sittón-Candanedo, I., Alonso, R.S., Rodríguez-González, S., García Coria, J.A., De La Prieta, F., Edge Computing Architectures in Industry 4.0: A General Survey and Comparison BT, in: *14th International Conference on Soft Computing Models in Industrial and Environmental Applications (SOCO 2019)*, pp. 121–131, 2020.
30. Balubaid, M. and Alsaadi, N., Achieving Sustainability in Manufacturing through Additive Manufacturing: An Analysis of Its Enablers. *Sustainability*, 15, 12, 9504, 2023. doi: 10.3390/su15129504.
31. Butt, J., Exploring the Interrelationship between Additive Manufacturing and Industry 4.0. *Designs*, 4, 2, 13, 2020. doi: 10.3390/designs4020013.
32. Karnik, N., Bora, U., Bhadri, K., Kadambi, P., Dhatrak, P., A comprehensive study on current and future trends towards the characteristics and enablers of industry 4.0. *J. Ind. Inf. Integr.*, 27, 100294, 2022. doi: https://doi.org/10.1016/j.jii.2021.100294.
33. Hernandez Korner, M.E., Lambán, M.P., Albajez, J.A., Santolaria, J., del C. Ng Corrales, L., Royo, J., Systematic Literature Review: Integration of Additive

Manufacturing and Industry 4.0. *Metals*, 10, 8, 1061, 2020. doi: 10.3390/met10081061.

34. Gaub, H., Customization of mass-produced parts by combining injection molding and additive manufacturing with Industry 4.0 technologies. *Reinf. Plast.*, 60, 6, 401–404, 2016. doi: https://doi.org/10.1016/j.repl.2015.09.004.
35. Parvanda, R. and Kala, P., Trends, opportunities, and challenges in the integration of the additive manufacturing with Industry 4.0. *Prog. Addit. Manuf.*, 8, 3, 587–614, 2023. doi: 10.1007/s40964-022-00351-1.
36. Mehrpouya, M., Dehghanghadikolaei, A., Fotovvati, B., Vosooghnia, A., Emamian, S.S., Gisario, A., The Potential of Additive Manufacturing in the Smart Factory Industrial 4.0: A Review. *Appl. Sci.*, 9, 18, 3865, 2019. doi: 10.3390/app9183865.
37. Iqbal, A., Zhao, G., Suhaimi, H., He, N., Hussain, G., Zhao, W., Readiness of subtractive and additive manufacturing and their sustainable amalgamation from the perspective of Industry 4.0: a comprehensive review. *Int. J. Adv. Manuf. Technol.*, 111, 9, 2475–2498, 2020. doi: 10.1007/s00170-020-06287-6.
38. Azarian, M., Yu, H., Solvang, W.D., Integrating Additive Manufacturing into a Virtual Industry 4.0 Factory BT, in: *Advanced Manufacturing and Automation X*, pp. 587–594, 2021.
39. Samal, B.B., Varshney, S.K., Kumar, C.S., Four-dimensional (4D) printing through FDM: Effect of infill density and bed temperature on shape memory properties in different thermo-mechanical programming conditions. *J. Mech. Sci. Technol.*, 38, 8, 4313–4319, 2024. doi: 10.1007/s12206-024-0727-3.
40. Samal, B.B., Jena, A., Varshney, S.K., Kumar, C.S., 4D Printing of Shape Memory Polymers: A Comparative Study of Programming Methodologies on Various Material Properties. *Smart Mater. Struct.*, 32, 7, 074003, May 2023. doi: 10.1088/1361-665X/acda6e.
41. Samal, B.B., Jena, A., Mishra, D., Controlled Shape Changing Components by Using 4D Printing Technology, in: *10th International Conference on Precision, Meso, Micro and Nano Engineering (COPEN 10)*, IIT Madras, pp. 78–81, 2017, no. October, Accessed: Feb. 05, 2021. [Online]. Available: http://www.copen.ac.in/proceedings/copen10/copen/19.revised_manuscript_262.pdf.
42. Samal, B.B., Jena, A., Varshney, S.K., Kumar, C.S., FDM 4D printing: A low-cost approach of shape programming and assessing the shape memory properties using angle measurement methods in hot water actuation testing apparatus. *Mater. Today Proc.*, 101, 7–14, 2024. doi: 10.1016/j.matpr.2023.02.038.
43. Samal, B.B., Jena, A., Varshney, S.K., Kumar, C.S., 4D printing: An experimental case study on processing of shape memory polymer by FDM/FFF for nature inspired structures, in: *Advances in Additive Manufacturing Artificial Intelligence, Nature-Inspired, and Biomanufacturing*, A. Kumar, R.K. Mittal, A. Haleem (Eds.), pp. 361–377, Elsevier, Amsterdam, 2023, doi: 10.1016/B978-0-323-91834-3.00019-3.

44. Khorasani, M., *et al.*, A review of Industry 4.0 and additive manufacturing synergy. *Rapid Prototyping J.*, 28, 8, 1462–1475, Jan. 2022. doi: 10.1108/RPJ-08-2021-0194.
45. Ben-Ner, A. and Siemsen, E., Decentralization and Localization of Production: The Organizational and Economic Consequences of Additive Manufacturing (3D Printing). *Calif. Manage. Rev.*, 59, 2, 5–23, 2017. doi: 10.1177/0008125617695284.
46. Zhang, L., *et al.*, Digital Twins for Additive Manufacturing: A State-of-the-Art Review. *Appl. Sci.*, 10, 23, 8350, 2020. doi: 10.3390/app10238350.
47. Leng, J., Wang, D., Shen, W., Li, X., Liu, Q., Chen, X., Digital twins-based smart manufacturing system design in Industry 4.0: A review. *J. Manuf. Syst.*, 60, 119–137, 2021. doi: https://doi.org/10.1016/j.jmsy.2021.05.011.
48. Olivotti, D., Dreyer, S., Lebek, B., Breitner, M.H., Creating the foundation for digital twins in the manufacturing industry: an integrated installed base management system. *Inf. Syst. e-Business Manage.*, 17, 1, 89–116, 2019. doi: 10.1007/s10257-018-0376-0.
49. Liu, X., *et al.*, A systematic review of digital twin about physical entities, virtual models, twin data, and applications. *Adv. Eng. Inf.*, 55, 101876, 2023. doi: https://doi.org/10.1016/j.aei.2023.101876.
50. Chigilipalli, B.K., Karri, T., Chetti, S.N., Bhiogade, G., Kottala, R.K., Cheepu, M., A Review on Recent Trends and Applications of IoT in Additive Manufacturing. *Appl. Syst. Innov.*, 6, 2, 50, 2023. doi: 10.3390/asi6020050.
51. Lammens, N., Gallego-Bordallo, F., Ni, J., Zinoviev, A., De Weer, T., Erdelyi, H., The role of simulation in the industrialization of Additive Manufacturing. *Procedia Struct. Integrity*, 34, 247–252, 2021. doi: https://doi.org/10.1016/j.prostr.2021.12.035.
52. Sajid, S., Haleem, A., Bahl, S., Javaid, M., Goyal, T., Mittal, M., Data science applications for predictive maintenance and materials science in context to Industry 4.0. *Mater. Today Proc.*, 45, 4898–4905, 2021. doi: https://doi.org/10.1016/j.matpr.2021.01.357.
53. Zonta, T., da Costa, C.A., da Rosa Righi, R., de Lima, M.J., da Trindade, E.S., Li, G.P., Predictive maintenance in the Industry 4.0: A systematic literature review. *Comput. Ind. Eng.*, 150, 106889, 2020. doi: https://doi.org/10.1016/j.cie.2020.106889.
54. Mukherjee, T. and DebRoy, T., A digital twin for rapid qualification of 3D printed metallic components. *Appl. Mater. Today*, 14, 59–65, 2019. doi: https://doi.org/10.1016/j.apmt.2018.11.003.
55. Junk, S. and Rothe, N., Lightweight design of automotive components using generative design with fiber-reinforced additive manufacturing. *Procedia CIRP*, 109, 119–124, 2022. doi: https://doi.org/10.1016/j.procir.2022.05.224.
56. Kumar, A., Ramadas, H., Samal, B.B., Kumar, C.S., Nath, A.K., Laser polishing of cobalt chrome alloy fabricated by laser powder bed fusion process: design of experiment-based approach for reducing surface roughness. *Int. J. Adv. Manuf. Technol.*, 1245–1264, 2024. doi: 10.1007/s00170-024-14101-w.

57. Kumar, A., Samal, B.B., Nath, A.K., Kumar, C.S., Laser surface modification of metal additive manufactured parts : A case study of *ex-situ* and *in-situ* methodology, in: *Surface Engineering*, pp. 121–144, CRC Press, Boca Raton, Florida, USA, 2022, doi: 10.1201/9781003319375-5.
58. Namin, A.T., *et al.*, Adoption of New Medical Technologies: The Case of Customized Individually Made Knee Implants. *Value Health*, 22, 4, 423–430, 2019. doi: https://doi.org/10.1016/j.jval.2019.01.008.
59. Ramola, M., Yadav, V., Jain, R., On the adoption of additive manufacturing in healthcare: a literature review. *J. Manuf. Technol. Manage.*, 30, 1, 48–69, Jan. 2019. doi: 10.1108/JMTM-03-2018-0094.
60. Gubin, A.V., Kuznetsov, V.P., Borzunov, D.Y., Koryukov, A.A., Reznik, A.V., Chevardin, A.Y., Challenges and Perspectives in the Use of Additive Technologies for Making Customized Implants for Traumatology and Orthopedics. *Biomed. Eng.*, 50, 4, 285–289, 2016. doi: 10.1007/s10527-016-9639-6.
61. Li, J., *et al.*, Hybrid additive manufacturing of 3D electronic systems. *J. Micromech. Microeng.*, 26, 10, 105005, Aug. 2016. doi: 10.1088/0960-1317/26/10/105005.
62. Korrayi, R.R., Jena, A., Marupalli, B.C.G., Samal, B.B., Varshney, S.K., Kumar, C.S., Four-Dimensional (4D) Microprinting, in: *Additive Manufacturing with Novel Materials*, pp. 429–457, John Wiley & Sons, Ltd, River Street, Hoboken, USA & Beverly, MA, USA, 2024, doi: https://doi.org/10.1002/9781394198085.ch14.
63. Lopera, L., Rodriguez, R., Yakout, M., Elbestawi, M., Emadi, A., Current and Potential Applications of Additive Manufacturing for Power Electronics. *IEEE Open J. Power Electron.*, 2, 33–42, 2021. doi: 10.1109/OJPEL.2021.3052541.
64. Kern, J., Additive Manufacturing, in: *The Digital Transformation of Logistics*, pp. 41–60, John Wiley & Sons, Ltd, Hoboken, New Jersey, USA, 2021, doi: https://doi.org/10.1002/9781119646495.ch4.
65. Bogers, M., Hadar, R., Bilberg, A., Additive manufacturing for consumer-centric business models: Implications for supply chains in consumer goods manufacturing. *Technol. Forecast. Soc. Change*, 102, 225–239, 2016. doi: https://doi.org/10.1016/j.techfore.2015.07.024.
66. Lacroix, R., Seifert, R.W., Timonina-Farkas, A., Benefiting from additive manufacturing for mass customization across the product life cycle. *Oper. Res. Perspect.*, 8, 100201, 2021. doi: https://doi.org/10.1016/j.orp.2021.100201.
67. Sun, C., Wang, Y., McMurtrey, M.D., Jerred, N.D., Liou, F., Li, J., Additive manufacturing for energy: A review. *Appl. Energy*, 282, 116041, 2021. doi: https://doi.org/10.1016/j.apenergy.2020.116041.
68. Verhoef, L.A., Budde, B.W., Chockalingam, C., García Nodar, B., van Wijk, A.J.M., The effect of additive manufacturing on global energy demand: An assessment using a bottom-up approach. *Energy Policy*, 112, 349–360, 2018. doi: https://doi.org/10.1016/j.enpol.2017.10.034.

69. Delgado Camacho, D., *et al.*, Applications of additive manufacturing in the construction industry – A forward-looking review. *Autom. Constr.*, 89, 110–119, 2018. doi: https://doi.org/10.1016/j.autcon.2017.12.031.
70. Paolini, A., Kollmannsberger, S., Rank, E., Additive manufacturing in construction: A review on processes, applications, and digital planning methods. *Addit. Manuf.*, 30, 100894, 2019. doi: https://doi.org/10.1016/j.addma.2019.100894.
71. Huang, R., *et al.*, Energy and emissions saving potential of additive manufacturing: the case of lightweight aircraft components. *J. Clean. Prod.*, 135, 1559–1570, 2016. doi: https://doi.org/10.1016/j.jclepro.2015.04.109.
72. Beiderbeck, D., Krüger, H., Minshall, T., The Future of Additive Manufacturing in Sports BT, in: *21st Century Sports: How Technologies Will Change Sports in the Digital Age*, S.L. Schmidt (Ed.), pp. 111–132, Springer International Publishing, Cham, 2020, doi: 10.1007/978-3-030-50801-2_7.
73. Novak, J., II and Novak, A.R., Is additive manufacturing improving performance in Sports? A systematic review. *Proc. Inst. Mech. Eng., Part P: J. Sports Eng. Technol.*, 235, 3, 163–175, 2021. doi: 10.1177/1754337120971521.
74. Zhou, K., Liu, T., Zhou, L., Industry 4.0: Towards future industrial opportunities and challenges, in: *2015 12th International Conference on Fuzzy Systems and Knowledge Discovery (FSKD)*, pp. 2147–2152, 2015, doi: 10.1109/FSKD.2015.7382284.
75. Tofail, S.A.M., Koumoulos, E.P., Bandyopadhyay, A., Bose, S., O'Donoghue, L., Charitidis, C., Additive manufacturing: scientific and technological challenges, market uptake and opportunities. *Mater. Today*, 21, 1, 22–37, 2018. doi: https://doi.org/10.1016/j.mattod.2017.07.001.

2

Additive Manufacturing Processing and Techniques: Focusing on Laser Powder Bed Fusion (L-PBF) and Its Various Post Processing Technologies

Abhishek Kumar[1], Bijaya Bikram Samal[2*], Ashish Kumar Nath[1], Shailendra Kumar Varshney[3] and Cheruvu Siva Kumar[1]

[1]Department of Mechanical Engineering, Indian Institute of Technology, Kharagpur, West Bengal, India

[2]Advanced Technology Development Centre, Indian Institute of Technology, Kharagpur, West Bengal, India

[3]Department of Electronics and Electrical Communication, Indian Institute of Technology, Kharagpur, West Bengal, India

Abstract

This chapter provides an extensive review of various additive manufacturing (AM) processes, with a primary focus on laser powder bed fusion (L-PBF). Through a comprehensive analysis of existing literature, we explore the significant advancements in AM technologies, specifically L-PBF, which has revolutionized the production of complex geometries with superior mechanical properties. This chapter also delves into critical process parameters such as laser power, scan speed, and powder characteristics, and their influence on surface roughness and overall part quality. Additionally, we discussed various post-processing techniques, particularly laser polishing, highlighting their role in enhancing surface integrity and reducing roughness. The interplay between surface texture, fatigue resistance, and corrosion behavior is also analyzed, with particular attention to the impact of heat treatments and surface metrology. The importance of optimizing process parameters and post-processing strategies to achieve higher performance and reliability in AM-produced parts across aerospace, automotive, and medical industries is

**Corresponding author*: bijaybikram3991@gmail.com

Bijaya Bikram Samal, Cheruvu Siva Kumar and Shailendra Kumar Varshney (eds.) 4D Printing Technology: Principles, Materials and Applications, (29–68) © 2025 Scrivener Publishing LLC

presented. Our findings contribute to the growing body of knowledge on AM, providing insights for future research and industrial applications.

***Keywords*:** Additive manufacturing, laser powder bed fusion, surface integrity, post-processing, laser polishing, surface roughness, fatigue resistance

2.1 Introduction

Additive manufacturing (AM), which is also commonly known as 3D printing, has emerged as a disruptive technology over conventional manufacturing, radically transforming the manner in which we think and create items [1]. The additive manufacturing (AM) technology allows designers and engineers to produce elaborate, bespoke components with an unprecedented level of accuracy and complexity. This is accomplished by stacking materials in accordance with digital blueprints. This technique marks a change from traditional subtractive manufacturing processes, which depend on cutting away material from a solid block [2]. Instead, it adds material just where it is required, thus eliminating waste and improving resource use. Additive manufacturing (AM), as outlined by ISO/ASTM 52900, refers to the method of constructing parts by incrementally adding material based on 3D model data, typically in a layer-by-layer fashion. This technology excels in producing small quantities of highly intricate components, where surface finishes may vary. During the process of building, temporary support structures are often required to stabilize the part, separate from the main geometry. Once the printing is complete, the removal of these supports and other post-processing steps are essential. Streamlining these tasks can lead to significant savings in both production time and costs [3]. The need for lightweight, tailored components is always expanding, and this unique technique provides considerable benefits across a wide range of sectors, including aerospace and automotive, healthcare, and consumer products, where the demand is constantly growing [4]. Additionally, additive manufacturing (AM) allows for quick prototyping, iterative design revisions, and production on demand, which drives innovation and accelerates the time-to-market for new goods. This is in addition to the design flexibility and efficiency that AM offers [5]. The revolutionary potential of additive manufacturing (AM) is set to revolutionize all sectors and open new horizons in manufacturing excellence as it continues to improve and combine with other sophisticated technologies such as artificial intelligence and robots [6].

AM has grown over the years from its beginnings with rapid prototyping to include a diversified range of processing methods with extensive industrial applications in a variety of industries, including aerospace, automotive, healthcare, and consumer products. AM originated in the 1980s, largely as a technique for rapid prototyping. During this time, pioneering methods including stereolithography (SLA) and selective laser sintering (SLS) were developed, setting the groundwork for future advances in the industry [7]. The 1990s saw substantial growth and refinement in AM technology, notably in terms of materials and processes. During this time, innovations led to the development of new additive manufacturing processes and to the diversity of accessible materials. Scott Crump created fused deposition modeling (FDM) in 1988, and it quickly gained popularity because of its ability to print using thermoplastics layer by layer. Recent advancements in metal additive manufacturing (AM) techniques, like direct metal laser sintering (DMLS) and electron beam melting (EBM), have made it possible to create metal components with highly complex geometries. These breakthroughs have expanded the potential for metal AM in industries such as aerospace, automotive, and healthcare, where intricate, high-performance parts are essential. The ability to produce such detailed metal structures is driving innovation and opening new doors for applications in these sectors [8]. The 2000s was a watershed moment for additive manufacturing, as the technique started to receive broad industrial use. Companies from a variety of industries began to see AM's potential for end-use component manufacturing, in addition to prototyping. During this time, developments aimed to improve the speed, accuracy, and scalability of AM techniques to fulfill the needs of industrial production. Customization became a primary emphasis, with AM allowing the creation of unique, individualized items based on individual tastes and needs [9]. The automobile and aerospace sectors, in particular, have adopted additive manufacturing for quick tooling, lightweighting, and geometric optimization. The 2010s saw the widespread adoption of additive manufacturing as a viable production technique in a variety of sectors. Continued breakthroughs in materials, processes, and machine capabilities fueled substantial innovation in the industry [10]. AM technologies have developed to provide better surface finishes, mechanical qualities, and manufacturing efficiency, making them suitable for a wider variety of applications. Hybrid AM methods, which blend additive and subtractive manufacturing techniques, originated as a way to capitalize on the capabilities of both methodologies [11]. Furthermore, advancements in software, automation, and post-processing methods have broadened AM's capabilities and uses, opening the door for its incorporation into Industry 4.0 projects and smart manufacturing ecosystems. Today, additive manufacturing is at the vanguard of Industry 4.0, driving digital transformation and redefining the face of contemporary production. AM's integration with digital technologies like artificial intelligence, machine learning, and the Internet of Things (IoT) allows for real-time monitoring, predictive maintenance, and autonomous manufacturing processes [12]. As AM evolves, continuous research and development activities are focused on enhancing material characteristics, scaling production, lowering prices, and solving issues like as process repeatability, quality control, and sustainability. With the ability to transform supply chains, allow on-demand production, and open up new design possibilities, additive manufacturing is positioned to play a critical role in the future of manufacturing [13].

In recent years, the processing of shape memory alloys (SMAs) via L-PBF has garnered attention for its potential in creating adaptive structures that can change the shape or properties in response to environmental stimuli [14]. AM of smart materials which can change its shape over time is termed as Four dimensional (4D) printing [15]. The ability to manufacture patient-specific implants on-demand using L-PBF technology not only enhances biological integration due to tailored surface roughness but also reduces lead times from design to surgery. Effective post-processing techniques are crucial in this context, as they improve the mechanical properties and functionality of SMAs, ensuring that these innovative materials perform as intended in medical applications.

2.2 Classification of Additive Manufacturing

ASTM classifies additive manufacturing into seven types: material extrusion (ME), which includes fused deposition modelling (FDM) and direct ink writing (DIW); Vat photopolymerization (VP), which includes stereolithography (SLA) and digital light processing (DLP); powder bed fusion (PBF) includes selective laser sintering (SLS) and selective laser melting (SLM); material jetting (MJ) includes poly-jet and ink-jet printing; binder jetting (BJ) includes metal binder jetting (MBJ) and sand binder jetting (SBJ); sheet lamination (SL) includes laminated object manufacturing (LOM) and ultrasonic additive manufacturing (UAM); and directed energy deposition (DED) includes laser metal deposition (LMD) and electron beam melting (EBM) [16]. Additive manufacturing (AM) methods can also be categorized differently depending on different attributes as provided in Table 2.1 [17, 18]. These classifications are provided in Figure 2.1.

Table 2.1 Type of AM process (ASTM classification).

AM category	Brief description
Binder Jetting	Liquid Binding agent is used to join powder particle
Directed Energy Deposition	Thermal Energy melt the feed material deposited
Material Extrusion	Material selectively dispensed from the nozzle/ orifice
Material Jetting	Selective deposition of the droplet of feed material
Powder Bed Fusion	Thermal energy selectively fuses powder particle
Sheet Lamination	Sheet of material is bonded and cut at specific contour
Vat Photopolymerization	Liquid photopolymer is selectively cured

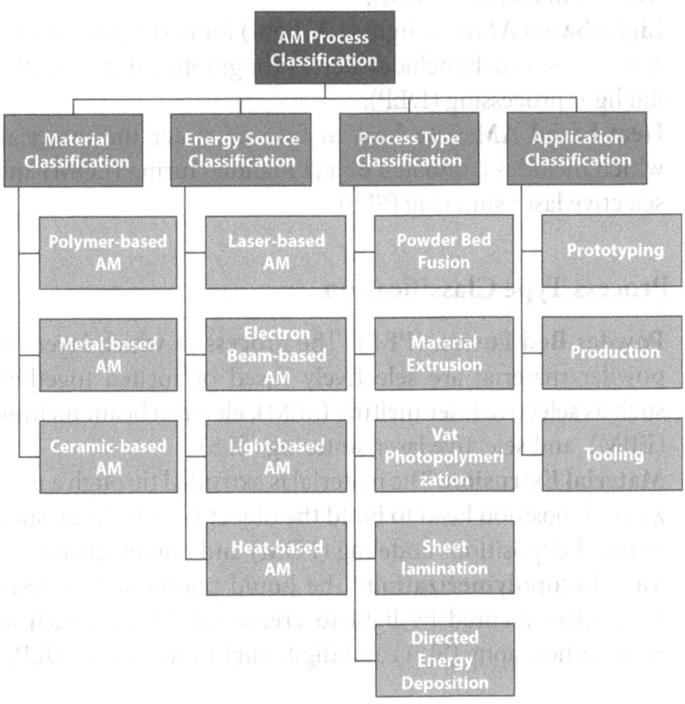

Figure 2.1 Classification of AM processes.

2.2.1 Material Classification

- **Polymer-based AM:** Use polymer materials: stereolithography (SLA); fused deposition modeling (FDM); and selective laser sintering (SLS).
- **Metal-based AM:** Use metals or metal alloys: selective laser melting (SLM); electron beam melting (EBM); and direct metal laser sintering (DMLS).
- **Ceramic-based AM:** Use ceramic materials: ceramic stereolithography (CSL) and selective laser sintering of ceramics.

2.2.2 Energy Source Classification

- **Laser-based AM:** Use laser energy for melting, sintering, or curing the materials, which includes selective laser melting (SLM), electron beam melting (EBM), and stereolithography (SLA).
- **Electron Beam-based AM:** Use an electron beam as the energy source for melting or sintering, which includes electron beam melting (EBM).
- **Light-based AM:** Use light (UV light) for curing photopolymer resins, which includes stereolithography (SLA) and digital light processing (DLP).
- **Heat-based AM:** Use heat to fuse or sinter the material, which includes laminated object manufacturing (LOM) and selective laser sintering (SLS).

2.2.3 Process Type Classification

- **Powder Bed Fusion (PBF):** The process in which layers of powder material are selectively fused or melted together, such as selective laser melting (SLM), electron beam melting (EBM), and selective laser sintering (SLS).
- **Material Extrusion:** The material is extruded through a nozzle or deposition head to build the object layer by layer, such as fused deposition modeling (FDM) and binder jetting.
- **Vat Photopolymerization:** The liquid photopolymer resin is selectively cured by light to create solid layers, such as Stereolithography (SLA) and digital light processing (DLP).

- **Sheet lamination:** The process where sheets of material are layered and bonded together, such as laminated object manufacturing (LOM).
- **Directed Energy Deposition (DED):** The process in which material is deposited or melted using a directed energy source, typically a laser or electron beam.

2.2.4 Application Classification

- **Prototyping:** It is used primarily for rapid prototyping and iterative design, allowing for the quick and cost-effective creation of physical models for testing and validation.
- **Production:** AM processes can produce end-use parts in various industries, such as aerospace, automotive, medical, and consumer products.
- **Tooling:** It fabricates molds, jigs, fixtures, and other tooling components to support traditional manufacturing processes.

2.3 LPBF

The L-PBF process comes under metal-based additive manufacturing, which involves metal or metal alloys. L-PBF, selective laser melting (SLM), and direct metal laser sintering (DMLS) are closely related additive manufacturing processes that utilize laser energy to fuse powder material selectively [19]. The working principle of this process can be divided into six significant subsystems as shown in Figure 2.2: Joining mechanism, which can be solid-based sintering, liquid phase sintering, or complete melting; powder material can be coated, mixed, pre-alloyed, polymer/wax, or metal; scanning system include galvanometer, flying optics, and mask; powder delivery system can be top-feeding or bottom-feeding; powder recoating systems include blades, rollers, slot feeders, and electrostatics [20]. An atmosphere can be inert, vacuumed, or open. The powder is applied to a build area in a thin, 30-100 mm-thick coating. In powder-based additive manufacturing, a recoating mechanism, such as a blade or counter-rotating roller, evenly distributes a fine layer of material powder over the build surface. A high-energy source, like a laser, then precisely directs heat onto the desired areas of the powder bed, causing the material to melt and fuse, forming solid connections between the particles of the current layer and the layer beneath it. Fiber lasers, typically ranging between 200 and 500 watts, are commonly used for metals due to their superior energy efficiency.

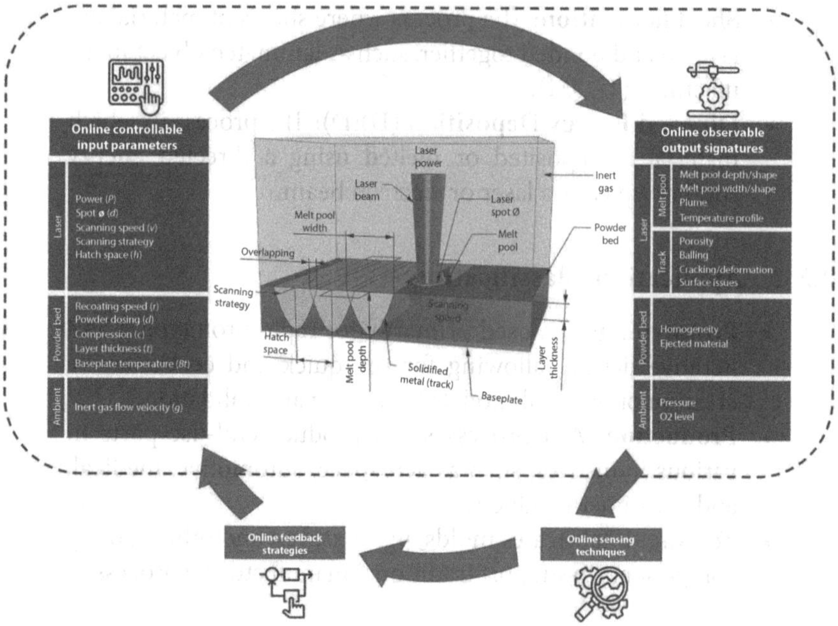

Figure 2.2 Schematic representation of L-PBF along with labels for some controllable input parameters and observable output signatures [23].

As the laser interacts with the material at the correct wavelength and energy level, the powder particles are heated to the point of melting, leading to their consolidation into a uniform, solid structure [21, 22].

2.3.1 L-PBF Process Details

- **CAD Model Preparation**: Creating a 3D CAD model of the desired components and defining the parts' geometry, dimensions, and other specifications.
- **Slicing**: The 3D CAD model is sliced into thin cross-sectional layers with layer specifications depending upon the materials and machine requirements. The appropriate support structure is also provided.
- **Machine Setup**: The build plate is installed, and proper leveling of the build plate is done to ensure the uniform spreading of the powder layer. Metal powder is loaded into the powder bed collector chamber.

- **Preheating and establishment of process condition:** The build plate is preheated to a temperature below the material melting point, thus helping to prevent thermal shock and giving uniform heat distribution. The proper flow of recirculating gas establishes lens cleaning and build process conditions.
- **Layer by Layer Deposition**: A powered laser scans the surface of each layer, tracing the cross-sectional geometry of the part to be built, thus selectively melting and fusing metal powder particles layer by layer.
- **Powder distribution and layer renewal**: After each full scan of each layer, a fresh layer of metal powder is uniformly spread over the previously melted and solidified layer.
- **Cooling and Solidification**: After each layer is scanned and the metal powder is melted and fused, rapid cooling and solidification form a solid part. It also affects the part's microstructure, mechanical properties, and surface finish.
- **Post Processing**: After the building process is over, the part undergoes post-processing, like removal of the support structure, sandblasting to remove excess powder, heat treatment, and performing additional machining, grinding, or polishing operations to achieve the desired surface finish and dimensional accuracy.

Some of the metallic feedstock for PBF are aluminum alloys, cobalt-chromium alloys, nickel alloys, gold, silver, stainless steel, tool steel, and titanium alloys, etc. The first DMLS machine was developed in 1995 by Electro-Optics (EOS). Significant advantages of DMLS include the production of highly complex and intricate geometries with a wide variety of metal materials. It also enabled the production of highly customized and personalized parts. It allows the integration of multiple components or features into a single printed part, and the ability to build complex parts with less material waste makes it a sustainable manufacturing process [24].

Major applications of the process include the aerospace and defense industry, medical and dental applications, the automotive industry, tooling and mold making, jewelry design, and other engineering and industrial sectors. In the aerospace and defense industries, processes mainly focus on lightweight structural components, brackets, mounts, satellite propulsion systems, avionics housings, customized sensor housing, antenna mounts, engine parts, turbine blades, fuel nozzles, and heat exchangers [25]. The medical and dental applications include patient-specific orthopedic implants (hip, knee, and spinal implants); customized dental implants, including crowns, bridges, and abutments; custom dental instruments, including dental drills, probes, and guides; limb and cranial prosthetics; specialized surgical instruments such as retractors, clamps, and guides; patient-specific guides and templates; [26, 27] Endoscopic and robotic components, Biocompatible components such as brackets, connectors, and housings for medical devices are patient-specific anatomical models as shown in Figure 2.3. Automotive industry applications include intake and exhaust manifolds; turbocharger components, including impellers, compressor housings, and turbine housings; lightweight suspension components, including control arms, knuckles, uprights, and tooling inserts; customized accessories, including emblems, logos, and interior trim components; and heat exchangers such as radiators, intercoolers, and oil coolers [28]. Tooling and mold-making applications include conformal cooling within molds, complex geometrical inserts, and cores. Other engineering applications include heat sinks for electronic devices, jigs, fixtures, and gauges. Thus, material compatibility, design flexibility, and the ability to create complex geometries make it valuable for various sectors, enabling improved performance, customization, and efficiency in manufacturing processes [29].

Although DMLS offers many advantages, it also has specific limitations and disadvantages that need to be addressed, such as higher surface

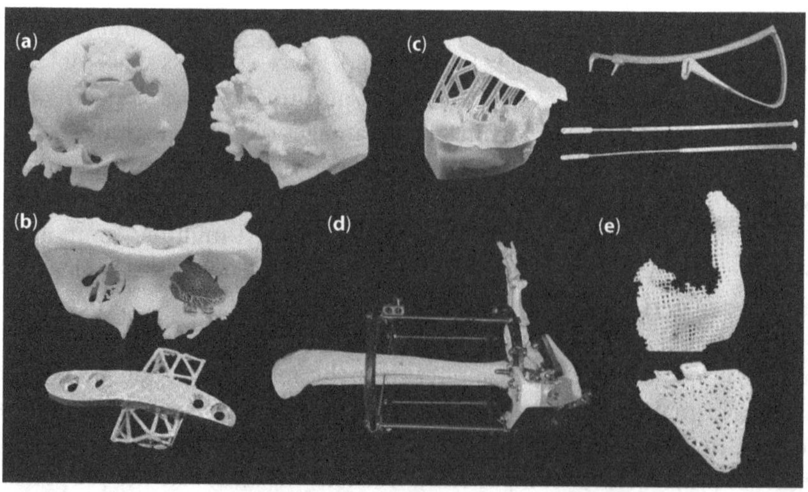

Figure 2.3 Additive manufacturing applications in biomedical engineering (a) medical models; (b) implants; (c) tools, instruments and parts for medical devices; (d) medical aids, supportive guides, splints and prostheses; (e) biomanufacturing [30].

roughness due to the staircase effect, a spattering of the satellite particles, the morphology of the solidified laser tracks, the interaction of the neighboring track forming complex geometry, and attachment of the residual powder particle, especially on the inclined and lateral parts [31]. Deviation in dimension due to thermal stress and shrinkage during the cooling process, removal of support material, which can be time-consuming, thereby leaving residual marks, limited material selection, expensive costs including material cost, machine cost, post-processing cost, process complexity, and specific design limitations in which powder particle and support structure removal pose significant challenges. Ongoing advancements in technology and materials can address some limitations, making this process more accessible and versatile [32].

2.3.2 Process Parameters

Quality in laser additive manufacturing is determined by several interconnected process parameters that must be optimized in conjunction,

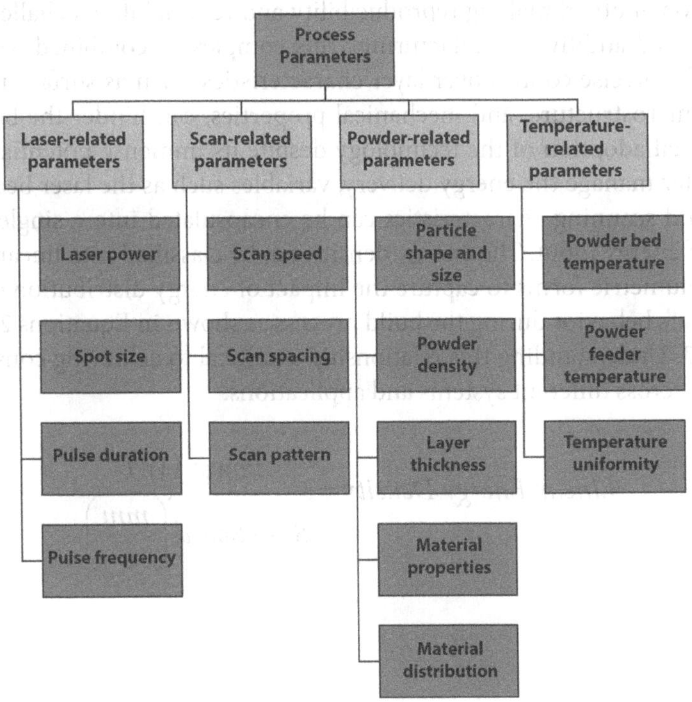

Figure 2.4 Process parameters involved in LPBF AM.

depending on the material and machine type. Powder bed fusion can be broadly classified into four types based on the process parameters as shown in Figure 2.4.

It should be noted that most of these characteristics are highly interconnected and have interactions. Research indicates that more than 25 process parameters influence the laser additive manufacturing process. Laser power, scan speed, beam diameter, hatching strategy, amount of powder discharge, layer thickness, gas flow rate, and orientation are the critical or customizable factors for a specific system [33]. The process parameters in laser additive manufacturing play a crucial role in determining product quality, including geometrical accuracy, manufacturing time, and surface finish. Factors such as the shape, size, and distribution of the powder directly influence laser absorption, powder bed density, heat transfer, and the efficiency of powder spreading. Key elements like powder bed temperature, laser power, scan speed, and scan spacing must be carefully balanced to optimize the melt pool size, ensure dimensional precision, enhance surface quality, and maintain a high build rate while achieving desirable mechanical properties [34]. However, these parameters are highly system-specific, meaning that settings optimized for one machine may not perform consistently on another, making reproducibility and repeatability a challenge in laser-based additive manufacturing. This complexity, combined with the need for precise control over layer characteristics, such as surface roughness, microstructure, and mechanical properties, can hinder the broader industrial adoption of the technology despite its immense potential [35]. To better manage the energy delivery, variables such as the laser beam, its size, and scanning characteristics can be encapsulated into a single measure of energy input. The energy density can be classified into linear, area, and volumetric forms to capture the impact of energy distribution on the material's behavior during the build process as shown in Equations 2.1, 2.2 and 2.3. Understanding this relationship is critical to achieving consistent results across different systems and applications.

$$\text{Linear Energy Density} = \frac{\text{Power}\,(W)}{\text{Scan Speed}\left(\frac{mm}{s}\right)} \tag{2.1}$$

$$Area\ Energy\ Density = \frac{Power\,(W)}{Scan\ Speed\left(\frac{mm}{s}\right) * Hatch\ Distance\,(mm)} \tag{2.2}$$

$$Volume\ Energy\ Density = \frac{Power\,(W)}{Scan\ Speed\left(\frac{mm}{s}\right) * Hatch\ Distance\,(mm) * Layer\ thickness\,(mm)} \tag{2.3}$$

While linear energy density (LED) and area energy density (AED) are commonly referenced in existing research, it is more appropriate to adopt volume energy density (VED) as a standard measure. VED provides a more accurate representation of how power is distributed in three-dimensional space to effectively melt materials.

2.3.3 Surface Roughness and Morphology of L-PBF

In laser powder bed fusion (L-PBF), surface roughness tends to be greater than in traditional manufacturing processes, with its variability affected by factors like the angle of inclination and local processing conditions. The distinctive surface textures and properties produced during L-PBF are directly connected to the unique nature of the additive manufacturing technique. Multiple design variables throughout the entire production process influence the degree of surface roughness. Before the build begins, factors such as powder particle size, distribution, the preheating of the build plate, and the specific conditions within the L-PBF chamber are crucial. During the actual manufacturing phase, critical parameters like laser power, scanning speed, hatch spacing, layer thickness, and the chosen scanning strategy significantly impact the quality of the surface. Finally, post-build treatments—ranging from heat treatments to various material removal techniques—also play a vital role in refining surface characteristics, highlighting the intricate challenges of achieving optimal finishes in L-PBF technology [36]. The surface characteristics of components produced via laser powder bed fusion (L-PBF) are influenced by a variety of factors, such as stair-stepping effects, spatter, the presence of satellite particles, melt pool dynamics, track uniformity, and recoil pressure effects. The orientation of the surface plays a critical role in how well residual particles are retained, leading to variations in roughness not only along the solidified tracks but also in the overlapping regions between them. In L-PBF, each layer solidifies as a result of multiple laser tracks that overlap according to predetermined parameters. Generally, the roughness is more accentuated at the overlaps between adjacent tracks compared to the central sections of the tracks themselves. Striking the right balance in overlap is essential; inadequate overlap can create noticeable peaks and valleys between tracks, while excessive overlap can prolong processing times and cause increased thermal accumulation, resulting in a higher likelihood of particle adhesion. Understanding and controlling these factors is crucial for optimizing surface quality in L-PBF applications [37, 38]. Surface quality in laser powder bed fusion (L-PBF) is significantly affected by several factors, including melt pool size, laser power settings, scanning speed, and the extent of track overlap. Excessive overlap can lead to increased thermal energy, which often exacerbates surface roughness. Conversely, insufficient overlap can result in uneven surfaces between tracks. Achieving an optimal balance among these parameters is essential for reducing surface roughness and enhancing the overall quality of L-PBF components. Understanding how each of these variables interacts allows for better control over the manufacturing process and the resultant surface finish [39].

2.3.4 Metallurgical Aspects and Mechanical Properties of L-PBF Parts

Surface treatments, such as post-printing polishing or coating operations, can improve surface smoothness, minimize roughness, and increase durability. Optimizing printing parameters such as layer thickness, temperature, and build orientation can aid in optimizing layer bonding and reduce the likelihood of weak interfaces. The L-PBF portion goes through a non-equilibrium phase of quick melting and crystallization during the laser melting process [40]. As a result, a completely different microstructure is obtained, far finer than a customarily manufactured product. Mechanical properties of L-PBF components fluctuate depending on the part build orientations, indicating that L-PBF parts are anisotropic. Because the mechanical qualities of any component are intimately related to its underlying microstructure, an L-PBF part exhibits mechanical properties comparable to its wrought equivalent under particular processing conditions [41]. Some mechanical and electrochemical properties, like fatigue and corrosion resistance, are affected by the surface features of the part. As-built L-PBF parts with high surface roughness and residual stresses have low fatigue life and corrosion characteristics. However, with suitable post-processing, varied mechanical properties, including fatigue life, can be obtained in L-PBF parts that are on par, if not better, than wrought counterparts.

2.4 Post Processing of Additive Manufactured Parts

Postprocessing techniques are crucial for achieving printed parts' desired final properties, functionality, and aesthetics. They are necessary to improve surface finish, precise dimensional accuracy, tight tolerance, and enhance mechanical properties. Postprocessing can be divided into four subprocesses: powder removal, recycling, conditioning, part removal and postprocessing for machining operations, finishing parts for improved aesthetics, and heat treatment to enhance mechanical properties [42]. Despite the increasing cost and build time, it is essential in additive manufacturing to create precise features and interfaces.

2.4.1 Powder Removal, Recycling and Conditioning

Post-additive manufacturing, fine powder particles often cling to the edges of components, especially around melted areas. To effectively manage this residue, techniques such as exposure to argon gas or a vacuum environment can facilitate the sieving and recycling of unused powder. The remaining powder generally consists of a mix of partially melted and sintered particles adjacent to the energy beam paths, as well as untouched powder. It is crucial to evaluate the properties of condensate powder, as its altered characteristics may negatively impact the integrity of the final product; thus, it is recommended that this material be sieved out and discarded [43].

2.4.2 Part Removal and Postprocessing for Machining Operations

When it comes to part extraction and subsequent machining, wire-type electrical discharge machining (wire EDM) is often preferred due to its adaptability with various conductive materials and its stress-free operational characteristics. However, it's noteworthy that metals fabricated through additive manufacturing (AM) typically have lower machinability compared to their conventionally manufactured counterparts. To overcome this limitation, alternative machining methods such as electrochemical machining (ECM), along with chemical and photo-chemical etching techniques, can be employed to enhance the surface quality of metal parts.

Furthermore, for joining AM-produced metal components, processes like welding and brazing offer reliable solutions, effectively expanding the utility and application of additive manufacturing in diverse industrial settings [44].

2.4.3 Finishing of Part for Improved Surface Finish and Aesthetic

Various finishing procedures can improve the surface finish of AM metal objects, including mechanical, radiation, and chemical treatments. Traditional machining, milling, grinding, and polishing remove surface irregularities with abrasive tools. Vibratory grinding reduces surface roughness, while abrasive fluidized bed machining selectively exposes rough surfaces to moving abrasives. Micro-ploughing or cutting with randomly oriented abrasives removes high peaks and improves surface finish [45].

2.4.3.1 Mechanical Surface Post-Processing

Polishing is a critical technique for enhancing the surface quality of additively manufactured parts, eliminating the need for conventional machining or milling processes. Among the mechanical surface treatment options available, methods such as magnetically driven abrasive polishing, hydrodynamic cavitation abrasive finishing, and ultrasonic cavitation abrasive finishing are particularly effective in minimizing surface roughness in laser powder bed fusion (L-PBF) materials, while also facilitating internal modifications. Furthermore, impact-based techniques leverage controlled plastic deformation to achieve a more consistent surface profile, though they may not always deliver a perfectly smooth finish. Buffing, another mechanical approach, utilizes a motorized flexible wheel to refine surfaces, providing versatility in achieving a range of finishes from coarse to fine across diverse materials. Together, these polishing methods play a vital role in improving the functional and aesthetic properties of additively manufactured components [46].

2.4.3.2 Chemical Treatments and Electrochemical Surface Modification

Chemical surface treatments play a vital role in improving the characteristics of components produced via laser powder bed fusion (L-PBF).

These treatments tackle surface roughness through an array of techniques, including etching, polishing, chemical brightening, traditional machining, and electrochemical polishing. Each method contributes uniquely to refining the surface finish, ultimately elevating the overall performance and aesthetic appeal of the final product. By employing these advanced surface treatment strategies, manufacturers can significantly enhance the quality and functionality of L-PBF components, making them more suitable for a wide range of applications. These techniques are particularly well-suited for intricate and complex geometries, where achieving a smooth surface finish poses significant challenges. They also provide the flexibility of localized treatment, enabling specific areas to be refined through the strategic application of masking. Among these approaches, electrochemical etching is noteworthy for its ability to selectively remove support structures from L-PBF parts. This technique capitalizes on the unique properties of the outer layer, which is generally less chemically stable and more susceptible to etching processes. By leveraging this characteristic, manufacturers can gain superior control over the support removal, ensuring that only targeted areas are affected while maintaining the structural integrity of the part. The result is not just an improved surface finish, but also a significant enhancement in the overall performance and aesthetic quality of additively manufactured components. Ultimately, the application of chemical treatments plays a vital role in advancing the capabilities and applications of 3D-printed parts in high-precision industries [47].

2.4.3.3 Laser-Based Surface Polishing

Laser-based surface post-processing techniques represent a cutting-edge approach to polishing additively manufactured components without direct contact. By employing focused laser pulses, these methods induce microscale melting that significantly enhances the quality of the surface produced by AM. This innovative process leverages capillary pressure and surface tension effects to effectively fill in valleys and crevices, thereby smoothing surface roughness while preserving material integrity, as there is no material loss during the treatment. Among these techniques, laser remelting stands out for its dual capability to improve surface densification and reduce roughness. This method can be applied either immediately after layer deposition or as a final finishing step to refine the contours of the part. Additionally, laser shock peening has emerged as a prominent surface modification strategy for laser powder bed fusion (L-PBF) materials. This technique involves the vaporization of a thin layer of materials to further generate very high amplitude of the shock waves that can penetrate the target material. The deformation that is created due to this, introduces beneficial compressive stresses into the surface, effectively closing near-surface pores and enhancing the overall structural integrity. Moreover, laser ablation techniques, particularly femtosecond laser micromachining, have demonstrated remarkable effectiveness in improving the surface quality of L-PBF materials. By utilizing ultra-short laser pulses, this method can precisely reduce surface roughness, resulting in smoother finishes that meet the demanding specifications of various high-performance applications. Overall, laser-based post-processing not only enhances the aesthetic and functional properties of additively manufactured parts but also opens new avenues for innovation in material design and engineering [48].

2.4.3.4 Surface Coating

Surface coatings are utilized on laser powder bed fusion (L-PBF) materials to achieve specific surface functionalities and control surface morphologies. These coatings are designed to enhance various characteristics, including tribological performance, corrosion resistance, fatigue strength, crack propagation resistance, and biological compatibility. By applying these tailored coatings, manufacturers can significantly improve the overall durability and effectiveness of additively manufactured components across diverse applications [49]. Bioactive coatings or nanopatterns can modulate bio-integration of additively manufactured materials. Studies show coatings efficiency in promoting bone formation and induced antibacterial properties, such as hydroxyapatite coatings and silver impregnated coatings [50]. Mechanical surface treatments are commonly used in industrial applications to remove thin layers, but their major demerit is that they produce high residual stresses towards the surface region. Other methods like chemical surface treatments can reduce these stresses, making them more suitable for high-hardness materials and very high intricate geometries. Methods like peening which may involve sand blasting on the surface, shock peening, shot peening and cavitation peening are good for removing the residual stresses and hence, are preferred in improving the fatigue strength of the LPBF parts [51]. In contrast to these methods, Laser based surface modification and its post processing techniques have advantages of creating smoother surface, and it can be done in situ during the LPBF process.

2.4.3.5 *Heat Treatment*

Heat treatment plays a pivotal role in enhancing the properties of materials produced through additive manufacturing, particularly for techniques like laser powder bed fusion (L-PBF). Various methods, such as stress relief annealing, recrystallization, low-temperature solution treatments, and aging, are employed to fine-tune the material's performance. These treatments are essential because the inherent properties of additively manufactured components differ markedly from those of traditional bulk materials. Even subtle shifts in parameters such as heating temperature, duration of treatment, sequence of processes, and cooling rates can lead to significant transformations in microstructure and mechanical behavior. Consequently, meticulous management of the heat treatment parameters is crucial for achieving optimal mechanical strength, microstructural integrity, and electrochemical stability. Moreover, properly executed heat treatments can effectively mitigate the negative influences of surface roughness on the fatigue resistance of L-PBF components. This not only enhances the longevity and reliability of these parts but also opens avenues for their application in demanding engineering contexts. By harnessing the potential of advanced heat treatment techniques, manufacturers can unlock the full capabilities of additively manufactured materials, ensuring they meet the rigorous demands of modern industries [52].

2.4.3.6 *Polishing*

Polishing is a post-processing technique that enhances a part's surface finish by removing imperfections and roughness [53]. It can be automated or hand-polished, depending on the part's complexity and production volume, and can be classified as contact or non-contact. Contact-type polishing involves physical contact between a polishing tool and workpiece surface, typically using pressure to remove material, improve surface finish, or achieve desired characteristics, and includes processes like mechanical polishing, grinding, buffing, and lapping.

- Surface preparation including cleaning of surface to remove any contaminants, oils and residues.
- Selection of abrasives based on specific requirements of the workpiece material and desired surface finish.
- Application of abrasive which can be manual or using polishing machine, in the form of abrasive powder, paste or slurries.

- Polishing action include movement in a controlled manner to create relative motion.
- Progression of grit sizes, which involves a series of grit sizes or abrasive grades that are used sequentially to refine the surface finish gradually.
- Lubrication and cooling are required to reduce friction, prevent overheating and debris removal.
- Inspection and repeat until the desired surface finish is obtained.

Material removal involves adjusting parameters like applied pressure, polishing speed, and duration to achieve desired material removal and surface quality. It offers versatility, precise control over surface finish, and removal of surface imperfections. However, it has limitations such as material removal affecting dimensional accuracy, difficulty achieving high surface flatness or parallelism, difficulty accessing complex areas, heat generation due to friction, and labor-intensive, especially for complex parts [54].

Mechanical polishing has limitations, leading to alternative techniques like chemical polishing, electropolishing, vibratory finishing, abrasive flow machining (AFM), and other non-contact methods. Chemical polishing is used for complex geometries, electropolishing eliminates imperfections, vibratory finishing is effective for delicate features, and AFM addresses complex geometries and hard-to-reach areas. Non-contact polishing is a method that uses energy or chemicals to modify the surface of a part without physical contact, preserving the workpiece's integrity and allowing precise control over parameters [44]. It is suitable for various materials like metals, ceramics, polymers, and composites and can be easily integrated into automated systems, increasing productivity and cost-effectiveness. However, it has limitations such as limited bulk material removal, high initial equipment setup costs, and the potential for surface roughness, microcracks, or heat-affected zones. Surface reflectivity, surface cleanliness, and material composition also impact its effectiveness. Additional process optimization and control may be required to achieve consistent results across different materials or surface conditions. Non-contact polishing may also have limitations when working with heat-sensitive materials prone to thermal damage or distortion. Both contact and non-contact polishing processes have advantages and disadvantages and are used in various industries and applications [55].

2.4.3.7 *Laser Polishing*

Laser polishing is a non-contact process that uses laser energy to melt and reflow material, creating a molten layer. This molten material flows and levels out, smoothing irregularities and filling gaps. The process relies on precise control of laser energy to achieve this, and heat input must be carefully regulated to prevent thermal damage. Process optimization ensures the desired surface finish while maintaining dimensional accuracy, including laser parameters and scanning strategies [56]. Laser polishing process with various parameters affecting the process is depicted in Figure 2.5.

2.4.3.8 *Mechanism of Laser Polishing*

Laser polishing is a technique that employs a high-powered laser beam to interact directly with the surface of a material, leading to localized melting and subsequent reflow of the material, as illustrated in Figure 2.6. This process can be broken down into several key stages:

2.4.3.8.1 Absorption of Laser Energy

High absorptivity materials absorb more energy, while low absorption may require surface modifications or coatings. Absorption mechanisms can be classified into electronic, vibrational, and free electron transitions. Electronic transitions convert laser energy into heat, while vibrational

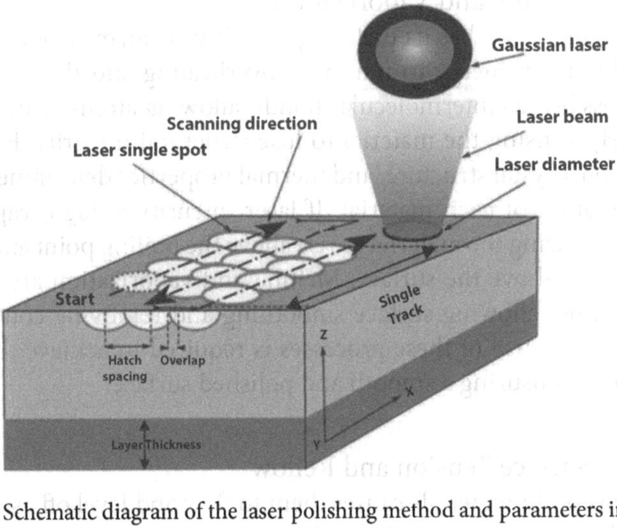

Figure 2.5 Schematic diagram of the laser polishing method and parameters involved [62].

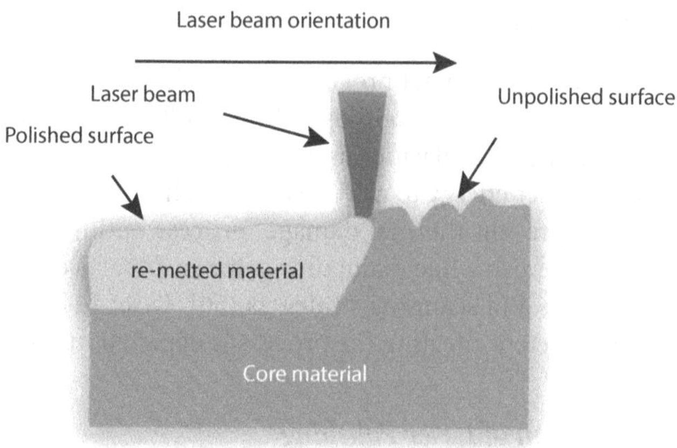

Figure 2.6 Mechanism of laser polishing [62].

transitions cause molecules or atoms to vibrate, converting energy into heat. Free electron transitions involve laser photons interacting with free electrons in materials, gaining energy and moving them to higher states. Material composition, structure, surface conditions, and laser beam wavelength affect energy absorption [57]. Different materials have different absorption properties, and proper laser parameter selection, including wavelength, power, pulse duration, and spot size, is crucial for efficient laser processing techniques.

2.4.3.8.2 Melting and Vaporization

When a material reaches its melting point, it transforms from a solid to a liquid using laser energy to minimize overheating and thermal damage. This process breaks intermolecular bonds, allowing atoms or molecules to move freely, causing the material to lose structural integrity. Factors like composition, crystal structure, and thermal properties determine the melting temperature of each material. If laser intensity is high, vaporization may occur, causing the material to rise above the boiling point and forming a vapor cloud above the surface. Melting and vaporization are crucial in laser polishing, allowing surface smoothing and removing contaminants [58]. Precise control of these processes is required to achieve the desired surface finish, ensuring a smooth and polished surface.

2.4.3.8.3 Surface Tension and Reflow

Surface tension in materials causes them to flow and level off as they melt, seeking a lower energy state. This reflows procedure smoothens the surface and removes surface abnormalities like scratches, pits, and roughness. Surface tension is a liquid property that causes the surface to behave like a stretched elastic membrane. In laser polishing, surface tension is essential, as the molten material absorbs laser energy and reaches its melting point. These forces reduce surface area, forcing molten material to flow and level out. Surface abnormalities are smoothened as the material moves from higher curvature to lower curvature areas [59]. Reflow is the process of moving and leveling molten material during laser polishing. It involves the material filling gaps, dips, and imperfections, redistributing itself to achieve a smoother surface—factors like viscosity, surface tension, and local surface shape influence reflow behavior. Laser parameters like power, scanning speed, and pulse duration must be optimized to prevent excessive material spreading. Surface tension and reflow are crucial for achieving a smooth surface. Material composition, surface energy, and thermal conditions influence the material's ability to flow and redistribute itself. Proper control and optimization of laser parameters are necessary for achieving the desired finish.

2.4.3.8.4 Marangoni Effect

The Marangoni effect is a fluid dynamics phenomenon that describes fluid flow due to surface tension gradients along a liquid's surface. It occurs when surface tension is not uniform across the liquid, causing it to move. Surface tension is a force that reduces a liquid's surface area, forming a distinct surface. A surface tension gradient is formed when temperature, concentration, or composition changes cause a difference in surface tension along the liquid surface. Cohesive forces on liquid molecules vary at different points on the surface, causing mass transfer. The difference in pressure along the liquid surface is caused by the concentration gradient, which causes the liquid to flow from areas with higher surface tension to areas with lower surface tension [60]. Understanding the Marangoni effect is crucial for studying and controlling fluid behavior in various applications. Factors affecting the Marangoni effect include the material's composition, surface energy, thermal properties, and laser parameters like power, pulse duration, and scanning speed.

2.4.3.8.5 Solidification and Surface Finishing

Laser polishing involves the rapid solidification of molten material due to heat dissipation and ambient temperature. The solidification rate impacts this process, which in turn affects the surface texture and polishing quality. The material then cools as the laser energy is removed, influenced by thermal properties, the environment, and heat dissipation mechanisms. The solidification front, the boundary between solid and liquid phases, moves through the molten material as it cools. Laser parameters can be adjusted to improve the surface finish, including intensity, pulse duration, scanning speed, and beam overlap. The energy input can be accurately regulated to minimize material removal, thermal damage, or distortion [61].

2.4.3.9 Type of Laser Polishing

Laser Polishing can be broadly classified into two types: Continuous, pulse, hybrid as shown in Table 2.2.

2.4.3.9.1 Continuous Laser Polishing

A continuous laser beam with a consistent power level is utilized in this technique to heat the material's surface. Localized melting is caused by the laser energy, and surface tension effects aid in the smoothening and reflow of the melted material, resulting in better surface quality [63]. Laser beam can be used in circular spot or in line focused.

2.4.3.9.2 Pulse Laser Polishing

Short-duration laser pulses are used in pulsed laser polishing to modify the surface of the workpiece. The pulsed laser energy causes rapid heating and melting of the material, which is then solidified and resolidified. It helps produce fine surface finishes and remove surface flaws.

2.4.3.9.3 Hybrid Laser Polishing

Hybrid laser polishing is a surface treatment method that combines laser polishing with other complementary methods to improve a workpiece's overall surface finish. It combines laser energy with other processes like mechanical polishing, chemical etching, or abrasive jet machining. Combining these techniques enables improved surface quality control, the removal of surface defects, and achieving desired surface characteristics (Temmler *et al.*, 2012).

2.4.3.10 Advantage, Limitation and Application of Laser Polishing

Laser polishing offers precise material removal and surface modification, making it ideal for intricate parts, fragile materials, or critical surfaces.

Table 2.2 Type of laser polishing: advantage and limitation.

Laser polishing type	Advantage	Limitation
Continuous Wave (CW)	High precision control, suitable for small to medium area, Fine surface finish	Thermal damage by high heat accumulation, Time consuming for complex part
Pulsed Wave (PW)	Suitable for small area, less thermal damage	Surface finish may not be able to reach desired level, Limited application on complex surface
Hybrid Wave	Enhanced material removal rate, Versatility for different materials	Process optimization required, Complexity of combining multiple techniques

It is non-contact, material-efficient, and environmentally friendly due to its nonabrasive nature and reduced need for post-polishing cleaning methods, reducing waste and conserving resources [64]. Laser polishing has limitations, including its effectiveness on flat or gently curved surfaces, consistency on intricate or heavily curved surfaces, and shallow roughness values. It's best for small to medium-sized objects, as large-scale components require precise control over laser beam movement and heat distribution, making the procedure complex and time-consuming. Additionally, part shape can affect its effectiveness, making uniform polishing difficult. A laser beam's speed and scanning influence processing time, making laser polishing more cost-effective in high-volume production. However, high-powered laser beams can be dangerous if not appropriately managed, necessitating safety measures. Laser polishing can be enhanced with other finishing techniques or optimized process parameters, with ongoing research and advancements aiming to expand its capabilities. Laser polishing is crucial in various industries, including aerospace, medical implants, dental prosthetics, automotive, electronic devices, and microelectromechanical systems. It improves aerodynamic performance, fatigue resistance, and surface finish while enhancing biocompatibility and patient comfort. It also enhances wear resistance [65].

2.5 Surface Metrology and Characterization

Surface metrology is the study and practice of measuring, characterizing, and analyzing surface properties and features. It assesses surface topography, texture, and geometric aspects at various scales, providing quantitative and qualitative data on surface quality, roughness, form, and other surface-related parameters. Surface metrology uses various measurement techniques, tools, and analysis procedures, including profilometers, atomic force microscopy (AFM), scanning electron microscopy (SEM), interferometry, and optical imaging [66, 67]. Surface parameters such as roughness, waviness, form error, texture, wear, flatness, and other geometric aspects are measured [68]. Surface metrology allows measuring and managing surface quality during production processes, ensuring specification adherence, and optimizing product performance. It aids in evaluating surface functioning, durability, and dependability and guiding surface treatment, finishing, and quality improvement activities. Data and information obtained are used for various purposes, including quality control, process optimization, surface characterization, wear analysis, failure investigation, material selection, and design validation [69]. Producers and researchers can comprehend surface qualities, identify deviations from essential requirements, and make informed decisions for improvement or problem resolution.

2.5.1 Surface Integrity and Topography

Surface integrity is the quality and condition of a surface resulting from manufacturing processes and material interactions, crucial for ensuring structural performance, dependability, and lifespan of engineered components. Surface properties, including roughness and residual stresses, can influence material corrosion resistance, and control can improve this resistance and protect against environmental degradation. Surface integrity is crucial for joining and bonding processes, as poor conditions can compromise the strength and integrity of welded joints, adhesive bonds, and surface coatings, leading to premature failure under cyclic loading conditions [70]. Surface topography is a three-dimensional representation of a surface's texture attributes, measured using profilometers, optical profilers, and atomic force microscopy. Optical profilers capture high-resolution 3D images, while AFM scans the surface with a sharp tip, characterizing micro-roughness and nanoscale shape. White light interferometry measures surface height variations by measuring the distance between a surface and an instrument using white light interference [71].

2.5.2 Surface Texture

Surface texture refers to a surface's physical attributes, including roughness and waviness, which can affect a material's utility, performance, and aesthetics. Areal surface texture is characterized by three-dimensional surfaces, measured and recorded at distinct coordinates. The ISO 25178 standard series outlines rules for measuring, characterizing, and evaluating surface texture in three dimensions [72]. The ISO 25178 series includes these parts: ISO 25178-1:2016 - Geometrical product specifications (GPS) - Surface texture: Areal - Part 1: Indication of surface texture; ISO 25178-2:2012 - Geometrical product specifications (GPS) - Surface texture: Areal - Part 2: Terms, definitions, and surface texture parameters; ISO 25178-3:2012 - Geometrical product specifications (GPS) - Surface texture: Areal - Part 3: Calibration and measurement requirements; ISO 25178-4:2010 - Geometrical product specifications (GPS) - Surface texture: Areal - Part 4: Rules and procedures for the assessment of surface texture; ISO 25178-6:2010 - Geometrical product specifications (GPS) - Surface texture: Areal - Part 6: Classification of methods for measuring surface texture [73].

The four components of surface texture are form, waviness, roughness, and microroughness.

2.5.2.1 Form

Form refers to a surface's overall contour, characterized by long wavelengths. It denotes deviations from the proper shape and must be removed when analyzing surface texture. Polynomial functions are used to calculate form, crucial in industries like automotive, aerospace, precision engineering, and metrology-assisted manufacturing processes.

2.5.2.2 Waviness

Waviness is larger-scale deviations or undulations on a surface, typically longer than surface roughness. It is a periodic or repetitive variation in the surface profile, occurring at a lower frequency than fine-scale irregularities. Waviness is crucial in applications like mating or sealing surfaces, optical components, and parts that must fit precisely. Excessive waviness can cause poor contact, leakage, and optical performance.

2.5.2.3 Roughness

Surface roughness refers to microscopic irregularities on an object's surface, affecting its texture, smoothness, and functionality. It is crucial in various industries like automotive, aerospace, electronics, medical devices, and precision engineering for quality control, process optimization, and performance. Controlled roughness is needed for adhesion, lubrication, or bonding. Surface roughness measurements are essential for aesthetics, reducing friction and wear, and promoting adhesion, lubrication, and surface bonding.

2.5.2.4 Micro Roughness

Microroughness, the finest component of surface texture, is influenced by the material's structure, microscopic features, and sampling noise. It significantly impacts a surface's appearance and functionality. Removing microroughness allows for a distinct separation of roughness, useful in practical applications. Microroughness is less important in larger-scale variations and functional characteristics of surfaces.

2.5.2.5 Filtering

Filtering is a technique used in metrology to separate surface topography components based on frequency or wavelength. It aims to isolate specific surface features and improve data analysis. The distribution of waviness and roughness components is based on a specified cut-off value, with waviness components assigned frequencies above the cut-off and roughness components assigned frequencies below [74]. The quality of separation depends on the filter type and cut-off value. Techniques like Gaussian, band-pass, and Fourier transforms can achieve desired separation. The standard values for cutoff lengths are distributed in a geometric series in ISO 16610-21 for profiles and ISO 16610-61 for surfaces. Standard values of 0.08 mm, 0.25 mm, 0.8 mm, 2.5 mm, and 8 mm are mentioned. These values provide various cutoff length options based on the desired level of separation between waviness and roughness components [75]. Custom cutoff length values can also be used if the standard options do not meet specific requirements. It allows the filtering process to be tailored to the specific characteristics of the surface being analyzed. Filtering a profile or surface involves deleting data from both sides, which can have negative consequences. It is common to subtract a length equal to the cutoff from both edges to focus on the appropriate wavelength range and avoid distortion. However, retaining the original data up to the edges may be necessary or desired, especially when analyzing surfaces with critical features [76].

ISO 16610 is a set of standards developed by the International Organization for Standardization (ISO) for surface profile data filtering and analysis. It includes guidelines, definitions, and specifications for various surface metrology topics, such as filtering techniques, characterization parameters, and analysis methods. The ISO 16610 series is divided into parts, each covering a different aspect of surface profile analysis, such as Gaussian filters, robust profile filters, and median filters. Different types of filters include gaussian, robust gaussian, spline, morphological, and band-pass filters. The gaussian filter and robust gaussian filter are most commonly used, which can be described in more detail as:

2.5.2.5.1 Gaussian Filter

The Gaussian filter is a popular digital filter in signal processing and image analysis, used to reduce noise while preserving essential features. It works by convolving a signal or image with a Gaussian kernel, a matrix with Gaussian weights. The filter's kernel size and standard deviation determine the amount of smoothing applied, with larger sizes enhancing smoothing effects. However, it can blur fine details and edges, introducing unwanted artifacts called ringing or blurring effects.

2.5.2.5.2 Robust Gaussian Filter

The Robust Gaussian filter is an advanced version of the traditional Gaussian filter that enhances data analysis capabilities and statistical accuracy. It uses an iterative statistical method to generate a mean line, minimizing the impact of outliers or noise. This method works well when outliers or non-Gaussian noise, resulting in smoother data analysis results. It can be applied to various data types, including profiles and surfaces, but has limitations such as increased computational time and complexity due to the iterative process.

2.5.3 Surface Texture Parameters

ISO 25178-1:2016 is a standard that specifies surface texture parameters and their corresponding values, which are crucial for accurately representing desired surface texture characteristics. These parameters are quantitative measurements or calculations that describe various surface topography aspects, including roughness, waviness, form, and other pertinent characteristics. The importance of specifying appropriate values or ranges for each parameter is emphasized in ISO 25178-1. Organizations can facilitate accurate interpretation, analysis, and evaluation of surface texture by clearly specifying parameter values or ranges, allowing for effective communication, and ensuring that desired surface characteristics are achieved in the final product. Areal parameters can be classified broadly according to ISO25178-2:2030, including height, functions, and related parameters, spatial, feature, hybrid, and miscellaneous parameters.

Profile method parameters can be classified broadly according to ISO 4287:1997, ISO 12085:1996, and ISO 13565-2:1996, including amplitude parameters, amplitude average parameters, spacing parameters, material ratio curves, and related parameters, surface parameters with stratified functional properties, and morph parameters. In summary, ISO 25178-1:2016 is a crucial standard for accurately representing surface texture characteristics and ensuring accurate interpretation, analysis, and evaluation of surface texture [77].

2.5.4 Significance of L-PBF Surface Roughness for Various Applications

Laser powder bed fusion (L-PBF) has several commercial uses, where surface roughness is a critical factor in determining performance. In the medical sector, the roughness of implants is key to ensuring proper integration with biological tissues. For components subjected to dynamic loads, such as in aerospace or automotive applications, rough surfaces can reduce fatigue life, increasing the risk of failure under repeated stress. Moreover, in thermal management systems, the roughness of internal cooling channels directly affects both heat transfer efficiency and fluid flow, making it a vital consideration for maintaining system effectiveness [78]. L-PBF technologies have unlocked new possibilities in several fields, particularly in medical, aerospace, and thermal management systems. In healthcare, L-PBF allows for the creation of implants with controlled surface roughness, improving biological compatibility and promoting tissue integration. The capability to design patient-specific implants tailored to unique anatomical structures further enhances treatment options. Additionally, in applications subject to dynamic loads, surface roughness plays a key role in reducing the risk of fatigue failure. For thermal applications, the roughness within internal cooling channels is vital for optimizing heat transfer efficiency and fluid flow, leading to improved overall performance [79]. Wear debris, commonly generated through frictional processes, is a critical factor in osteolysis, a condition where bone is broken down by osteoclasts without proper regeneration. These particles can set off a series of harmful biological responses. The typical surface roughness of L-PBF components is in the micrometer range, which may help in promoting cell adhesion and bone integration (osseointegration). However, this same roughness can also contribute to particle wear and detachment, leading to inflammatory reactions. Once implanted, L-PBF devices encounter a complex biological environment, where proteins quickly adhere to the surface, forming a matrix composed of blood components like plasma and serum, followed by platelet adhesion. The fatigue performance of dynamically loaded L-PBF structures is strongly linked to surface roughness, as rougher surfaces generally reduce fatigue life. While rough surfaces can aid in promoting bone growth for implants, they may also accelerate fatigue failure in components exposed to repeated stress. In studies on the fatigue behavior of L-PBF-produced Ti6Al4V parts built at a 45-degree angle, it was observed that cracks tend to initiate on rougher, downward-facing surfaces, highlighting the importance of surface texture in fatigue resistance [80].

The local surface roughness of components produced through L-PBF has a profound impact on thermal conduction, as it alters both the area available for conductive heat transfer and the turbulence of the fluid flow. Enhanced turbulence improves the mixing of fluids, which in turn boosts convective heat transfer efficiency. However, the increased surface roughness can also elevate the friction factor beyond that seen in traditional designs. As a result, L-PBF presents a novel manufacturing approach for hydraulic and thermofluidic systems, particularly suited for designs that demand intricate geometries and small production runs. The importance of surface roughness in fluid applications lies in its influence on the effective cross-sectional area of fluid channels, as well as its effects on particle removal and thermal conductivity. Factors such as the choice of material, the design of the channel geometry, and the orientation of the powder bed during the L-PBF process collectively determine the surface roughness, ultimately affecting the overall performance of these fluidic systems [81].

2.6 Conclusions

Additive manufacturing (AM) represents a transformative shift in the manufacturing landscape, fundamentally redefining the design and production of components. Its capability to produce intricate, customized parts with minimal waste distinguishes it from traditional methods, offering unparalleled design freedom and streamlined production processes.

This is particularly evident in high-stakes industries such as aerospace, automotive, and healthcare, where precision and efficiency are paramount.

However, to fully realize the benefits of AM, effective post-processing techniques are essential for optimizing the properties and performance of additively manufactured parts. This review has illuminated a range of subprocesses—including powder removal, recycling, finishing methods, machining enhancements, and heat treatments—that collectively refine the final attributes of printed components. Each technique, whether mechanical, chemical, or laser-based, addresses critical challenges related to surface roughness, dimensional accuracy, and mechanical strength, underscoring the complexity of materials and geometries inherent in AM.

As the adoption of AM technologies expands, mastering post-processing strategies will be vital to meet stringent performance and aesthetic standards. Ongoing advancements in these techniques are anticipated to enhance efficiency, enabling greater customization and seamless integration into production workflows. Furthermore, the interplay between AM and emerging smart technologies, such as artificial intelligence and the Internet of Things, promises to revolutionize supply chains and manufacturing strategies, facilitating on-demand production and reduced lead times.

Ultimately, the continuous evolution of post-processing methods not only enhances the functionality and appeal of additive manufactured parts but also propels innovation across multiple sectors. Future research should concentrate on optimizing these processes to lower costs and processing times, thereby fostering broader applications and accelerating the adoption of AM technologies. The expansive potential of AM not only redefines existing manufacturing paradigms but also paves the way for a future characterized by efficiency, customization, and sustainability, heralding a new era in production that transcends traditional limitations.

Acknowledgments

BBS, AK, and AJ acknowledge the Ministry of Human Resource Development, Government of India, for the graduate fellowship. All authors are grateful to the research scholars and staff of Direct Digital Manufacturing Lab, Department of Mechanical Engineering, and Fiber Optics Systems Lab, Department of Electronics and Electrical Communication Engineering, Indian Institute of Technology, Kharagpur, for their immense help throughout.

References

1. Gibson, I., *et al.*, *Additive manufacturing technologies*, vol. 17, Springer, Switzerland, 2021.
2. Gibson, I., Rosen, D.W., Stucker, B., Development of Additive Manufacturing Technology, in: *Additive Manufacturing Technologies: Rapid Prototyping to Direct Digital Manufacturing*, pp. 36–58, Springer US, Boston, MA, 2010, doi: 10.1007/978-1-4419-1120-9_2.
3. ISO/ASTM 52900, Standard Terminology for Additive Manufacturing General Principles Part 1: Terminology, in: *American Society for Testing of Materials*, 2015.
4. Bourell, D.L., Leu, M., Rosen, D., *Roadmap for Additive Manufacturing: Identifying the Future of Freeform Processing*, University of Texas at Austin Labratory for Freeform Fabrication Advanced Manufacturing Center, Austin, Texas, USA, 2009, [Online]. Available: https://books.google.co.in/books?id=6IwzzwEACAAJ.
5. Kern, J., Additive Manufacturing, in: *The Digital Transformation of Logistics*, pp. 41–60, John Wiley & Sons, Ltd, Hoboken, New Jersey, USA, 2021, doi: https://doi.org/10.1002/9781119646495.ch4.
6. Wong, K.V. and Hernandez, A., A Review of Additive Manufacturing. *ISRN Mech. Eng.*, 2012, 208760, 2012. doi: 10.5402/2012/208760.
7. Attaran, M., The rise of 3-D printing: The advantages of additive manufacturing over traditional manufacturing. *Bus. Horiz.*, 60, 5, 677–688, 2017. doi: https://doi.org/10.1016/j.bushor.2017.05.011.
8. Guo, N. and Leu, M.C., Additive manufacturing: technology, applications and research needs. *Front. Mech. Eng.*, 8, 3, 215–243, 2013. doi: 10.1007/s11465-013-0248-8.
9. Kruth, J.P., Material Incress Manufacturing by Rapid Prototyping Techniques. *CIRP Ann.*, 40, 2, 603–614, 1991. doi: https://doi.org/10.1016/S0007-8506(07)61136-6.
10. F., O.J., G., J.D., P., J.R.B., Petrovic, V., Gonzalez, J.V.H., Griñan, L.P., Additive layered manufacturing: sectors of industrial application shown through case studies. *Int. J. Prod. Res.*, 49, 4, 1061–1079, 2011. doi: 10.1080/00207540903479786.
11. Zhai, Y., Lados, D.A., LaGoy, J.L., Additive Manufacturing: Making Imagination the Major Limitation. *JOM*, 66, 5, 808–816, 2014. doi: 10.1007/s11837-014-0886-2.
12. Azarian, M., Yu, H., Solvang, W.D., Integrating Additive Manufacturing into a Virtual Industry 4.0 Factory BT, in: *Advanced Manufacturing and Automation X*, pp. 587–594, 2021.
13. Haleem, A. and Javaid, M., Additive Manufacturing Applications in Industry 4.0: A Review. *J. Ind. Integr. Manage.*, 04, 04, 1930001, Aug. 2019. doi: 10.1142/S2424862219300011.

14. Kubášová, K., *et al.*, A Review on Additive Manufacturing Methods for NiTi Shape Memory Alloy Production. *Materials*, 17, 6, 1248, 2024. doi: 10.3390/ma17061248.
15. Samal, B.B., Varshney, S.K., Kumar, C.S., Four-dimensional (4D) printing through FDM: Effect of infill density and bed temperature on shape memory properties in different thermo-mechanical programming conditions. *J. Mech. Sci. Technol.*, 38, 8, 4313–4319, 2024. doi: 10.1007/s12206-024-0727-3.
16. Ribeiro, I., *et al.*, Framework for Life Cycle Sustainability Assessment of Additive Manufacturing. *Sustainability*, 12, 3, 929, 2020. doi: 10.3390/su12030929.
17. Hague, S.M.R. and Saleh, N., Material and design considerations for rapid manufacturing. *Int. J. Prod. Res.*, 42, 22, 4691–4708, 2004. doi: 10.1080/00207840410001733940.
18. Shanthar, R., Chen, K., Abeykoon, C., Powder-Based Additive Manufacturing: A Critical Review of Materials, Methods, Opportunities, and Challenges. *Adv. Eng. Mater.*, 25, 19, 2300375, 2023. doi: https://doi.org/10.1002/adem.202300375.
19. King, W.E., *et al.*, Observation of keyhole-mode laser melting in laser powder-bed fusion additive manufacturing. *J. Mater. Process. Technol.*, 214, 12, 2915–2925, 2014. doi: https://doi.org/10.1016/j.jmatprotec.2014.06.005.
20. Gu, D.D. and Shen, Y.F., Development and characterisation of direct laser sintering multicomponent Cu based metal powder. *Powder Metall.*, 49, 3, 258–264, 2006. doi: 10.1179/174329006X95662.
21. de Moura Nobre, R., *et al.*, Role of laser powder bed fusion process parameters in crystallographic texture of additive manufactured Nb–48Ti alloy. *J. Mater. Res. Technol.*, 14, 484–495, 2021. doi: https://doi.org/10.1016/j.jmrt.2021.06.054.
22. Bourell, D., Frazier, W., Kuhn, H., Seifi, M., *ASM Handbook: Volume 24 Additive Manufacturing Processes*, vol. 24, ASM International, Materials Park, Ohio, USA, 2020.
23. Lupi, F., Pacini, A., Lanzetta, M., Laser powder bed additive manufacturing: A review on the four drivers for an online control. *J. Manuf. Processes*, 103, 413–429, 2023. doi: https://doi.org/10.1016/j.jmapro.2023.08.022.
24. Sghaier, T.A.M., Sahlaoui, H., Mabrouki, T., Sallem, H., Rech, J., Selective Laser Melting of Stainless-Steel: A Review of Process, Microstructure, Mechanical Properties and Post-Processing treatments. *Int. J. Mater. Form.*, 16, 4, 41, 2023. doi: 10.1007/s12289-023-01769-w.
25. Liu, S. and Shin, Y.C., Additive manufacturing of Ti6Al4V alloy: A review. *Mater. Des.*, 164, 107552, 2019. doi: https://doi.org/10.1016/j.matdes.2018.107552.
26. van Noort, R., The future of dental devices is digital. *Dent. Mater.*, 28, 1, 3–12, 2012. doi: https://doi.org/10.1016/j.dental.2011.10.014.
27. Chia, H.N. and Wu, B.M., Recent advances in 3D printing of biomaterials. *J. Biol. Eng.*, 9, 1, 4, 2015. doi: 10.1186/s13036-015-0001-4.

28. Charles, A., Hofer, A., Elkaseer, A., Scholz, S.G., Additive Manufacturing in the Automotive Industry and the Potential for Driving the Green and Electric Transition, in: *Sustainable Design and Manufacturing*, pp. 339–346, 2022.
29. Al Rashid, A., Khan, S.A., Al-Ghamdi, S.G., Koç, M., Additive manufacturing: Technology, applications, markets, and opportunities for the built environment. *Autom. Constr.*, 118, 103268, 2020. doi: https://doi.org/10.1016/j.autcon.2020.103268.
30. Salmi, M., Additive Manufacturing Processes in Medical Applications. *Materials*, 14, 1, 191, 2021. doi: 10.3390/ma14010191.
31. Moeinfar, K., Khodabakhshi, F., Kashani-bozorg, S.F., Mohammadi, M., Gerlich, A.P., A review on metallurgical aspects of laser additive manufacturing (LAM): Stainless steels, nickel superalloys, and titanium alloys. *J. Mater. Res. Technol.*, 16, 1029–1068, 2022. doi: https://doi.org/10.1016/j.jmrt.2021.12.039.
32. Yasa, E., Kruth, J.-P., Deckers, J., Manufacturing by combining Selective Laser Melting and Selective Laser Erosion/laser re-melting. *CIRP Ann.*, 60, 1, 263–266, 2011. doi: https://doi.org/10.1016/j.cirp.2011.03.063.
33. Gong, G., *et al.*, Research status of laser additive manufacturing for metal: a review. *J. Mater. Res. Technol.*, 15, 855–884, 2021. doi: https://doi.org/10.1016/j.jmrt.2021.08.050.
34. Oliveira, J.P., LaLonde, A.D., Ma, J., Processing parameters in laser powder bed fusion metal additive manufacturing. *Mater. Des.*, 193, 108762, 2020. doi: https://doi.org/10.1016/j.matdes.2020.108762.
35. Khairallah, S.A., Anderson, A.T., Rubenchik, A., King, W.E., Laser powder-bed fusion additive manufacturing: Physics of complex melt flow and formation mechanisms of pores, spatter, and denudation zones. *Acta Mater.*, 108, 36–45, 2016. doi: https://doi.org/10.1016/j.actamat.2016.02.014.
36. Sheshadri, R., *et al.*, Experimental investigation of selective laser melting parameters for higher surface quality and microhardness properties: taguchi and super ranking concept approaches. *J. Mater. Res. Technol.*, 14, 2586–2600, 2021. doi: https://doi.org/10.1016/j.jmrt.2021.07.144.
37. Spierings, A.B., Herres, N., Levy, G., Influence of the particle size distribution on surface quality and mechanical properties in AM steel parts. *Rapid Prototyping J.*, 17, 3, 195–202, Jan. 2011. doi: 10.1108/13552541111124770.
38. Kumar, A., Samal, B.B., Nath, A.K., Kumar, C.S., Laser surface modification of metal additive manufactured parts : A case study of *ex-situ* and *in-situ* methodology, in: *Surface Engineering*, pp. 121–144, CRC Press, Boca Raton, Florida, USA, 2022, doi: 10.1201/9781003319375-5.
39. Kumar, A., Ramadas, H., Samal, B.B., Kumar, C.S., Nath, A.K., Laser polishing of cobalt chrome alloy fabricated by laser powder bed fusion process: design of experiment-based approach for reducing surface roughness, in: *The International Journal of Advanced Manufacturing Technology*, 1245–1264, 2024, doi: 10.1007/s00170-024-14101-w.

40. Yasa, E., Deckers, J., Kruth, J., The investigation of the influence of laser re-melting on density, surface quality and microstructure of selective laser melting parts. *Rapid Prototyping J.*, 17, 5, 312–327, Jan. 2011. doi: 10.1108/13552541111156450.
41. Calignano, F., Manfredi, D., Ambrosio, E.P., Iuliano, L., Fino, P., Influence of process parameters on surface roughness of aluminum parts produced by DMLS. *Int. J. Adv. Manuf. Technol.*, 67, 9, 2743–2751, 2013. doi: 10.1007/s00170-012-4688-9.
42. Boschetto, A., Bottini, L., Macera, L., Veniali, F., Post-Processing of Complex SLM Parts by Barrel Finishing. *Appl. Sci.*, 10, 4, 1382, 2020. doi: 10.3390/app10041382.
43. Lanzutti, A. and Marin, E., The Challenges and Advances in Recycling/Re-Using Powder for Metal 3D Printing: A Comprehensive Review. *Metals*, 14, 8, 886, 2024. doi: 10.3390/met14080886.
44. Maleki, E., Bagherifard, S., Bandini, M., Guagliano, M., Surface post-treatments for metal additive manufacturing: Progress, challenges, and opportunities. *Addit. Manuf.*, 37, 101619, 2021. doi: https://doi.org/10.1016/j.addma.2020.101619.
45. Hassanin, H., Elshaer, A., Benhadj-Djilali, R., Modica, F., Fassi, I., Surface Finish Improvement of Additive Manufactured Metal Parts, in: *Micro and Precision Manufacturing*, K. Gupta (Ed.), pp. 145–164, Springer International Publishing, Cham, 2018, doi: 10.1007/978-3-319-68801-5_7.
46. Shiyas, K.A. and Ramanujam, R., A review on post processing techniques of additively manufactured metal parts for improving the material properties. *Mater. Today Proc.*, 46, 1429–1436, 2021. doi: https://doi.org/10.1016/j.matpr.2021.03.016.
47. Chaghazardi, Z. and Wüthrich, R., Review—Electropolishing of Additive Manufactured Metal Parts. *J. Electrochem. Soc.*, 169, 4, 43510, Apr. 2022. doi: 10.1149/1945-7111/ac6450.
48. Rosa, B., Mognol, P., Hascoët, J., Laser polishing of additive laser manufacturing surfaces. *J. Laser Appl.*, 27, S2, S29102, 2015. doi: 10.2351/1.4906385.
49. Jolly, A., Vitry, V., Azar, G.T., Guaraldo, T.T., Cobley, A.J., Surface Defect Mitigation of Additively Manufactured Parts Using Surfactant-Mediated Electroless Nickel Coatings. *Materials*, 17, 2, 406, 2024. doi: 10.3390/ma17020406.
50. Wu, Y., Lu, Y., Zhao, M., Bosiakov, S., Li, L., A Critical Review of Additive Manufacturing Techniques and Associated Biomaterials Used in Bone Tissue Engineering. *Polymers*, 14, 10, 2117, 2022. doi: 10.3390/polym14102117.
51. Peng, X., Kong, L., Fuh, J.Y.H., Wang, H., A Review of Post-Processing Technologies in Additive Manufacturing. *J. Manuf. Mater. Process.*, 5, 2, 38, 2021. doi: 10.3390/jmmp5020038.
52. Junker, D., Hentschel, O., Schmidt, M., Merklein, M., Investigation of Heat Treatment Strategies for Additively-Manufactured Tools of X37CrMoV5-1. *Metals*, 8, 10, 2018. doi: 10.3390/met8100854.

53. Kumbhar, N.N. and Mulay, A.V., Post Processing Methods used to Improve Surface Finish of Products which are Manufactured by Additive Manufacturing Technologies: A Review. *J. Inst. Eng. India Ser. C*, 99, 4, 481–487, 2018. doi: 10.1007/s40032-016-0340-z.
54. Boban, J., Ahmed, A., Jithinraj, E.K., Rahman, M.A., Rahman, M., Polishing of additive manufactured metallic components: retrospect on existing methods and future prospects. *Int. J. Adv. Manuf. Technol.*, 121, 1, 83–125, 2022. doi: 10.1007/s00170-022-09382-y.
55. Lei, J., Ge, Y., Liu, T., Wei, Z., Effects of Heat Treatment on the Microstructure and Mechanical Properties of Selective Laser Melting 316L Stainless Steel. *Shock Vib.*, 2021, 1, 6547213, 2021. doi: https://doi.org/10.1155/2021/6547213.
56. Kuisat, F., Lasagni, F., Lasagni, A.F., Smoothing additive manufactured parts using ns-pulsed laser radiation. *Prog. Addit. Manuf.*, 6, 2, 297–306, 2021. doi: 10.1007/s40964-021-00168-4.
57. Dutta Majumdar, J. and Manna, I., Laser processing of materials. *Sadhana*, 28, 3, 495–562, 2003. doi: 10.1007/BF02706446.
58. Hassan, A.F., El-Nicklawy, M.M., El-Adawi, M.K., Nasr, E.M., Hemida, A.A., El-Ghaffar, O.A.A., Heating effects induced by a pulsed laser in a semi-infinite target in view of the theory of linear systems. *Opt. Laser Technol.*, 28, 5, 337–343, 1996. doi: https://doi.org/10.1016/0030-3992(95)00083-6.
59. Ma, C., Zhao, J., Cao, C., Lin, T.-C., Li, X., Fundamental Study on Laser Interactions With Nanoparticles-Reinforced Metals—Part II: Effect of Nanoparticles on Surface Tension, Viscosity, and Laser Melting. *J. Manuf. Sci. Eng.*, 138, 12, 121002, 2016. doi: 10.1115/1.4033446.
60. Hibiya, T. and Ozawa, S., Marangoni Flow and Surface Tension of High Temperature Melts, in: *High-Temperature Measurements of Materials*, H. Fukuyama and X.X.X.Y. Waseda (Eds.), pp. 39–59, Springer Berlin Heidelberg, Berlin, Heidelberg, 2009, doi: 10.1007/978-3-540-85918-5_3.
61. Kumar, A., Saha, S., Kumar, C.S., Nath, A.K., Laser surface re-melting of additive manufactured samples with a line focused beam. *Mater. Today Proc.*, 26, 1221–1225, 2020. doi: https://doi.org/10.1016/j.matpr.2020.02.245.
62. Ali, M., Almotari, A., Algamal, A., Qattawi, A., Recent Advancements in Post Processing of Additively Manufactured Metals Using Laser Polishing. *J. Manuf. Mater. Process.*, 7, 3, 115, 2023. doi: 10.3390/jmmp7030115.
63. Krishnan, A. and Fang, F., Review on mechanism and process of surface polishing using lasers. *Front. Mech. Eng.*, 14, 3, 299–319, 2019. doi: 10.1007/s11465-019-0535-0.
64. Hong, J., Zhang, L., Zhang, Z., Huang, H., Yan, J., A comparative study on mechanical polishing and laser polishing of laser-nitrided Zr-based metallic glass surface. *Int. J. Adv. Manuf. Technol.*, 124, 3, 959–971, 2023. doi: 10.1007/s00170-022-10529-0.
65. Gisario, A., Barletta, M., Veniali, F., Laser polishing: a review of a constantly growing technology in the surface finishing of components made by additive

manufacturing. *Int. J. Adv. Manuf. Technol.*, 120, 3, 1433–1472, 2022. doi: 10.1007/s00170-022-08840-x.
66. Quinten, M., Optical Surface Metrology – Physical Basics, in: *A Practical Guide to Surface Metrology*, pp. 67–93, Springer International Publishing, Cham, 2019, doi: 10.1007/978-3-030-29454-0_4.
67. Quinten, M., Optical Surface Metrology: Methods, in: *A Practical Guide to Surface Metrology*, pp. 95–198, Springer International Publishing, Cham, 2019, doi: 10.1007/978-3-030-29454-0_5.
68. Quinten, M., Introduction to Surfaces and Surface Metrology, in: *A Practical Guide to Surface Metrology*, pp. 1–41, Springer International Publishing, Cham, 2019, doi: 10.1007/978-3-030-29454-0_1.
69. Archenti, A., Gao, W., Donmez, A., Savio, E., Irino, N., Integrated metrology for advanced manufacturing. *CIRP Ann.*, 73, 2, 639–665, 2024. doi: https://doi.org/10.1016/j.cirp.2024.05.003.
70. Thakur, A. and Gangopadhyay, S., State-of-the-art in surface integrity in machining of nickel-based super alloys. *Int. J. Mach. Tools Manuf.*, 100, 25–54, 2016. doi: https://doi.org/10.1016/j.ijmachtools.2015.10.001.
71. Greenwood, J.A., Three-dimensional surface topography: Stout & Blunt, Penton Press ISBN No. 1-8571-8026-7. *Tribol. Int.*, 37, 3, 289–290, 2004. doi: https://doi.org/10.1016/j.triboint.2003.10.001.
72. Franco, L.A. and Sinatora, A., 3D surface parameters (ISO 25178-2): Actual meaning of Spk and its relationship to Vmp. *Precis. Eng.*, 40, 106–111, 2015. doi: 10.1016/j.precisioneng.2014.10.011.
73. Tian, Y., Weckenmann, A., Hausotte, T., Schuler, A., He, B., *Measurement strategies in optical 3-D surface measurement with focus variation*, IMEKO-International Measurement Federation Secretariat, Krakow - Kielce, Poland, 2013.
74. Raja, J., Muralikrishnan, B., Fu, S., Recent advances in separation of roughness, waviness and form. *Precis. Eng.*, 26, 2, 222–235, 2002. doi: https://doi.org/10.1016/S0141-6359(02)00103-4.
75. Dobrzański, P. and Pawlus, P., A study of filtering techniques for areal surface topography assessment. *Proc. Inst. Mech. Eng., Part B: J. Eng. Manuf.*, 225, 11, 2096–2107, 2011. doi: 10.1177/0954405410397235.
76. Shao, Y., Xu, F., Chen, J., Lu, J., Du, S., Engineering surface topography analysis using an extended discrete modal decomposition. *J. Manuf. Processes*, 90, 367–390, 2023. doi: https://doi.org/10.1016/j.jmapro.2023.02.005.
77. I. O. for Standardization, ISO 4287-1997 Geometrical product specifications (GPS)-Surface texture: Profile method-Terms, definitions and surface texture parameters. *Int. Stand. ISO*, 1998, 25, 2007.
78. Ciochon, A., Kennedy, J., Leiba, R., Flanagan, L., Culleton, M., The impact of surface roughness on an additively manufactured acoustic material: An experimental and numerical investigation. *J. Sound Vib.*, 546, 117434, 2023. doi: https://doi.org/10.1016/j.jsv.2022.117434.

79. Lebea, L., Ngwangwa, H.M., Desai, D., Nemavhola, F., Experimental investigation into the effect of surface roughness and mechanical properties of 3D-printed titanium Ti-64 ELI after heat treatment. *Int. J. Mech. Mater. Eng.*, 16, 1, 16, 2021. doi: 10.1186/s40712-021-00138-2.
80. Heimbrook, A. and Gall, K., Effect of surface topography on the fatigue behavior of additively manufactured Ti6Al4V and CoCr alloys. *Mater. Sci. Eng. A*, 909, 146821, 2024. doi: https://doi.org/10.1016/j.msea.2024.146821.
81. Jackson, M. and Dring, K., A review of advances in processing and metallurgy of titanium alloys. *Mater. Sci. Technol.*, 22, 8, 881–887, 2006. doi: 10.1179/174328406X111147.

3

The Rise of Smart Materials: Recent Developments

Mariel Amparo Fernandez Aramayo[1,2,3], Mohd Rehan[2,3], Bijaya Bikram Samal[4*], Cheruvu Siva Kumar[5] and Shailendra Kumar Varshney[6]

[1]*Polytechnic School, Chemical Engineering Department, University of São Paulo, São Paulo, Brazil*

[2]*School of Engineering, Mackenzie Presbyterian University, São Paulo, Brazil*

[3]*MackGraphe-Mackenzie Institute for Research in Graphene and Nanotechnologies, Mackenzie Presbyterian University, São Paulo, Brazil*

[4]*Advanced Technology Development Centre, Indian Institute of Technology, Kharagpur, West Bengal, India*

[5]*Department of Mechanical Engineering, Indian Institute of Technology, Kharagpur, West Bengal, India*

[6]*Department of Electronics and Electrical Communication Engineering, Indian Institute of Technology, Kharagpur, West Bengal, India*

Abstract

Recently 4D printing has drawn wide attention from both academia and industry for the ability to create dynamic parts and structures using various smart materials with the ability to alter shape or qualities over time in reaction to environmental stimuli such as heat, moisture, light, or electrical current. This cutting-edge printing method harnesses the capabilities of smart materials offering immense potential for creating complex structures that can adapt, reconfigure, self-morph, self-heal, or self-assemble after printing. Recent developments in new kinds of smart materials and their composites such as shape memory polymers, magnetic shape memory alloys, hydrogels, electro-responsive polymers, liquid crystal elastomers, and smart inorganic polymers have fueled the development of 4D printing technology. The utilization of various smart materials in 4D printing is based on its specific applications and requirements. The incorporation of smart materials into 4D printing allows for the creation of objects that actively respond to stimuli, allowing for dynamic behavior, improved performance, and novel solutions to complicated challenges.

**Corresponding author*: bijaybikram3991@gmail.com

Bijaya Bikram Samal, Cheruvu Siva Kumar and Shailendra Kumar Varshney (eds.) 4D Printing Technology: Principles, Materials and Applications, (69–92) © 2025 Scrivener Publishing LLC

Keywords: 4D printing, smart materials, shape memory materials, piezoelectric material, photo responsive materials, hydrogels

3.1 Introduction

Materials have long served as the foundational building blocks of our universe, offering essential attributes such as structure, functionality, and aesthetic appeal. However, with the relentless pace of technological advancement, the demand for innovative materials capable of meeting the evolving needs of diverse applications has intensified. In response to this demand, a remarkable new class of materials known as smart materials has emerged at the forefront of scientific and engineering innovation. This class of materials is gaining widespread recognition and acclaim for its unparalleled potential to revolutionize numerous fields by exhibiting unique properties and dynamic responses to external stimuli. These ground-breaking materials go beyond their passive functions, actively reacting to light, temperature, pressure, and even electrical fields, displaying striking alterations in conductivity, color, and shape [1].

The transformative potential of 4D printing is catalyzing a paradigm shift in the realm of materials research and manufacturing. Building upon the principles of 3D printing, which constructs three-dimensional objects layer by layer, 4D printing introduces a revolutionary fourth dimension: time. This groundbreaking advancement imbues objects with the remarkable ability to dynamically transform, adapt, and evolve in response to external stimuli over time, transcending the static and rigid nature of traditional manufacturing processes [2]. Unlike conventional manufacturing methods that yield static structures, 4D printing heralds a new era of adaptive and responsive materials. By harnessing the intrinsic capabilities of smart materials, such as shape memory polymers and hydrogels, 4D printing enables the creation of dynamic structures capable of self-assembly, shape-shifting, and environmental responsiveness. This unprecedented level of adaptability holds immense promise across a spectrum of applications, from self-optimizing architectural structures that adapt to changing environmental conditions to biomedical implants that seamlessly conform to the dynamic physiological landscape of the human body.

As researchers delve deeper into the possibilities offered by 4D printing, a vast landscape of innovation and discovery unfolds. The exploration of novel materials, advanced fabrication techniques, and intricate design methodologies paves the way for unprecedented advancements in diverse industries. From aerospace engineering to healthcare, the potential applications of 4D printing are boundless, offering transformative solutions to complex challenges and reshaping the fabric of materials research and manufacturing as we know it [3–5].

With the use of self-transformable or smart materials, this new technology allows for the production of dynamic components and structures that are a significant shift from the traditional manufacturing [6, 7]. These materials are capable to change their shape, properties, or functionality in response to environmental stimuli such as heat, moisture, light, or electrical current (as observed in Figure 3.1) and, for those reasons their study in 4D printing has attracted considerable interest in academia and industry [8, 9].

Smart materials represent a groundbreaking frontier with transformative potential across a spectrum of industries. In healthcare, these materials hold the promise of revolutionizing medical practices through innovations such as self-adjusting stents and precision drug capsules, ushering in a new era of personalized and responsive treatments. Concurrently, in the realm of architecture, smart materials empower structures to dynamically respond to environmental changes, enabling them to adapt their shape in real-time to optimize energy efficiency and withstand extreme weather conditions. This convergence of advanced materials science and innovative design principles marks a paradigm shift in how we approach challenges and opportunities across diverse sectors, driving forward the boundaries of technological innovation and sustainable development [6, 10, 11]. In the aerospace industry, the integration of smart materials holds immense promise for revolutionizing aircraft design and performance. By incorporating lighter and more efficient components that dynamically adapt during flight, these advancements have the potential to significantly enhance aerodynamics and safety standards. Furthermore, the utilization

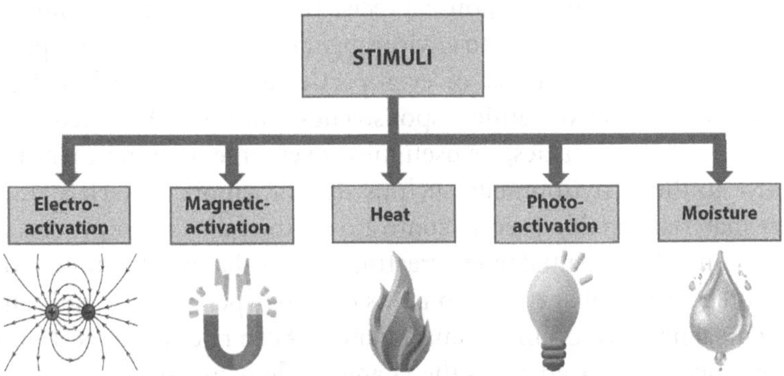

Figure 3.1 Common environmental stimuli for smart materials used in mainstream 4D printing.

of smart materials in aerospace applications has the added benefit of reducing fuel consumption, aligning with the industry's ongoing efforts to improve sustainability and environmental impact. This convergence of cutting-edge materials science with aerospace engineering represents a pivotal advancement with far-reaching implications for the future of air travel and aerospace technology [7, 11, 12].

In this chapter, we embark on an in-depth exploration of the revolutionary advancements in 4D printing, delving into the enthralling domain of smart materials for fourth-dimensional design. As we navigate through this dynamic landscape, we unveil the transformative capabilities of smart materials, propelling objects beyond conventional static forms to autonomously adapt and evolve in response to environmental stimuli. Our comprehensive discussion spans fundamental concepts, diverse applications, and profound implications of smart materials across an array of sectors, including healthcare, architecture, consumer products, and engineering domains such as robotics and aeronautics. Through meticulous analysis and insightful exploration, we illuminate the profound impact of smart materials in shaping the future of design, engineering, and innovation across diverse industries.

3.2 Importance of Smart Materials in 4D Printing

Smart materials in 4D printing enable the development of objects that actively respond to stimuli, allowing for dynamic behavior, improved performance, and innovative solutions to complicated problems. As a result, smart materials are essential in the field of 4D printing because they have the ability to change the structure and functionality of printed objects [9, 13, 14]. The importance of smart materials in 4D printing can be highlighted by their dynamic response, precise control, and sustainability.

Smart materials can react to environmental stimuli such as temperature fluctuations, humidity levels, or even particular triggers such as light or magnetic fields. This dynamic responsiveness enables 4D printed items to alter shape, characteristics, or usefulness over time, making them highly adaptive to their environment, such us shape memory materials [15, 16]. Smart materials enable precise control over the fabrication process. This level of control is essential when creating objects that need to undergo certain modifications at specific moments or under particular conditions. In the medical field, for example, smart materials can release medication with pinpoint accuracy or adapt to the changing demands of a patient's body [6, 10]. Smart materials can enhance resource usage while also reducing waste and environmental effect. Furthermore, by creating objects that can adapt and change, the lifespan and usefulness of products can be extended, adding to long-term sustainability initiatives [12, 17].

3.3 Types of Smart Materials

This cutting-edge printing method harnesses the potential of smart materials, offering immense possibilities for creating complex structures capable of adapting, reconfiguring, self-morphing, self-healing, or self-assembling after printing [13, 14, 18]. Recent advancements in novel types of smart materials and their composites have propelled the growth of 4D printing technology. Smart materials, such as shape memory polymers, magnetic shape memory alloys, hydrogels, electro-responsive polymers, liquid crystal elastomers, and smart inorganic polymers, have played a pivotal role in driving the advancement of 4D printing technology [18, 19].

Shape memory materials are able to memorize their original shape and return to it through the application of an external stimulus such as heat or UV radiation [8, 20, 21]. Piezoelectric materials generate an electric charge under mechanical stress and change their shape when an electric field is applied [22, 23]. Magnetostrictive materials are substances that can change their shape or dimensions when exposed to a magnetic field, exhibiting a magnetostrictive effect [24, 25]. Thermo responsive materials can change their shape, properties, or behavior in response to changes in temperature over time [26–28]. Photo responsive materials respond to light by changing their properties, exhibiting deformation, color change, or chemical reactions [21, 28]. Chromoactive materials can change their color or appearance in response to temperature, light, or environmental conditions [29]. Rheological fluids change their viscosity or rheology in response to temperature, magnetic fields, or mechanical stress [30, 31]. Self-healing materials are able to recover the structural integrity or functionality of the printed objects after sustaining damage or wear and tear. They have the ability to repair themselves utilizing heat or light as triggers [15, 19]. Hydrogels can absorb and retain water, leading to changes in their shape, size, or properties [26, 32, 33]. Liquid crystal elastomers combine the properties of both liquid crystals and elastomers. They can change their shape, stiffness, and optical properties in response to heat or light [34, 35].

In the upcoming sections, each of these smart materials is described in detail by exploring its properties, mechanisms and applications. Each material possesses distinct properties and capabilities that will be thoroughly examined to provide a comprehensive understanding of their significance in the field of advanced manufacturing by 4D printing.

3.3.1 Shape Memory Materials

3.3.1.1 Concepts of Shape Memory Effects

The shape memory effect was first discovered in the 1930s by the scientist Arne Ölander while studying gold-cadmium alloys. He observed that these materials could return to their original shape after being deformed when exposed to temperature variations. In the 1980s, researchers began to develop shape memory polymers (SMPs), which are different from shape memory alloys but share the ability to return to a predefined shape upon activation. SMPs broadened the variety of possible applications, particularly in industries requiring lightweight and flexible materials. In recent years, the combination of shape memory materials with 4D printing technology has gained attention, enabling for the creation of objects and structures that can change their shape or function in response to environmental stimuli over time [8, 20, 29].

Shape memory materials can return to their original shape after an external stimulus, such as heat or UV radiation and can be divided in shape memory polymers and shape memory alloys [31, 36]. Figure 3.2 depicts the recovery mechanism, which follows a four-step cycle. First, the structure is deformed by an external load at a high temperature; second, the temperature is reduced while the external load is maintained; third, the structure is unloaded at the low temperature and the desired shape is obtained; and fourth, the original shape can be recovered by reheating the structure [36, 37].

3.3.1.1.1 Types of Shape Memory Effects

Shape memory materials display different shape memory effects, each with its own distinct set of properties and applications. The most common types of shape memory effects are: one-way shape memory effect, two-way shape

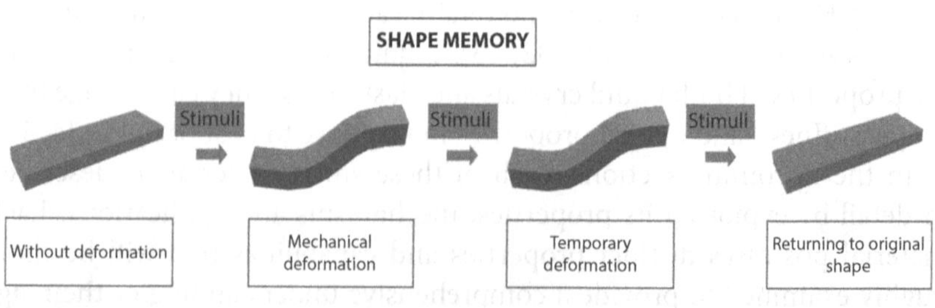

Figure 3.2 Shape memory material mechanism.

memory effect, triple-shape memory effect, tunable multi-shape effect, reversible plasticity shape memory effect and pseudoelasticity [37–41].

The material in the one-way shape memory effect has the capacity to recover to a single memorized state in response to a specific stimulus, most commonly a change in temperature. It is suitable for one-time application; otherwise, a process is needed to re-create the temporary shape. This issue can be overcome by using two-way shape memory effect, which can change the shape reversibly. As a result, two-way shape memory combines the reversibility of shape altering material with the programmability of shape memory material [39–43]. Figure 3.3 shows the difference between one-way and two-way shape memory effect [39].

Tripleshape memory effect is an extension of the traditional two-shape memory effect. Here, it introduces a third shape, providing additional functionality and versatility. Tunable multi-shape effect allows the transition among multiple different shapes in a material, and this effect offers the capability to adjust or "tune" these shapes according to specific requirements or conditions. The material in the reversible plasticity shape memory effect is capable of reversible change in shape, specifically from a deformed or temporary state to its original or permanent state, with no long-term deformation or damage on its structure [39–43]. Pseudoelastic effect (PE) or superelasticity, is another type of shape memory effect, however, it is not as desirable as the other two shape memory effects. Here, the material returns to its original shape completely when the load is removed.

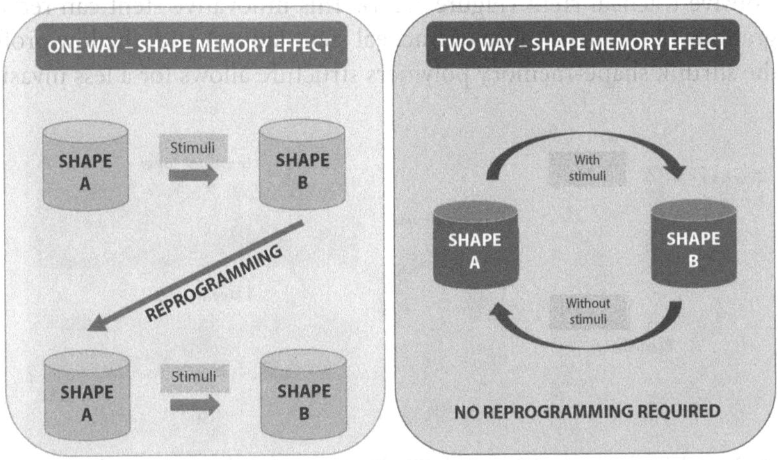

Figure 3.3 Illustration of the difference between one-way and two-way shape memory effect.

This functionality is similar to that of an elastic solid, therefore, this effect is valuable in applications where mechanical actuation is required, such as in orthodontic wires or eyeglass frames [20, 38].

3.3.1.2 *Shape Memory Polymers*

When the polymer is heated above its transition temperature, a deformation force is applied on it, then, when the material is cooled under the influence of this deformation force, its shape becomes fixed. This phenomenon occurs because when an external deformation force is applied, the conformations of the polymeric chain change, leading to a state of reduced entropy (also referred to as a temporary shape) that corresponds to a macroscopic change in shape. When the polymer is cooled, the segments of the molecular chain lose their molecular mobility, fixing the macroscopic structural shape. Upon reheating the polymer, the chains return to their state of maximum entropy, resulting in the recovery of the permanent shape [41, 44].

As the SMP undergoes a thermo-mechanical programming process for the shape memory effect, additive manufacturing like fused deposition modeling, which is based on the thermal processing can be utilized to create shape changing components. The pre-strain stored during the process gets released to provide the shape change thus providing an economical method for programming [45].

This technology has been applied in the fabrication of medical devices. Zarek *et al.* (2017) reported the fabrication of a 4D printed thermo-responsive tracheal stent (Figure 3.4 a). This innovative stent can reduce migrations (a frequent cause of tracheal stent failure) and, the low profile of the shrunk shape-memory polymers structure allows for a less invasive

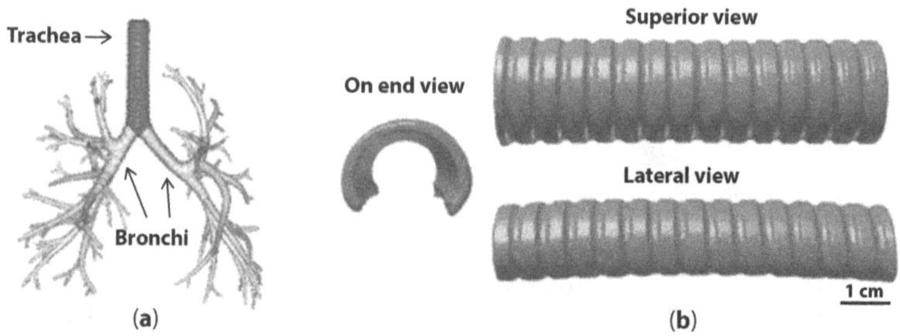

Figure 3.4 (a) Tracheal stent 4D printed with shape memory polymer, (b) tracheal stent after being implanted in the body, the stent evolved from an open duct to a closed one [46].

deployment. This medical device can better suit the natural arcade pattern and cartilaginous rings, ensuring stability and preventing unwanted migrations. The shrunken shape of the stent during insertion, minimizes potential harm. Once inserted into the body, the tracheal stent can expand to its permanent shape in response to a local increase in temperature, providing optimal functionality (Figure 3.4 b) [46].

3.3.1.3 Shape Memory Alloys

Shape memory alloys (SMAs) are a type of smart material that can "memorize" a given shape and return to it when exposed to specific environmental stimuli, most notably temperature changes. This unique behavior is due to a thermoelastic martensitic transformation, which occurs as the material transitions between two phases: austenite (high-temperature phase) and martensite (low-temperature phase). When the material is cooled, it transforms from austenite to martensite, and when heated, it reverts from martensite back to austenite. The alloys respond to two main types of stimuli: thermoresponsive and magnetoresponsive [15, 47]. This specific property has resulted in a wide range of applications in a variety of sectors. In the context of aeronautics, these alloys offer the advantage of being lightweight, which is crucial for aircraft because greater weight means higher fuel consumption [12, 39]. The most common SMAs are composed of nickel and titanium, known as Nitinol, which experiences 4% of reduction in length during this transition [48, 49].

Applications in medical equipment include self-expanding stents used in minimally invasive procedures to maintain blood arteries open, as well as orthodontic wires used to exert continuous pressure on teeth to aid orthodontic changes. These materials are employed in the aviation and automobile industries as antennas, solar arrays, and engine components such as temperatureresponsive valves, respectively. In civil engineering, they are also utilized to construct buildings that can absorb vibrations during earthquakes [6, 50, 51].

3.3.1.3.1 Thermoresponsive Shape Memory Alloys

Thermoresponsive shape memory alloys (TSMAs) are a subgroup of shape memory alloys that exhibit shape memory effects in response to changes in temperature. The choice of alloy composition determines the specific temperature range at which the TSMA exhibits its shape memory effect. These alloys can change their shape or mechanical properties when heated or cooled within a specific temperature range and, this effect is reversible and repeatable, requiring precise temperature control. Nickel-titanium (Nitinol), copperaluminumnickel and other alloys are employed in a variety of applications, including actuators in robotics [52]. They can change their components shape and create adaptive structures. In the case of consumer products, they can be used to develop self-adjusting eyeglass frames that respond to temperature cues [53].

3.3.1.3.2 Magnetic Shape Memory Alloys

Magnetic shape memory alloys can change their shape and mechanical properties in response to external magnetic fields, returning to their original shape once the magnetic field is removed. These smart materials are employed in the fabrication of microscale actuators and sensors, which can convert magnetic energy into mechanical movement making them valuable for applications in robotics, biomedical devices, and adaptive structures. Their unique properties and repeatable shape transformations make them promising for aerospace and defense applications. Ongoing research focuses on optimizing composition, enhancing durability, and integrating them into multifunctional systems to improve performance [11, 19, 47].

3.3.2 Piezoelectric Materials

This type of smart materials is capable of generating an electric charge when subjected to mechanical stress or deformation. Additionally, they can change their shape in response to an applied electric field. Piezoelectric materials are used in sensors, actuators, and energy harvesting systems. For example, they can transform mechanical vibrations or deformations into electrical energy, making them suitable for harvesting energy from sources such as footfalls or vibrations in machinery [13, 22, 23, 30].

Grinberg *et al.* (2019) developed a sensor-based piezoelectric material by dispersing inorganic ferroelectric submicron particles in a polymer matrix. The authors demonstrated the potential application of this sensor in a 4D printed knee prosthesis. This sensor simplifies the detection of the current temperature and the residual life of the prothesis by applying mechanical strain drive on it, as shown in Figure 3.5. This development holds promise for future 4D printing technologies, especially in smart implants and prosthetics [54].

3.3.3 Magnetostrictive Materials

Similar to piezoelectric materials, magnetostrictive materials can change their shape or dimensions when subjected to a magnetic field, exhibiting a

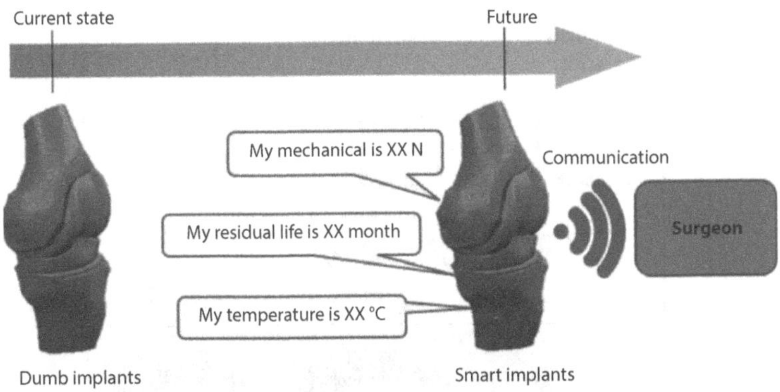

Figure 3.5 Concept of 4D printing technology for a 4D printed knee prosthesis based on a sensor-based piezoelectric material [54].

magnetostrictive effect. Although they can be effective, it's important to note that they exhibit significant hysteresis losses, which need to be accounted for when utilizing these materials in sensor applications. Therefore, these materials could enable soft robots change their shape and function for various tasks [13, 25].

Zhang *et al.* (2015) showed that magnetic field recovery differs from direct heating recovery, in which structures are programmed at specified transition temperatures and typically restored at the same temperatures. In the context of 3D printed origami (as shown in Figure 3.6 a), variable output powers can be employed to trigger particularly specified changes in shape. The unfolding behavior of the smart box is triggered by placing it in a magnetic coil subjected to magnetic field heating at different output powers. As depicted in Figure 3.6 b), the three-dimensional box was unfolded back from the temporary shape to a flat sheet step by step. This result can be applied in implant manufacturing as well as in biomedical science [55].

3.3.4 Thermoresponsive Materials

Thermoresponsive materials can change their shape, properties, or behavior in response to changes in temperature over time. These materials can experience transition between two or more different states within a specified temperature range, such as a temporary deformed state and their original or permanent state. Thermoresponsive materials, when employed in 4D printing, can provide dynamic and programmable shape-changing capabilities over time. For example, in drug delivery systems, drugs can change their shape or mode of release in response to temperature variations

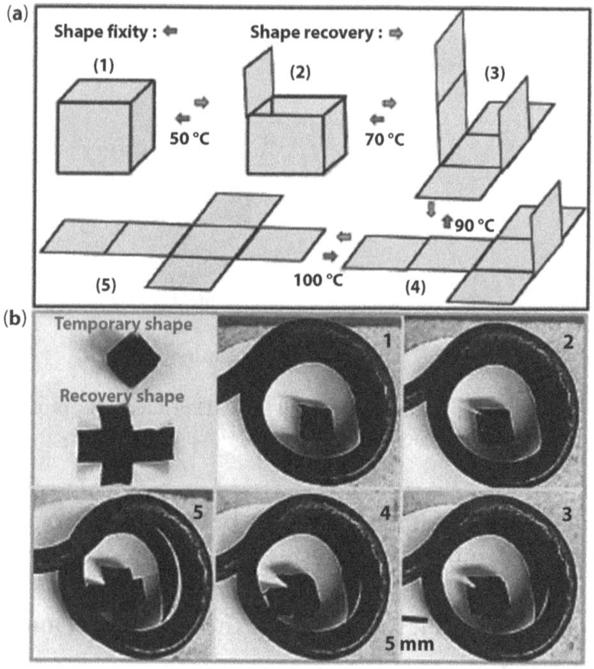

Figure 3.6 3D shape recovery performance of (a) the designed shape recovery process of the origami box, (b) magnetic field triggered shape recovery in the same origami box [55].

within the body. Similarly, in the realm of 4D-printed clothing or textiles, these intelligent materials are employed to adjust their porosity or shape to regulate temperature or moisture, enhancing wearer comfort. This versatility is particularly valuable in sportswear, outdoor gear, and medical textiles [26–28, 48].

Zarek *et al.* (2016) showed a morphological change of a 3D printed flower due to the temperature response of the polymer material, as depicted in Figure 3.7. The petals are capable to bloom or close when subjected to heat, with the top row showing the permanent shape and the bottom row displaying the temporary shape. Heating above the melting temperature enables the shape memory thermo-responsive structures to return to their permanent form [56].

3.3.5 Photo Responsive Materials

Photo responsive materials, also known as light-responsive materials, can experience reversible changes in their properties or behaviors when

Figure 3.7 Photographs of thermo-responsive 3D printed flower. The top row is the permanent shape of the structures. The bottom row is the temporary shape. On heating above the melting temperature, the shape memory thermo-responsive structures recover their permanent shape [56].

exposed to light. When these materials are used for 4D printing, they can be programmed to change shape, color, stiffness, or other properties characteristics when illuminated with specific wavelengths of light [21, 28].

Yang *et al.* (2017) fabricated a 3D-printed, photo-responsive sunflower inspired by the heliotropism of sunflowers (Figure 3.8 a). When the temperature of the sunflower climbs over Tg (30°C), its petals change from rigid to an elastomeric and pliant state, below that Tg, any externally induced deformation can be fixed (Figure 3.8 b). When illuminated for a period of time, the petal shape gradually returns to its original state, resulting in a change in shape from the closed to the open state. The sunflower fully flowered after 280 seconds of illumination, displaying light-triggered, plant-like shape-memory behavior (Figure 3.8 c) [21].

3.3.6 Chromoactive Materials

Chromoactive materials, characterized by their ability to undergo reversible changes in color or optical properties in response to external stimuli, have emerged as a significant area of research in recent years. These materials exhibit diverse chromoactive behaviors, including photochromism, thermochromism, electrochromism, and others, making them valuable for various applications across different fields [57].

Understanding the structure-property relationships governing chromoactive behavior is essential for the design and development of novel materials with tailored properties. Chromoactive materials exhibit a fascinating array of behaviors, categorized based on the triggering stimulus. Photochromism, as the name suggests, involves light-induced color

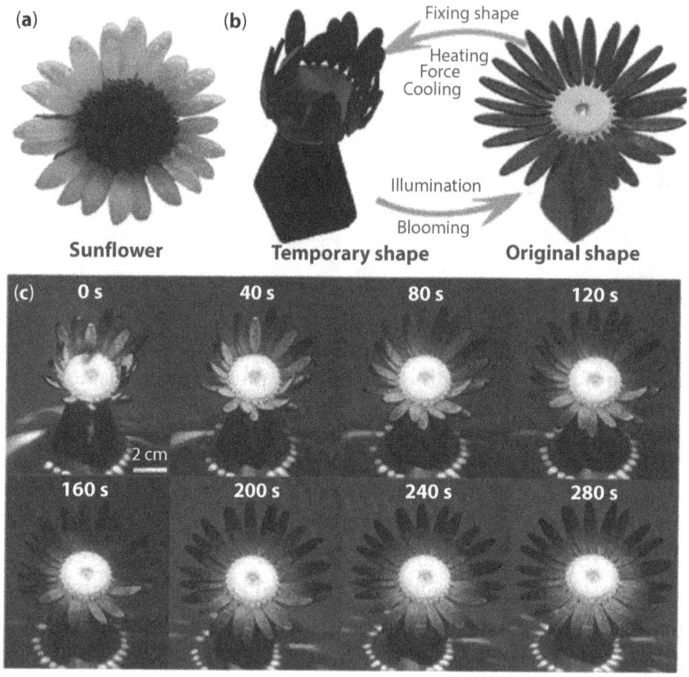

Figure 3.8 (a) The real image of a sunflower, (b) photo-triggered shape memory behavior of a 3D printed sunflower from closed to opened state like the blooming of flowers and, (c) the different deformation of 3D printed sunflower with different illumination time [21].

changes. Familiar examples include light-sensitive sunglasses that darken in bright sunlight. Thermochromism, on the other hand, responds to temperature fluctuations, like the color-changing cups that reveal hidden messages when filled with hot beverages. Electrochromism utilizes electrical fields to manipulate color, as seen in smart windows that adjust transparency on demand [58]. Additionally, piezochromism responds to pressure, halochromism to changes in acidity, and mechanochromism to mechanical stress, each demonstrating the versatility of this material class.

The underlying mechanisms fueling these color transformations are as intricate as their effects. Photochromic materials often rely on photoisomerization, where light exposure triggers rearrangement of molecular structures, altering light absorption and hence, color. Thermochromic materials may involve temperature-dependent changes in crystal structure or molecular interactions, impacting light scattering or absorption. Electrochromic materials typically involve redox reactions triggered by electric fields, altering the oxidation state of molecules and their light absorption properties.

Understanding these structure-property relationships is crucial for tailoring chromoactive materials with specific responses and functionalities.

3.3.7 Rheological Fluids

Once limited to the use in shock absorbers and lubricants, rheological fluids are now becoming more and more popular in the field of smart materials. They are distinguished from static materials and responsive systems by their capacity to dynamically alter viscosity in response to external stimuli such as electric fields, shear stress, or magnetic fields. Just a few examples of the possibilities represented by these smart fluids include prosthesis that modify resistance for smooth movement, structures that respond to wind pressure, and car suspensions that adjust in real-time to the terrain. The applications are numerous and wide-ranging; from magnetorheological fluids that stiffen under magnetic fields [59] for vibration control to electrorheological fluids that solidify with electric stimuli for adaptive brakes. Although issues like cost and reaction time still exist, developments in microfluidics and nanomaterials demonstrate ample motivation for further research and development to make it more useful for various applications [60].

3.3.8 Self-Healing Materials

The amazing ability of self-healing materials to fix damage on their own, restoring functioning and prolonging their lifespan, is impressive. This skill can be demonstrated through a variety of ways, each with unique advantages and disadvantages. The process of intrinsic healing depends on a material's innate qualities, such as reversible chemical bonds or microcapsules that hold healing agents and release them when the material is damaged [61]. Heat, light, or electrical fields are examples of external stimuli that are used in extrinsic healing to start repair processes. Notably, bio-inspired self-healing uses bacteria or enzymes to repair fractures and fissures, emulating nature's own restorative processes [62]. The widespread use of self-healing materials is hampered by a number of issues, despite their enormous potential. One major challenge is yet to provide effective healing across many forms of injuries. Crucial study areas also include maximizing healing speed and guaranteeing full return of functionality. However, intriguing solutions are provided by developments in computer modeling and nanotechnology. For example, adding nanoparticles can improve the effectiveness of healing, and using computational techniques can help create materials with specific self-healing characteristics [63].

3.3.9 Hydrogels

A well-known family of smart materials called hydrogels are three-dimensional networks composed of hydrophilic polymers interwoven with water. Their special physicochemical qualities and ability to react to outside stimuli have made them highly desirable. These "water-laden gels" are extremely sought-after in a variety of scientific fields because they surpass the constraints of conventional materials and demonstrate extraordinary versatility and usefulness. Because hydrogels are naturally sensitive to their surroundings, they react quickly to changes in temperature, pH, light, mechanical stress, and biomolecules [64]. The interaction between the entrapped water and the polymer network causes changes in mechanical characteristics, swelling behavior, and even drug release rates, which accounts for this responsiveness. For example, thermoresponsive hydrogels can expand or contract in response to temperature changes; this property makes them useful for drug delivery systems as they can activate at specific body temperatures [65]. Similar to this, pH-responsive hydrogels allow for the regulated release of medicinal chemicals into specific tissues by changing their structure in response to acidity. The possibilities include mechanoresponsive hydrogels that react to pressure or stress for use in sensors and actuators, and photo responsive hydrogels that release encapsulated biomolecules when exposed to light. Hydrogels are not just responsive; they are versatile. Hydrogels with adjustable qualities, from soft and tissue-like to stiff and mechanically robust, can be made by researchers by carefully choosing and adjusting the polymer mix and crosslinking density [66]. This extraordinary tunability creates opportunities for a wide range of uses. Biocompatible hydrogels function as scaffolds for tissue engineering and regeneration, providing support and signals for cellular growth in an analogous manner to the extracellular matrix. Furthermore, because they can encapsulate and release biomolecules, they are perfect for drug delivery systems that allow for regulated and sitespecific therapeutic targeting. Hydrogels are also used in soft robotics, environmental remediation, and biosensors, demonstrating their versatility in a range of industries [67].

3.3.10 Liquid Crystal Elastomers (LCEs)

LCEs have a unique structure that combines the flexibility of polymer networks with the ordered organization of liquid crystals. Their capacity to experience significant and reversible deformations in response to stimuli like heat, light, or electricity results from this synergy. The fundamental process is based on how the rod-shaped liquid crystal molecules are arranged inside the polymer chains. These molecules vary their orientation in response to outside stimuli, which significantly alters the size and structure of the elastomer [68]. There are many possible uses for LCEs due to their varied actuation mechanisms. Thermally sensitive LCEs have potential applications in biomimicry, self-deployable structures, and artificial muscles since they can change their shape significantly in response to temperature changes [69]. Light-triggered photo-responsive LCEs present intriguing opportunities for light-driven valves, adaptive optics, and microfluidic devices [70]. Tunable lenses, haptic feedback devices, and energy-efficient displays are made possible by electro responsive LCEs that can be controlled by electric fields. LCEs' intrinsic biocompatibility and biomimetic qualities add to their allure. Because of their structural similarity to muscle tissue, they are perfect for creating bioinspired robotics and 4D-printed tissues [71]. Moreover, their capacity to react to minute alterations in their surroundings corresponds with the adaptive tactics found in living systems, providing opportunities for biomimetic sensors and materials that possess the ability to mend themselves. LCEs have enormous promise, but they also have major drawbacks. There is still need for improvement in the areas of cost-effectiveness, response speed and output force restrictions, and fabrication complexity. Ongoing studies in computer modeling, nanoengineering, and polymer chemistry, however, present encouraging alternatives. To fully realize the potential of these adaptable materials, advances in stimuli-responsive control mechanisms and microscale production are essential.

3.4 Conclusion

The integration of smart materials into 4D printing marks a paradigm shift in the world of design and manufacturing. By harnessing the inherent capabilities of these responsive materials, from shape memory polymers to hydrogels, 4D printing transcends the limitations of static objects and empowers the creation of dynamic, adaptive structures.

This transformative technology boasts numerous advantages:

- Dynamic Response: Objects can actively respond to stimuli like temperature, light, or pressure, enabling self-assembly, shape-shifting, and environmental responsiveness.
- Tailored Functionality: Diverse smart materials offer a spectrum of properties, catering to specific application needs in healthcare, architecture, aerospace, and beyond.

- Sustainability: Self-healing capabilities and adaptable shapes extend product lifespan and resource usage, aligning with green initiatives.

The landscape of available smart materials is constantly evolving, fueled by ongoing research and innovation. The future holds immense potential for groundbreaking advancements in diverse fields:

- Healthcare: Personalized medicine with self-adjusting drug capsules and responsive implants.
- Architecture: Buildings that dynamically adapt to weather conditions and energy demands.
- Aerospace: Lighter, more efficient aircraft components that optimize performance and safety.

Ultimately, the fusion of smart materials with 4D printing opens a captivating frontier of possibilities. As we delve deeper into this realm, we unlock the potential to reshape diverse industries and engineer solutions to complex challenges, paving the way for a smarter and more sustainable future.

References

1. Zhao, Q., Qi, H.J., Xie, T., Recent progress in shape memory polymer: New behavior, enabling materials, and mechanistic understanding. *Prog. Polym. Sci.*, 49–50, 79–120, 2015. https://doi.org/10.1016/j.progpolymsci.2015.04.001.
2. Samal, B.B., Jena, A., Mishra, D., Controlled Shape Changing Components by Using 4D Printing Technology, in: *Proceedings of 10th International Conference on Precision, Meso, Micro and Nano Engineering (COPEN 10)*, IIT Madras. IIT Matras, pp. 78–81, 2017, http://www.copen.ac.in/proceedings/copen10/copen/19.revised_manuscript_262.pdf.
3. Joharji, L., Mishra, R.B., Alam, F., Tytov, S., Al-Modaf, F., El-Atab, N., 4D printing: A detailed review of materials, techniques, and applications. *Microelectron. Eng.*, 265, 1–21, 2022. https://doi.org/10.1016/j.mee.2022.111874.
4. Singholi, A.K.S. and Sharma, A., Review: Recent advancement and research possibilities in 4D printing technology. *Materwiss Werksttech*, 51, 1332–1340, 2020. https://doi.org/10.1002/mawe.202000008.
5. Samal, B.B., Jena, A., Varshney, S.K., Kumar, C.S., 4D printing of shape memory polymers: A comparative study of programming methodologies on

various material properties. *Smart Mater. Struct.*, 32, 1–17, 2023. https://doi.org/10.1088/1361-665X/acda6e.

6. Sahafnejad-Mohammadi, I., Karamimoghadam, M., Zolfagharian, A., Akrami, M., Bodaghi, M., 4D printing technology in medical engineering: a narrative review. *J. Braz. Soc. Mech. Sci. Eng.*, 44, 1–26, 2022. https://doi.org/10.1007/s40430-022-03514-x.
7. Bajpai, A., Baigent, A., Raghav, S., Brádaigh, C., Koutsos, V., Radacsi, N., 4D printing: Materials, technologies, and future applications in the biomedical field. *Sustainability (Switzerland)*, 12, 1–32, 2020. https://doi.org/10.3390/su122410628.
8. Alshebly, Y.S., Nafea, M., Mohamed Ali, M.S., Almurib, H.A.F., Review on recent advances in 4D printing of shape memory polymers. *Eur. Polym. J.*, 159, 1–20, 2021. https://doi.org/10.1016/j.eurpolymj.2021.110708.
9. Cataldi, P., Liu, M., Bissett, M., Kinloch, I.A., A Review on Printing of Responsive Smart and 4D Structures Using 2D Materials. *Adv. Mater. Technol.*, 7, 1–21, 2022. https://doi.org/10.1002/admt.202200025.
10. Javaid, M. and Haleem, A., 4D printing applications in medical field: A brief review. *Clin. Epidemiol. Glob. Health*, 7, 317–321, 2019. https://doi.org/10.1016/j.cegh.2018.09.007.
11. Javaid, M. and Haleem, A., Exploring smart material applications for covid-19 pandemic using 4d printing technology. *J. Ind. Integr. Manage.*, 5, 481–494, 2020. https://doi.org/10.1142/S2424862220500219.
12. Maraveas, C., Bayer, I.S., Bartzanas, T., 4D printing: Perspectives for the production of sustainable plastics for agriculture. *Biotechnol. Adv.*, 54, 1–23, 2022. https://doi.org/10.1016/j.biotechadv.2021.107785.
13. Mondal, K. and Tripathy, P.K., Preparation of smart materials by additive manufacturing technologies: A review. *Materials*, 14, 1–24, 2021. https://doi.org/10.3390/ma14216442.
14. Palmero, E.M. and Bollero, A., 3D and 4D Printing of Functional and Smart Composite Materials, in: *Encyclopedia of Materials: Composites*, pp. 402–419, Elsevier, Amsterdam, Netherlands, 2021, https://doi.org/10.1016/B978-0-12-819724-0.00008-2.
15. Farid, M.I., Wu, W., Liu, X., Wang, P., Additive manufacturing landscape and materials perspective in 4D printing. *Int. J. Adv. Manuf. Technol.*, 115, 2973–2988, 2021. https://doi.org/10.1007/s00170-021-07233-w/Published.
16. Subeshan, B., Baddam, Y., Asmatulu, E., Current progress of 4D-printing technology. *Prog. Addit. Manuf.*, 6, 495–516, 2021. https://doi.org/10.1007/s40964-021-00182-6.
17. Georgantzinos, S.K., Giannopoulos, G.I., Bakalis, P.A., Additive manufacturing for effective smart structures: The idea of 6d printing. *J. Compos. Sci.*, 5, 1–11, 2021. https://doi.org/10.3390/jcs5050119.

18. Zhang, Z., Demir, K.G., Gu, G.X., Developments in 4D-printing: a review on current smart materials, technologies, and applications. *Int. J. Smart Nano Mater.*, 10, 205–224, 2019. https://doi.org/10.1080/19475411.2019.1591541.
19. Haleem, A., Javaid, M., Singh, R.P., Suman, R., Significant roles of 4D printing using smart materials in the field of manufacturing. *Adv. Ind. Eng. Polym. Res.*, 4, 301–311, 2021. https://doi.org/10.1016/j.aiepr.2021.05.001.
20. Subash, A. and Kandasubramanian, B., 4D printing of shape memory polymers. *Eur. Polym. J.*, 134, 1–17, 2020. https://doi.org/10.1016/j.eurpolymj.2020.109771.
21. Yang, H., Leow, W.R., Wang, T., Wang, J., Yu, J., He, K., Qi, D., Wan, C., Chen, X., 3D Printed Photoresponsive Devices Based on Shape Memory Composites. *Adv. Mater.*, 29, 1–7, 2017. https://doi.org/10.1002/adma.201701627.
22. Kim, H., Torres, F., Wu, Y., Villagran, D., Lin, Y., Tseng, T.-L., Integrated 3D printing and corona poling process of PVDF piezoelectric films for pressure sensor application. *Smart Mater. Struct.*, 26, 1–10, 2017. https://doi.org/10.1088/1361-665x/aa738e.
23. Bodkhe, S., Turcot, G., Gosselin, F.P., Therriault, D., One-Step Solvent Evaporation-Assisted 3D Printing of Piezoelectric PVDF Nanocomposite Structures. *ACS Appl. Mater. Interfaces*, 9, 20833–20842, 2017. DOI:10.1021/acsami.7b04095.
24. Peterson, G.I., Larsen, M.B., Ganter, M.A., Storti, D.W., Boydston, A.J., 3D-printed mechanochromic materials. *ACS Appl. Mater. Interfaces*, 7, 577–583, 2015. https://doi.org/10.1021/am506745m.
25. Hristoforou, E. and Ktena, A., Magnetostriction and magnetostrictive materials for sensing applications. *J. Magn. Magn. Mater.*, 316, 372–378, 2007. https://doi.org/10.1016/j.jmmm.2007.03.025.
26. Han, D., Lu, Z., Chester, S.A., Lee, H., Micro 3D Printing of a Temperature-Responsive Hydrogel Using Projection Micro-Stereolithography. *Sci. Rep.*, 8, 1–10, 2018. https://doi.org/10.1038/s41598-018-20385-2.
27. Taylor, M.J., Tomlins, P., Sahota, T.S., Thermoresponsive gels. *Gels*, 3, 1–31, 2017. https://doi.org/10.3390/gels3010004.
28. Kuksenok, O. and Balazs, A.C., Stimuli-responsive behavior of composites integrating thermo-responsive gels with photo-responsive fibers. *Mater. Horiz.*, 3, 53–62, 2016. https://doi.org/10.1039/c5mh00212e.
29. Rognoli, V. and Ferraro, V., *ICS Materials Interactiveconnected, and smart materials*, FrancoAngeli, Milano, Italy, 2021.
30. Nkomo, N.Z., A Review of 4D Printing Technology and Future Trends, in: *Eleventh South African Conference on Computational and Applied Mechanics*, pp. 1–11, 2018.
31. Sossou, G., Demoly, F., Montavon, G., Gomes, S., Design for 4D printing: Rapidly exploring the design space around smart materials, in: *Procedia CIRP*, pp. 120–125, Elsevier B.V., Nantes, France, 2018, https://doi.org/10.1016/j.procir.2018.02.032.

32. Ahmed, E.M., Hydrogel: Preparation, characterization, and applications: A review. *J. Adv. Res.*, 6, 105–121, 2015. https://doi.org/10.1016/j.jare.2013.07.006.
33. Champeau, M., Heinze, D.A., Viana, T.N., de Souza, E.R., Chinellato, A.C., Titotto, S., 4D Printing of Hydrogels: A Review. *Adv. Funct. Mater.*, 30, 1–22, 2020. https://doi.org/10.1002/adfm.201910606.
34. Wang, Z. and Cai, S., Recent progress in dynamic covalent chemistries for liquid crystal elastomers. *J. Mater. Chem. B*, 8, 6610–6623, 2020. https://doi.org/10.1039/d0tb00754d.
35. Ula, S.W., Traugutt, N.A., Volpe, R.H., Patel, R.R., Yu, K., Yakacki, C.M., Liquid crystal elastomers: An introduction and review of emerging technologies. *Liq. Cryst. Rev.*, 6, 78–107, 2018. https://doi.org/10.1080/21680396.2018.1530155.
36. Huang, W.M., Ding, Z., Wang, C.C., Wei, J., Zhao, Y., Purnawali, H., Shape memory materials. *Mater. Today*, 13, 54–61, 2010. https://doi.org/10.1016/S1369-7021(10)70128-0.
37. Momeni, F., Mehdi Hassani, S.M., Liu, N.,.X., Ni, J., A review of 4D printing. *Mater. Des.*, 122, 42–79, 2017. https://doi.org/10.1016/j.matdes.2017.02.068.
38. Lee, A.Y., An, J., Chua, C.K., Two-Way 4D Printing: A Review on the Reversibility of 3D-Printed Shape Memory Materials. *Engineering*, 3, 663–674, 2017. https://doi.org/10.1016/J.ENG.2017.05.014.
39. Hager, M.D., Bode, S., Weber, C., Schubert, U.S., Shape memory polymers: Past, present and future developments. *Prog. Polym. Sci.*, 49–50, 3–33, 2015. https://doi.org/10.1016/j.progpolymsci.2015.04.002.
40. Zhou, J. and Sheiko, S.S., Reversible shape-shifting in polymeric materials. *J. Polym. Sci. B Polym. Phys.*, 54, 1365–1380, 2016. https://doi.org/10.1002/polb.24014.
41. Xie, T., Recent advances in polymer shape memory. *Polym. (Guildf)*, 52, 4985–5000, 2011. https://doi.org/10.1016/j.polymer.2011.08.003.
42. Wu, X., Huang, W.M., Zhao, Y., Ding, Z., Tang, C., Zhang, J., Mechanisms of the shape memory effect in polymeric materials. *Polym. (Basel)*, 5, 1169–1202, 2013. https://doi.org/10.3390/polym5041169.
43. Tcharkhtchi, A., Abdallah-Elhirtsi, S., Ebrahimi, K., Fitoussi, J., Shirinbayan, M., Farzaneh, S., Some new concepts of shape memory effect of polymers. *Polym. (Basel)*, 6, 1144–1163, 2014. https://doi.org/10.3390/polym6041144.
44. Samal, B.B., Jena, A., Varshney, S.K., Kumar, C.S., 4D printing: An experimental case study on processing of shape memory polymer by FDM/FFF for nature inspired structures, in: *Advances in Additive Manufacturing Artificial Intelligence, Nature-Inspired, and Biomanufacturing*, pp. 361–377, Elsevier, Amsterdam, Netherlands, 2023, https://doi.org/10.1016/b978-0-323-91834-3.00019-3.
45. Samal, B.B., Jena, A., Varshney, S.K., Kumar, C.S., FDM 4D printing: A low-cost approach of shape programming and assessing the shape memory properties using angle measurement methods in hot water actuation testing

apparatus. *Mater. Today Proc.*, 101, 7–14, 2024. https://doi.org/10.1016/j.matpr.2023.02.038.
46. Zarek, M., Mansour, N., Shapira, S., Cohn, D., 4D Printing of Shape Memory-Based Personalized Endoluminal Medical Devices. *Macromol. Rapid Commun.*, 38, 1–6, 2017. https://doi.org/10.1002/marc.201600628.
47. Fornell, J., Magnetic Shape Memory Composites, in: *Encyclopedia of Materials: Composites*, pp. 487–495, Elsevier, Amsterdam, Netherlands, 2021, https://doi.org/10.1016/B978-0-12-803581-8.11867-3.
48. Biswas, M.C., Chakraborty, S., Bhattacharjee, A., Mohammed, Z., 4D Printing of Shape Memory Materials for Textiles: Mechanism, Mathematical Modeling, and Challenges. *Adv. Funct. Mater.*, 31, 1–25, 2021. https://doi.org/10.1002/adfm.202100257.
49. Kim, S., Hawkes, E., Cho, K., Jolda, M., Foley, J., Wood, R., Micro artificial muscle fiber using NiTi spring for soft robotics, in: *IEEE/RSJ International Conference on Intelligent Robots and Systems*, IEEE, pp. 1–7, 2009.
50. Liu, Y., Du, H., Liu, L., Leng, J., Shape memory polymers and their composites in aerospace applications: A review. *Smart Mater. Struct.*, 23, 1–22, 2014. https://doi.org/10.1088/0964-1726/23/2/023001.
51. Sokolowski, W.M. and Tan, S.C., Advanced self-deployable structures for space applications. *J. Spacecr. Rockets*, 44, 750–754, 2007. https://doi.org/10.2514/1.22854.
52. Jena, A., Samal, B.B., Kumar, C.S., Varshney, S.K., Analysis of electro-thermo-mechanical behavior of thin film Ni50-Ti50and Ni40-Ti50-Cu10 shape memory alloys for application in thermal actuators. *Mater. Today Proc.*, 47, 4578–4583, 2021. https://doi.org/10.1016/j.matpr.2021.05.448.
53. Leng, J.S., Lan, X., Liu, Y.J., Du, S.Y., Huang, W.M., Liu, N., Phee, S.J., Yuan, Q., Electrical conductivity of thermoresponsive shape-memory polymer with embedded micron sized Ni powder chains. *Appl. Phys. Lett.*, 92, 1–4, 2008. https://doi.org/10.1063/1.2829388.
54. Grinberg, D., Siddique, S., Le, M.Q., Liang, R., Capsal, J.F., Cottinet, P.J., 4D Printing based piezoelectric composite for medical applications. *J. Polym. Sci. B Polym. Phys.*, 57, 109–115, 2019. https://doi.org/10.1002/polb.24763.
55. Zhang, F.H., Zhang, Z.C., Luo, C.J., Lin, I.T., Liu, Y., Leng, J., Smoukov, S.K., Remote, fast actuation of programmable multiple shape memory composites by magnetic fields. *J. Mater. Chem. C Mater.*, 3, 11290–11293, 2015. https://doi.org/10.1039/c5tc02464a.
56. Zarek, M., Layani, M., Eliazar, S., Mansour, N., Cooperstein, I., Shukrun, E., Szlar, A., Cohn, D., Magdassi, S., 4D printing shape memory polymers for dynamic jewellery and fashionwear. *Virtual Phys. Prototyp.*, 11, 4, 263–270, 2016. https://doi.org/10.1080/17452759.2016.1244085.
57. Tian, H. and Zhang, J. (Eds.), *Photochromic Materials: Preparation, Properties and Applications*, Wiley-VCH, Weinheim, Germany, 2016, https://doi.org/10.1002/9783527683734.

58. Gu, C., Jia, A.B., Zhang, Y.M., Zhang, S.X.A., Emerging Electrochromic Materials and Devices for Future Displays. *Chem. Rev.*, 122, 14679–14721, 2022. https://doi.org/10.1021/acs.chemrev.1c01055.
59. Filipcsei, G., Csetneki, I., Szilágyi, A., Zrínyi, M., Magnetic field-responsive smart polymer composites. *Adv. Polym. Sci.*, 206, 137–189, 2007. https://doi.org/10.1007/12_2006_104.
60. Kumar, J.S., Paul, P.S., Raghunathan, G., Alex, D.G., A review of challenges and solutions in the preparation and use of magnetorheological fluids. *Int. J. Mech. Mater. Eng.*, 14, 13, 2019. https://doi.org/10.1186/s40712-019-0109-2.
61. Aramayo, M.A.F. and Aoki, I.V., Synthesis of Innovative Epoxy Resin and Polyamine Hardener Microcapsules and their Age Monitoring by Confocal Raman Imaging. *J. Appl. Polym. Sci.*, e55342, 141, 19, 2024. https://doi.org/10.1002/app.55342.
62. Qi, M., Yang, R., Wang, Z., Liu, Y., Zhang, Q., He, B., Li, K., Yang, Q., Wei, L., Pan, C., Chen, M., Bioinspired Self-healing Soft Electronics. *Adv. Funct. Mater.*, 33, 2214479, 2023. https://doi.org/10.1002/adfm.202214479.
63. Akram, M., Adetunji, C.O., Ahamed, I., Sohail, A., Ghaffar, I., Michael, O.S., Anwar, H., Muhibi, M.A., Adetunji, J.B., Laila, U., Olaniyan, M., Self-Healing Polymers, in: *Self-Healing Smart Materials and Allied Applicaitons*, pp. 511–530, Scrivener Publishing LLC, River Street, Hoboken, USA & Beverly, MA, USA, 2021, https://doi.org/10.1002/9781119710219.ch21.
64. Buwalda, S.J., Vermonden, T., Hennink, W.E., Hydrogels for Therapeutic Delivery: Current Developments and Future Directions. *Biomacromolecules*, 18, 316–330, 2017. https://doi.org/10.1021/acs.biomac.6b01604.
65. Gupta, P., Vermani, K., Garg, S., Hydrogels: from controlled release to pH-responsive drug delivery. *Res. Focus*, 7, 569–579, 2002. https://doi.org/10.1016/S1359-6446(02)02255-9.
66. Koetting, M.C., Peters, J.T., Steichen, S.D., Peppas, N.A., Stimulus-responsive hydrogels: Theory, modern advances, and applications. *Mater. Sci. Eng. R Rep.*, 93, 1–49, 2015. https://doi.org/10.1016/j.mser.2015.04.001.
67. Buwalda, S.J., Boere, K.W.M., Dijkstra, P.J., Feijen, J., Vermonden, T., Hennink, W.E., Hydrogels in a historical perspective: From simple networks to smart materials. *J. Controlled Release*, 190, 254–273, 2014. https://doi.org/10.1016/j.jconrel.2014.03.052.
68. Bobnar, M., Derets, N., Umerova, S., Domenici, V., Novak, N., Lavrič, M., Cordoyiannis, G., Zalar, B., Rešetič, A., Polymer-dispersed liquid crystal elastomers as moldable shape-programmable material. *Nat. Commun.*, 14, 1–11, 2023. https://doi.org/10.1038/s41467-023-36426-y.
69. Zhang, J., Guo, Y., Hu, W., Sitti, M., Wirelessly Actuated Thermo- and Magneto-Responsive Soft Bimorph Materials with Programmable Shape-Morphing. *Adv. Mater.*, 33, 1–10, 2021. https://doi.org/10.1002/adma.202100336.

70. Ohm, C., Brehmer, M., Zentel, R., Applications of Liquid Crystalline Elastomers, in: *Liquid Crystal Elastomers: Materials and Applications. Advances in Polymer Science*, pp. 49–94, Springer-Verlag, Berlin Heidelberg, 2012, https://doi.org/10.1007/12_2011_164.
71. Ambulo, C.P., Burroughs, J.J., Boothby, J.M., Kim, H., Shankar, M.R., Ware, T.H., Four-dimensional Printing of Liquid Crystal Elastomers. *ACS Appl. Mater. Interfaces*, 9, 37332–37339, 2017. https://doi.org/10.1021/acsami.7b11851.

4

From 3D Printing to 4D Printing: Adding Time Dimension

Bijaya Bikram Samal[1*], Debadutta Mishra[2], Marwan Nafea[3], Anita Jena[1], Shailendra Kumar Varshney[4] and Cheruvu Siva Kumar[5]

[1]*Advanced Technology Development Centre, Indian Institute of Technology, Kharagpur, West Bengal, India*
[2]*Department of Production Engineering, Veer Surendra Sai University of Technology, Burla, Sambalpur, Odisha, India*
[3]*Department of Electrical and Electronic Engineering, Faculty of Science and Engineering, University of Nottingham Malaysia, Semenyih, Selangor, Malaysia*
[4]*Department of Electronics and Electrical Communication, Indian Institute of Technology, Kharagpur, West Bengal, India*
[5]*Department of Mechanical Engineering, Indian Institute of Technology, Kharagpur, West Bengal, India*

Abstract
The field of additive manufacturing has evolved significantly from traditional 3D printing to the emerging domain of 4D printing, where time becomes a critical functional dimension. This paper explores the transition from 3D to 4D printing, focusing on the incorporation of dynamic behaviors in printed materials that respond to external fillip such as temperature, light, humidity, and magnetic fields. It highlights the historical development of 4D printing, driven by advancements in smart materials such as shape memory polymers, hydrogels, and liquid crystal elastomers. The study identifies key challenges in this transition, including material limitations, scalability issues, and regulatory concerns. Despite these obstacles, 4D printing holds immense potential, particularly in sectors like aerospace, healthcare, and architecture, where adaptable and self-transforming materials can unlock unprecedented functionality. The paper also examines future trends, such as the development of multi-functional materials, the integration of artificial intelligence, and the expansion of 4D printing applications in biomedicine and

**Corresponding author*: bijaybikram3991@gmail.com

Bijaya Bikram Samal, Cheruvu Siva Kumar and Shailendra Kumar Varshney (eds.) 4D Printing Technology: Principles, Materials and Applications, (93–124) © 2025 Scrivener Publishing LLC

sustainable manufacturing. As research continues, 4D printing is poised to transform the manufacturing industry by enabling the development of intelligent, programmable structures capable of dynamically adapting to changing environments.

***Keywords*:** Additive manufacturing, 3D printing, 4D printing, history of 4D printing, decade of 4D printing, 3D to 4D printing, laws of 4D printing

4.1 Introduction

From the fundamental achievements of three-dimensional (3D) printing to the cutting-edge developments of 4D printing, the area of additive manufacturing (AM) has seen a remarkable transformation. Since its modest origins, 3D printing, also known as AM, has changed conventional manufacturing methods by enabling 3D things to be created layer by layer directly from its digital models, mainly generated using computer (CAD) [1]. This innovative technology has revolutionized the manufacturing industries and found applications in a variety of sectors, including aerospace, automotive, healthcare, and consumer products. Its advantages for fast/rapid prototyping, mass customization, on-demand manufacturing, rapid tooling has transformed how things are designed, created, produced, and manufactured [2]. However, as the possibilities of 3D printing expanded, researchers tried to push the limitations even farther by investigating processes to include dynamic behavior in the printed part. This resulted in the starting of 4D printing. Skylar Tibbits from MIT self-assembly Lab was the first to coin this word "4D printing" in a TED talk [3]. It is a ground-breaking breakthrough that incorporates time into the qualities of printed things. Unlike traditional 3D printing, which produces static objects, 4D printing enables printed materials to alter their shape, adapt, or self-assemble when exposed to environmental stimuli such as temperature, humidity, or light [4]. This paradigm shift in additive manufacturing offers up a whole new universe of possibilities, including the ability to create dynamic structures, adaptable materials, and smart gadgets with previously unheard-of functionality [5]. By seamlessly incorporating the dimension of time into the printing process, 4D printing technology has evolved drastically, transforming industries such as responsive architecture, programmable materials, biomedical devices, and more. The transitions from 1D to 4D printing is well depicted in Figure 4.1. As we approach the peak of this technological revolution, the future of design, engineering, and manufacturing is being redefined, opening the way for unparalleled levels of creativity, efficiency, and sustainability.

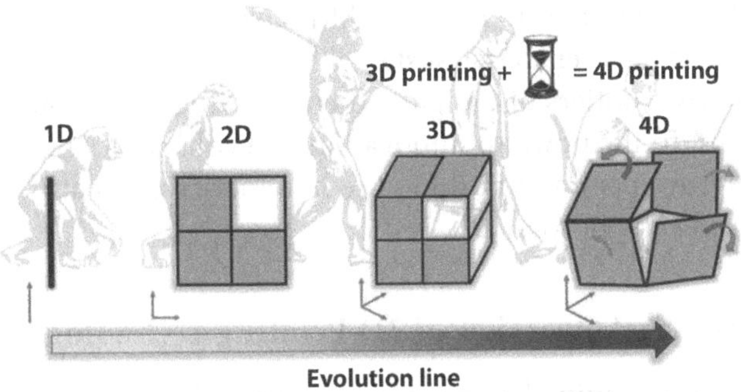

Figure 4.1 Evolution of 4D printing from 1D, 2D, and 3D [1].

4.2 Additive Manufacturing Evolution

AM, often referred to as 3D printing, has undergone a remarkable evolution since its inception, transforming from a niche technology to a disruptive force across various industries. In the early days of additive manufacturing, innovation was just beginning to take shape. Researchers were exploring uncharted territory, and applications were few and far between. However, groundbreaking technologies like stereolithography (SLA) and selective laser sintering (SLS) emerged as the pioneers of this new field, establishing a foundation that would lead to the remarkable advancements we see today. These early technologies set the stage for a revolution, allowing manufacturing to evolve beyond traditional methods into something far more dynamic and flexible [6]. As patents expired in the early 2000s, the democratization of AM accelerated, leading to the proliferation of desktop 3D printers and spurring innovation within the maker community. This period marked a shift towards consumer-grade AM and fueled experimentation with new materials and printing techniques. Concurrently, industries began to recognize the potential of AM for rapid prototyping, customization, and on-demand production, leading to its integration into manufacturing processes across aerospace, automotive, healthcare, and beyond. The evolution of AM has been driven by technological breakthroughs in materials science, hardware development, and software innovation [7]. AM now spans a broad spectrum of techniques, including fused deposition modeling (FDM), stereolithography (SLA), and digital light processing (DLP), each catering to specific needs and offering distinct benefits. Recent breakthroughs have significantly extended AM's capabilities, enabling the production of highly complex shapes, multi-material components, and functional parts with remarkable precision and speed [8].

As technology continues to evolve, AM holds the promise of revolutionizing supply chains, enabling sustainable manufacturing practices, and unlocking new frontiers of design and engineering. The evolution of AM is a testament to the power of innovation and collaboration, driving continuous improvement and pushing the boundaries of what is possible in manufacturing and beyond [9].

4.2.1 Historical Overview of 3D Printing Technology

The story of 3D printing begins in an unexpected place: the realm of science fiction. In 1945, a writer named Murray Leinster penned a captivating short story titled Things Pass By, where he imagined a remarkable machine that could transform sketches into real, tangible objects. This imaginative concept may have seemed far-fetched at that time, yet it laid the groundwork for a future where such technology could become a reality [10]. Fast forward to 1980, where a researcher named Hideo Kodama worked diligently at the Nagoya Municipal Industrial Research Institute in Japan. Inspired by the possibilities hinted at in Leinster's fiction, Kodama set out to turn the dream of 3D printing into a practical endeavor. He developed two innovative methods to create three-dimensional objects from a photo-curable thermosetting polymer, pushing the boundaries of what was technically possible. However, despite his groundbreaking contributions, the world wasn't quite ready for Kodama's vision. His patent for an XYZ plotter was ultimately abandoned, due to lack of funding and insufficient interest at the time. Yet, his pioneering efforts did not go unnoticed; they opened the door for future innovators who would build upon his ideas and propel the technology forward. Little did Kodama know that the seed he planted would grow into a revolution, forever changing the landscape of manufacturing and design, transforming the once-imagined dream of 3D printing into a vibrant reality. While purely imaginative at that time, this idea laid the groundwork for future technological breakthroughs [11].

The next major breakthrough came in 1984, when Chuck Hull invented stereolithography (SLA), a process that used ultraviolet light to harden layers of a liquid photopolymer, creating solid objects one layer at a time. Hull's invention was groundbreaking not only because of the technology itself but also because it introduced critical components like the STL file format and slicing techniques, which would become foundational for the 3D printing industry. Hull's work led to the founding of 3D Systems in 1986, and in 1988, the company released the SLA-1, the world's first commercial rapid prototyping system [12].

Meanwhile, in 1988, S. Scott Crump, while experimenting with a glue gun at home, developed the fused deposition modeling (FDM) process, which extruded heated plastic filament to form objects layer by layer [13]. Crump's process democratized 3D printing by offering a simpler, more cost-effective solution compared to stereolithography. This innovation ultimately led him to establish Stratasys, a company that remains a leading force in 3D printing even today.

During the 1990s, academic institutions like MIT, Stanford, and Carnegie Mellon University began pushing the boundaries of additive manufacturing. Researchers developed new methods for material deposition, including micro casting and material spraying, which allowed for more complex geometries. At MIT, Emanuel Sachs developed a powder-based printing technique that was later commercialized [14].

In 1995, the Fraunhofer Institute introduced selective laser melting (SLM), a game-changing development that enabled the production of metal parts with additive manufacturing. This innovation opened new doors for industries that required high-strength, functional parts, particularly in aerospace and automotive sectors [15].

As the technology advanced, 3D printing began expanding beyond its industrial and research roots. By the early 2000s, fields like architecture and medicine started exploring additive manufacturing for customized designs and medical implants. Around this time, Adrian Bowyer at the University of Bath launched the RepRap project, aiming to create a self-replicating 3D printer that could print most of its own components. This initiative, combined with open-source platforms like Fab@Home [16], led by Evan Malone and Hod Lipson, ignited a surge of interest in affordable, accessible 3D printers.

The year 2009 marked a pivotal moment in the evolution of desktop 3D printing with the expiration of several crucial patents, notably those pertaining to Fused Deposition Modeling (FDM) technology. This development catalyzed a surge of creativity, leading to the emergence of numerous companies and startups dedicated to producing affordable 3D printers for consumers. As a result, the landscape of additive manufacturing transformed dramatically, lowering entry barriers and broadening the spectrum of materials and applications available for this innovative technology. This wave of progress opened new doors, allowing hobbyists, educators, and entrepreneurs alike to explore the possibilities of 3D printing in ways previously thought unattainable [17].

By the 2010s, 3D printing had matured into a vital tool in industries such as aerospace, where companies like Boeing, Airbus, and General Electric began utilizing additive manufacturing to create lighter, more efficient

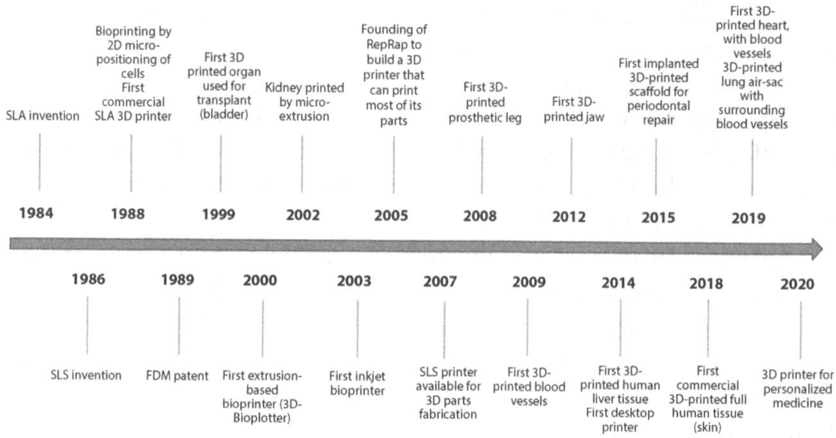

Figure 4.2 Important developments in AM [18].

parts. GE's introduction of 3D-printed fuel nozzles for its LEAP engine demonstrated the transformative potential of the technology, reducing part count and weight while enhancing performance.

Today, 3D printing continues to evolve, with advancements in materials, techniques, and accessibility shaping its future. What began as a science fiction dream is now a cornerstone of modern manufacturing, pushing the boundaries of what's possible in design, engineering, and production. Various important developments in the field of AM are shown in Figure 4.2.

Till the year 2010, 3D printers have made significant improvements in their efficiency and accuracy with lots of applications in the areas of aerospace industry to create parts that would improve the fuel efficiencies of the jet engines. Companies like Airbus, Boeing and General Electric started switching over to AM processes not just for prototyping but to use them as final functional parts for reduced weight, better efficiencies, reduced scraps etc. [19].

4.3 A Decade of 4D Printing (2013-2023)

4D printing history and evolution is based on the foundation of the 3D printing technology and its development. It is also highly dependent on the development in the field of smart materials. The history of the concept of 4D printing started with a TED talk given in 2013 [3], by MIT processor, designer, architect Skylar Tibbits of the Self-Assembly Lab. The concept of 4D printing, came from the ability of self-assembly in micro and nano scale by various systems like protein. Skylar while trying to make programmable materials, came with the concept of using AM with this kind of materials and ideas to create a 4D printed part [4]. In the same year, with collaborations from Autodesk and Stratasys, they printed the first 4D printed part. Some of the major works in the field of 4D printing from 2013 till now is shown in Figure 4.3.

In 2014, researchers embarked on an innovative study that revealed the potential to create a variety of shapes with intricate geometries capable of evolving independently over time. The first model showcased a curve that transformed dynamically across a two-dimensional plane, while the second design illustrated a grid that shifted into a sinusoidal wave and subsequently morphed into a hyperbolic surface. The third model combined both stretching and folding actions within a single structure, highlighting its versatility. The investigation delved into how the interplay of material properties and geometric designs could facilitate self-transformation. It introduced hydrophilic materials engineered for precise temporal deformations and devised a bending mechanism that allows for local coordination, thus enabling effective shape morphing in 4D printed objects. This research not only advanced the field of 4D printing but also opened new avenues for functional applications where structures can adapt and respond to their environments in real time [20]. Another group of researchers expanded the 4D printing concept by exploring the use of active origami, using 2D flat sheets with hinges made from polymer fibers. The hinges fold into 3D

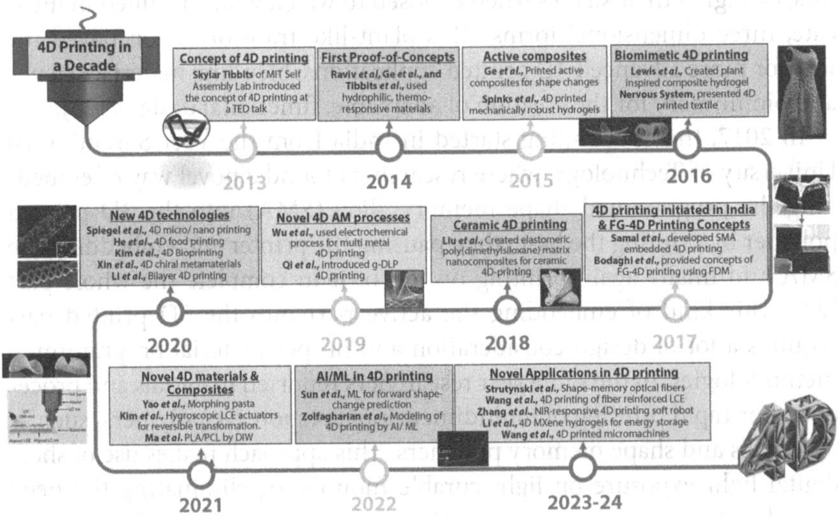

Figure 4.3 Most important developments in 4D printing in the last decade.

structures when heat is applied, and a theoretical model is developed for further applications [21].

By 2015, researchers began to explore the applications of 4D printing across various industries, including aerospace, healthcare, and architecture. In aerospace, for example, researchers envisioned 4D-printed components that could adapt to changing conditions during flight, leading to more efficient and resilient aircraft designs. Similarly, in healthcare, 4D-printed medical implants showed promise in adapting to the patient's body over time, improving comfort and effectiveness. A new type of 4D printing called active composite is developed in the same year [22]. Design of the composite is possible to be directly printed by utilizing a multi material AM and applying post-process as thermo-mechanical programming.

In 2016, the world saw significant breakthroughs in 4D printing, fueled by innovations in both printing technologies and the development of responsive smart materials. Scientists worked to refine the accuracy and control of 4D printing processes, making it possible to create structures that were not only more complex but also highly customizable. These advancements allowed for the fabrication of objects with dynamic, shape-shifting capabilities, driven by external stimuli. A key innovation involved the use of composite hydrogels, where cellulose fibrils were aligned in specific orientations to control how these materials would respond to environmental changes. By applying minimal theoretical frameworks, researchers were able to program these structures to transform in predictable ways. Inspired by the natural world, these hydrogel-based systems mimicked plant behaviors, changing their shapes when exposed to water, which resulted in intricate, three-dimensional forms. This plant-like transformation paved the way for more advanced 4D printed applications, blending material science and biomimicry for the creation of adaptable, functional designs [23].

In 2017, the 4D printing started in India from the Veer Surendra Sai University of Technology, where researchers found a novel way of embedding the programmed shape memory alloy (SMA) into the 3D printed polymer by printing the base, then pausing the printer and embedding the SMA and finally again printing on top of it to complete the whole part [24]. This kind of embedding the active part into the 3D printed part requires a lot of design consideration and proper material programming methodologies. In the same year researchers reported materials and procedures for rapid printing of multidimensional responsive polymers, such as hydrogels and shape memory polymers. This approach makes use of short digital light exposure on light-curable monomers, eliminating the need for a layer-by-layer procedure in the vertical dimension and sequential pixel manipulation in the planar dimensions [25]. A study highlights how

FDM has been employed to develop adaptive and shape changing/morphing metamaterials that integrate functional properties directly into their design. This approach leverages the thermo-mechanical characteristics of shape memory polymers (SMPs) and the layer-by-layer manufacturing technique of FDM to engineer materials capable of self-expansion behavior. By carefully controlling fabrication parameters such as printing speed and temperature, researchers have found how these variables influence the material's ability to program shape transformations during printing. In addition, other factors like infill density and bed temperature play a crucial role in enhancing the shape memory effect of these materials. Higher infill densities improve the structural rigidity, while the print bed temperature affects the activation and responsiveness of the shape change. This research opens new doors for creating programmable metamaterials that can adapt their structure in response to external stimuli, offering promising potential for future applications in areas requiring smart, self-transforming components [26, 27].

In 2018, researchers developed elastomeric poly(dimethylsiloxane) matrix nanocomposites (NCs) that can be printed, distorted, and changed into silicon oxycarbide matrix NCs, enabling the creation of intricate ceramic origami and 4D-printed structures. The microlattice has a compressive strength of 547 MPa and is time-efficient for origami and 4D ceramic printing [28]. In 2018, researchers from India also developed the strongest 4D printed part considering the FDM and SMA-SMP combination showing a compressive strength of 150 MPa [29]. Another explores AM of direct magnetic SMA utilizing Ni-Mn-Ga powders. Three types of powders were used: spark eroded in liquid nitrogen, spark eroded in liquid argon, and ball milled. The 3D printing method, known as powder bed binder jetting (3DP), allows for easy control of component porosity and the production of complex shapes. The process, which involves binder jetting, curing, and sintering, results in spring-like, three-dimensional net-shaped porous structures showing 4D effect. A composite ink system combining poly(dimethyl siloxane) (PDMS) and iron nanoparticles has been developed for 4D printing. This system allows for the creation of magnetically responsive 3D structures with a fast response time under external magnetic fields. The ink's elastomer component and soft magnetic Fe particle allow for programmable and actuated 3D structures, making it suitable for reversible 4D printing and enabling the creation of functional devices [30]. The article introduces triple SMPs using 4D printing technology for creating adaptive structures for mechanical/bio-medical devices. It uses hot-cold programming and FDM to characterize SMPs properties at low and high temperatures. It demonstrates the feasibility of dual and triple SMPs with self-bending features [31]. A new 4D printing technology for composites with embedded continuous fibers allows for programmable deformation with high accuracy. The deformation is caused by the difference in thermal expansion coefficients between fibers and flexible matrix. The technology also allows for deformation of deployable surfaces, with a 6% error between theoretical and experimental values [32].

In 2019, in a pioneering work, researchers presented a single-vat grayscale digital light processing (g-DLP) 3D printing method, which can produce functionally graded materials with high resolution and mechanical gradient [33]. It showcases direct fabrication of complex 2D/3D lattices, negative Poisson's ratio metamaterial, presurgical models, composites for 4D printing, and anti-counterfeiting 3D printing. The study explores the use of digital light processing AM (DLP-AM) to control chemical composition along all three axes of an object, offering advantages in stress-focusing designs for mechanoresponsive materials, simulated tissue models, controlled gradation in over molded parts, dynamic optical materials, and autonomous actuators [34]. The DLP-based approach enables metamaterial and spatially controlled compositions based on input wavelength.

In 2020, notable developments include a bilayer hydrogel strategy using tetra-(4-pyridylphenyl) ethylene (TPE-4Py) is designed to create stimuli-responsive hydrogels with fluorescence color, brightness, and shape changes. The active layer is made of poly(acrylamide-r-sodium 4-styrenesulfonate) (PAS) and TPE-4Py. Protonation of TPE-4Py causes fluorescence changes and deformation, making them attractive for 3D/4D printing, soft robots, and smart wearable devices [35]. Another study presents a 4D bioprinting system using a shape-controllable cell culture environment for *in vivo* applications. The method uses digital light processing (DLP) 3D bioprinting technology and photopolymerizable silk fibroin bioink. Hydrogels with high water holding capacity are suitable for 4D bioprinting due to their swelling properties. DLP has faster printing speed, higher cell viability, and better structural integrity than other methods [36]. For the very first time, a progress report was published that explores emerging materials for 4D printing at the microscale, focusing on hydrogels, liquid crystalline, and composite materials. It discusses challenges, design barriers, and performance in 4D microprinting. Strategies are proposed to overcome limitations and advance applications in biomedicine, micro robotics, and optics [37, 38]. A miniature biomimetic 4D printing of pH-responsive hydrogel is demonstrated using femtosecond laser direct writing, achieving microscale dimensions and sub second response speed. The technology allows shape transformation with multiple degrees of freedom, demonstrating potential applications in micromanipulation, single-cell analysis, and drug delivery [39]. Another study developed chiral metamaterials with large deformation, programmable, reconfigurable, and tunable properties through 4D printing. The chiral metamaterials can simulate specific tissues/organs' σ–λ curves and can be converted to another biomaterial through programming [40]. 4D printing of food research started from, the study where purple sweet potato flake as a source of anthocyanins for 3D printing. Anthocyanins, a bioactive substance with antioxidant capacity, can undergo color changes under specific environmental stimuli like pH. The gelatinized purple sweet potato powder was chosen for 3D printing, aiming to create ready-to-eat 4D foods with spontaneous color change under pH, providing valuable information in 4D food printing [41]. In another study high speed 4D printing is explored. Development of tBA-co-DEGDA 4D-printable shape memory polymer (SMP) with nano silica particles is done for DLP printing, highlighting their excellent mechanical strength, high strain, and shape memory properties, potentially enhancing printing speed for 3D and 4D printing processes [42].

In 2021, An excellent study by the researchers showed how morphing technologies can have greater impact on our day-to-day life. They showed that the diffusion-based mechanism enables transient morphing in parametric surface grooves, transforming flat objects into designed shapes. This mechanism can be realized using low-cost manufacturing methods and can induce temporary swelling or deswelling. Tuning the grooving pattern can achieve zero and nonzero Gaussian curvature geometries, potentially improving food manufacturing efficiency and sustainable packaging [43]. Various innovative applications of the 4D printing were developed. One study presents a new hydration method for creating a hygroscopic liquid crystal elastomer (LCE) actuator, enabling reversible shape transformation in response to changes in humidity [44]. This is the first time such an actuator has been created using DIW based 3D printing, expanding complex LCE geometries and enabling the fabrication of various types of actuators. The researchers have developed a new 3D printing material, poly (glycerol dodecanoate) acrylate (PGDA), by modifying the biodegradable poly (glycerol dodecanoate) (PGD) [45]. PGDA, a photo-curable PGD, has been used to create 3D structures with a large fixity ratio, recovery ratio, cycling stability, and short recovery time. The Young's moduli of the structures are comparable to soft biological tissues, reducing mechanical mismatch. The printed structures were tested *in vitro* and *in vivo* for potential biomedical applications. More and more composites and polymer blends have been developed for better printability and high shape memory properties. Researchers explored the use of DIW printing to fabricate polylactic acid (PLA)/polycaprolactone (PCL) composites [46]. The study examines the impact of PCL concentration on thermodynamic behaviors, structures, and mechanical properties. The composites were also analyzed for shape memory properties. The research suggests potential applications in bio-inspired robotics and biomedical devices. Another study presented a method for 2D-to-3D shape-shifting that overcomes limitations of in-plane anisotropies in traditional methods [47]. It introduces a simple design for hobbyist 3D printers to 4D-print on curved surfaces, using computational models to simulate shape-shifting behavior. The approach is demonstrated through the fabrication of reconfigurable materials and devices.

In 2022, A lot of development in new material development, innovative applications, artificial intelligence and new printing methodologies are developed for 4D printing applications. A new method has been developed for manufacturing MXene hydrogels using PEDOT:PSS as the crosslinker. This technology allows for the creation of a series of MXene hydrogels with diverse structures and properties, including Nb2CTx, Ti3C2Tx, and Mo2Ti2C3Tx hydrogels. The hydrogels are customizable and feature excellent hydrophilic properties, large specific surface areas, and high electrical conductivities. The technology also allows for low-temperature operation and high energy and power densities, making them potential energy-storage devices [48]. A new study presents an efficient approach for forward shape-change prediction and inverse material design for voxelized active composite beams using recurrent neural network-based machine learning and an EA. This method quickly achieves optimal designs for multiple target shapes and simplifies design-fabrication [49]. Another study presents 4D printed hydrogel capsules for smart controlled drug release, featuring a core-shell structure with a model drug and a cross-linked poly(N-isopropylacrylamide) hydrogel [48]. Temperature sensitivity, shell thickness, and environmental conditions are investigated in detail. Another research explored the fabrication of customized Ni-Ti jewels using four-dimensional printing [48]. The rings exhibit one-way shape memory effect and two-way shape memory effect, allowing them to adapt to environmental changes, improving comfort and fit. A novel 4D-printed gripper is also designed by utilizing actuators with various geometric patterns, such as grid, honeycomb, triangular, and rectilinear configurations. The design focuses on evaluating how these different patterns impact the functionality of the actuators, which are divided into two main sections: the upper and lower parts. The actuators are fabricated using ten layers of PLA, with consistent line and border widths across the structure. Notably, the lower section of the actuator is printed at a reduced speed to enhance control and precision in the movement, contributing to the gripper's overall adaptability and responsiveness in dynamic tasks [50]. The same research team introduced a novel approach using FDM technology to create PLA-based large-stroke XY micro positioning stages, significantly streamlining the fabrication process by reducing complexity, production time, and cost. These innovative micro stages demonstrate effective positioning capabilities when activated by temperature, marking a key advancement in 4D printing. Notably, this work is the first to develop large-stroke XY micro positioning stages made from PLA that do not require any manual intervention for shape programming, offering an automated and efficient solution for precise movements [51]. These stages are capable of functioning underwater without the risk of rust or the need for electricity. Finite element analysis is employed to model and simulate their actuation. One study investigated the use of a shape-memory PLA/PCL blend in 4D printing to develop tracheal scaffolds for medical purposes. The inclusion of PCL enhanced the blend's flexibility and its ability to maintain a temporary shape after deformation. The PLA70/PCL30 blend exhibited an ideal shape fixity of 91% and a recovery rate of 90%, resulting in a thermally activated tracheal scaffold that conformed perfectly to the lumen [52]. Another outstanding research presents a novel hybrid additive manufacturing system that 4D prints freestanding structures. It is a novel approach to 3D printing liquid crystalline elastomers (LCEs) involving an innovative single-step process. This technique integrates a top-down digital light processing (DLP) system with a direct ink writing (DIW) setup, combined with an *in-situ* laser curing mechanism. As the LCE ink is extruded and stretched by the moving nozzle, the laser cures it simultaneously, ensuring structural integrity. The DLP system, meanwhile, can provide optional supports, either structural or removable, to assist during the printing process. This method effectively overcomes the challenge of printing soft and flexible LCE materials in mid-air while preserving their desired form [53].

Till 2023, 4D printing has become a widely researched topic among AM scientists worldwide. A study explored the development of a mechanically robust, magneto-responsive SMP with 3D printability and excellent shape memory behavior using PLA/TPU polymer blends and Fe_3O_4 particles. The study evaluates printability, morphology, thermal, rheology, and mechanical properties, and designs two smart structures for 3D printing [54]. In another study, the effects of printing speed, temperature, passive-to-active layer ratio, layer height, and height of polylactic acid (PLA) actuators on internal strain. Other researchers developed a low cost testing setup [55] and also studied the shape memory properties for differently programmed samples and showed a comparison between both of them [56]. Recent developments also includes utilizing these programming for creating shape changing metamaterial structures [57]. This approach can be applied to thermal additive manufacturing processes, expanding materials used for 4D printing. The study uses fused filament fabrication (FFF) technology to thermally stimulate 2D planar structures. Engineers developed a 4D-printed electrode for small peripheral nerve interfacing, allowing self-folding and actuation. The device demonstrated reliable *in vivo* interfacing, successful stimulation, and easy extraction without damaging the nerve [58]. Laser-induced thermal stress can control 3D printed structures, creating a new 4D printing method using metallic 4D printing technology, enhancing structural transformation and making 3D structure production faster and more straightforward [59]. 4D printing strategy enables intricate ceramic structures through direct ink writing technology, allowing bilayer ceramic platelets to self-shape after sintering, offering a bottom-up demanding design approach [60]. The integration of polymer-derived ceramics (PDCs) and additive manufacturing techniques improves geometrical flexibility and mechanical properties. The first ceramic 4D printing system was developed in 2018, but its practical application is limited due to time-intensive transformations, low precision, and mechanical contact stimuli requirements. Shape memory ceramics (SMCs) have limitations due to material universality, geometrical flexibility, structural dimensions, and shape memory capability. A novel paradigm involving one-step shape/ material transformation is needed for further development [61]. Apart from metals and ceramics, regenerative living 4D printing allows for the transformation of a 3D-printed seed template into distinct geometric and mechanical structures through controlled localized growth and chemical regeneration [62].

4.4 Understanding the Fourth Dimension: Time in Printing

The idea of 4D represents a significant augmentation of our knowledge of reality, adding time to the conventional spatial dimensions of length, breadth, and height. This addition shifts our experience of the world from a static picture to a dynamic continuum in which events occur throughout time. In this environment, time acts as a channel for change, creating a temporal framework within which all activities develop and grow. Albert Einstein notably established the notion of time as a dimension in his theory of relativity, which combined space and time into a single entity known as spacetime [63].

Einstein envisioned spacetime as a four-dimensional continuum in which the three dimensions of space are inextricably linked to the dimension of time. Objects in the cosmos move through space and time, with their motion and interactions defined by the curvature of spacetime. The existence of mass and energy alters the fabric of spacetime, resulting in gravitational fields that impact the trajectory of things passing through it. This beautiful framework of spacetime gives a coherent account of the universe, helping us to explain phenomena ranging from planet motion to light bending by enormous celestial bodies.

In the field of manufacturing and technology, 4D printing is a game-changing breakthrough that uses spacetime principles to construct dynamic, adaptable structures. Unlike typical 3D printing, which creates static items layer by layer, 4D printing adds a new level of functionality by using materials that can react to environmental stimuli such as temperature, humidity, or light over time. This capacity allows printed things to morph, self-assemble, or modify their shape and attributes independently, bringing up new possibilities in engineering, architecture, and biological applications [64].

4D printing promises to transform manufacturing processes by leveraging the power of spacetime, allowing for the fabrication of smart materials and structures that can actively adapt to their surroundings and change over time [65].

4.4.1 Conceptualization of Time as a Dimension in Additive Manufacturing

The conceptualization of time as a dimension in AM represents a profound paradigm shift that has the potential to redefine the way we design, fabricate, and interact with objects. This innovative approach, known as 4D printing, builds upon the foundation laid by traditional 3D printing techniques but introduces a transformative element by incorporating time as the fourth dimension. While conventional 3D printing has enabled the creation of three-dimensional objects layer by layer from digital models, the addition of time in 4D printing introduces a dynamic aspect to the fabrication process. This dynamic aspect allows printed objects to respond to external stimuli or triggers over time, leading to shape changes, self-assembly, or other transformations that were previously unattainable with static fabrication methods [66]. The major differences between the conventional 3D printing or AM and 4D printing is well depicted in Figure 4.4, and summarized in Table 4.1.

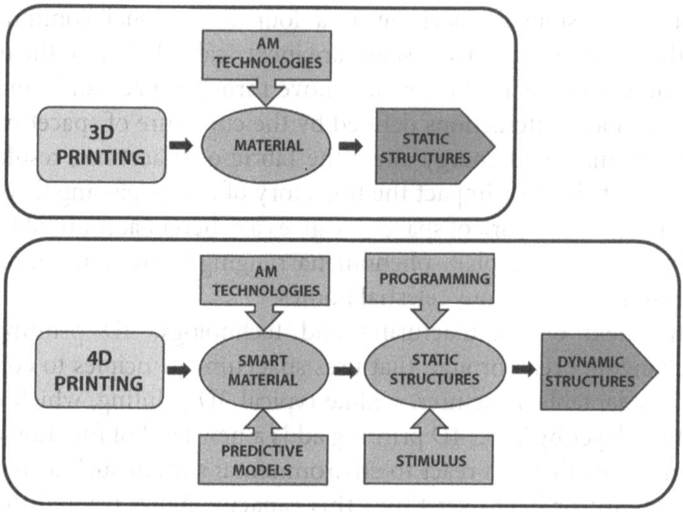

Figure 4.4 Difference between 3D and 4D printing technology.

Table 4.1 Differentiation between 3D and 4D printing considering various aspects.

Aspects	3D printing	4D printing
Core principle	Static 3D structures from digital models.	Adds time dimension, enabling shape change over time.
Materials used	Plastics (HIPS, ABS), metals, ceramics, resins.	Smart materials (SMPs, hydrogels, LCEs, SMAs).
Stimuli-response	No response after printing.	Responds to stimuli like temperature, moisture, light, pH.
Triggering stimuli	None post-printing.	Thermal, hydration, light, pH, magnetic stimuli.
Design customization	High geometric customization for static objects.	High customization for both geometry and time-based behavior.

(Continued)

Table 4.1 Differentiation between 3D and 4D printing considering various aspects. (*Continued*)

Aspects	**3D printing**	**4D printing**
Control over properties	Limited to geometry, surface texture, and structure.	Extends to time-dependent behavior, like morphing and stiffness.
Fabrication process	FDM, SLA, SLS, DLP, etc.	Same methods with programmable materials for time behavior.
Post-processing	May need curing or polishing.	Requires stimuli-based tuning for shape-shifting behavior, like heat
Material behavior	Fixed after printing.	Changes over time in response to stimuli.
Material programming	No post-printing programming.	Preprogrammed to react based on stimuli.
Applications	Prototyping, aerospace, automotive, medical, consumer goods, static implants, light weight structures	Adaptive devices, self-transforming implants adapting to body conditions, Shape-morphing wings or com-ponents.
Sustainability	Varies by process, not always recyclable.	Focus on biodegradable, self-healing materials.
Cost of Materials	Lower for common materials	Higher cost for advanced smart materials.
Emerging trends	Multi-material printing, faster speeds, improved resolution.	AI integration, multi-stimuli-responsive materials, bio-inspired designs.
Scalability	Scalable, but limited by print size and post-processing.	Challenging due to high material costs and complex programming.
Regulatory concerns	Standard regulations for static objects.	New safety regulations needed for shape-shifting components.

The integration of time as a dimension in AM opens up a wide array of possibilities for creating intelligent and adaptive materials that can exhibit behaviors such as self-repair, self-assembly, or shape-shifting in response to specific environmental conditions or user-defined parameters. This concept is inspired by natural phenomena observed in living organisms, where structures and materials can adapt and evolve over time to optimize their performance or functionality [67].

The journey towards conceptualizing time as a dimension in AM began with the foundational innovations of 3D printing, which revolutionized traditional manufacturing processes by enabling the direct fabrication of complex geometries from digital models. Initially developed for rapid prototyping purposes, 3D printing quickly expanded its applications across various industries, including aerospace, automotive, healthcare, and consumer goods. The ability to create customized parts on-demand with minimal material waste and lead times has made 3D printing a preferred manufacturing method for many companies seeking to streamline their production processes and enhance product design flexibility [67].

As advancements in 3D printing progressed, researchers and engineers began to push the boundaries by introducing an additional dimension to the process, giving rise to 4D printing. This breakthrough opens up new possibilities for fabricating dynamic structures, adaptive materials, and intelligent devices with innovative and versatile functionalities, marking a significant leap in additive manufacturing technology [68].

A major benefit of 4D printing lies in its capability to produce objects that autonomously respond to environmental changes or user inputs, without requiring external control. This self-activating behavior is made possible by incorporating smart materials, which have inherent properties that allow them to undergo specific transformations when exposed to stimuli like temperature shifts, humidity, light, or magnetic fields. By embedding these responsive materials into the printed structure and programming them to react to set triggers, engineers can design objects capable of altering their shape, flexibility, color, or other characteristics over time, enabling a new level of adaptive and interactive functionality [69].

The concept of time as a dimension in AM also opens up new avenues for research and innovation in materials science, as researchers seek to develop novel materials with tailored properties and functionalities for specific applications. Advances in material science, such as the development of shape memory polymers, hydrogels, and stimuli-responsive composites, have paved the way for the creation of intelligent materials that can undergo reversible changes in shape, volume, or mechanical properties in response to external stimuli. By combining these advanced materials with additive manufacturing techniques, researchers can fabricate complex structures with built-in functionality that goes beyond what is achievable with traditional manufacturing methods.

In addition to its applications in materials science and engineering, 4D printing technology has already seen high applicability in biomedical field. By using the shape memory properties of 4D-printed materials, researchers can create biomedical implants, drug delivery systems, and tissue engineering scaffolds that can adapt and evolve in response to changes in the patient's body or the surrounding environment. For example, 4D-printed implants could be designed to change shape or release drugs in response to changes in temperature, pH, or mechanical stress, providing personalized and adaptive treatment options for patients with complex medical conditions.

4D printing also has high potential applications in the field of robotics and autonomous intelligent systems by enabling the fabrication of self-assembling or self-moving structures and self-reconfigurable robots. By incorporating smart materials and programmable actuators into the design of robotic systems, engineers can create micro machines that can change their shape or size, or properties to adapt in real time for different operating environments. This concept, known as "robotic metamorphosis," could lead to the development of versatile and adaptable robots that can perform a wide range of tasks in dynamic and unpredictable environments.

4.5 Laws of 4D Printing

The "Laws of 4D Printing" outline three foundational principles that governs the stimuli responsive behavior depending upon time in 4D-printed structures [70] as shown in Figure 4.5. These laws provide a comprehensive framework for understanding how 4D-printed materials evolve over time, adding the crucial fourth dimension of time to traditional 3D printing. The foundational principle of multi-material 4D printing is that nearly all shape-morphing behaviors stem from the differential expansion between stimuli-responsive (active) and non-responsive (passive) materials. This difference in expansion drives intricate deformations like bending, twisting, and coiling. Through strategic placement of active and passive materials during the fabrication process, designers can pre-program precise and predictable shape changes that occur in response to external stimuli, enabling highly controlled transformations.

The second law classifies the physical phenomena driving these shape changes into four main categories: mass diffusion, thermal expansion,

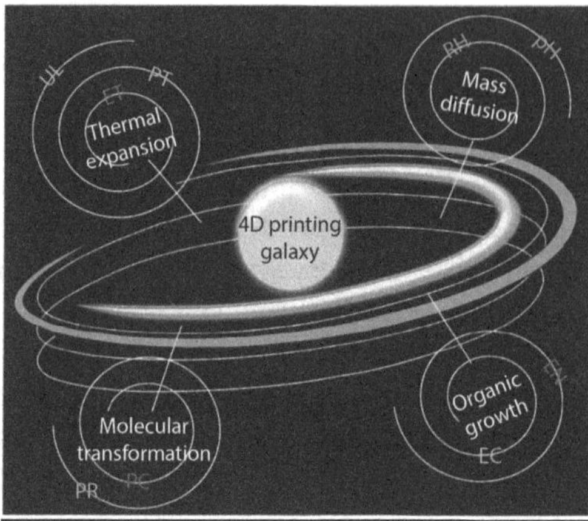

1st law of 4D printing: Almost all shape-morphing behaviors of multi-material 4D structures originate from "relative expansion" between active and passive materials

2nd law of 4D printing: The shape-morphing behaviors of almost all multi-material 4D structures have four types of physics (4D printing galaxy)

Unified formula of the 2nd law:

C_j depends on different factors
τ_j depends on different factors

$$\left\{\begin{array}{l} \varepsilon_{\text{mass diffusion}}(t) = C_1[1 - \exp(-t/\tau_1)] \\ \varepsilon_{\text{thermal expansion/contraction}}(t) = C_2[1 - \exp(-t/\tau_2)] \\ \varepsilon_{\text{molecular transformation}}(t) = C_3[1 - \exp(-t/\tau_3)] \\ \varepsilon_{\text{organic growth}}(t) = C_4[1 - \exp(-t/\tau_4)] \end{array}\right\} \Rightarrow C_j\left[1 - \exp\left(-\frac{t}{\tau_j}\right)\right]$$

Formula of the 3rd law:

$$k(t) = \frac{H_I\left(1 - e^{\frac{-t}{\tau_I}}\right) + H_{II}\left(1 - e^{\frac{-t}{\tau_{II}}}\right)}{\frac{h}{2} + \frac{2(E_1 I_1 + E_2 I_2)}{h}\left(\frac{1}{E_1 a_1} + \frac{1}{E_2 a_2}\right)}$$

3rd law of 4D printing: Time-dependent shape-morphing behavior of almost all multi-material 4D structures is governed by two "types" of time constants

Figure 4.5 Three laws of 4D printing [70].

molecular transformation, and organic growth. In mass diffusion, stimuli like moisture or solvent absorption cause a transport of matter, leading to expansion or contraction. Thermal expansion occurs when a change in temperature alters the distances between atoms or molecules, inducing expansion or contraction. Molecular transformation involves the realignment or reconfiguration of molecules (e.g., light-induced cis-trans conversion), which leads to structural changes without a mass or temperature change. Finally, organic growth refers to the biological expansion of living tissues, which can be integrated into 4D bioprinting applications. Each of these categories exhibits a time-dependent response that follows a bi-exponential model, capturing the dynamics of short- and long-term behavior.

The third law introduces two distinct time constants that govern the time-dependent shape transformations in 4D-printed structures. The first time constant, sI, relates to the mismatch-driven stress at the interface between active and passive materials. This internal stress plays a critical role in determining how quickly the materials deform in response to external stimuli. The second time constant, sII, describes the material's response to fillips, such as heat, light, or moisture. Together, these two-time constants capture the non-linear, time-dependent behavior of 4D-printed structures. The overall shape transformation is modeled by a unified bi-exponential formula, which accurately predicts how 4D structures evolve over time.

4.6 Challenges and Opportunities in the Transition to 4D Printing from 3D

The transition from 3D to 4D printing introduces new challenges as well as significant opportunities. The key distinction between the two lies in the added temporal dimension, where objects can transform over time in response to external stimuli. However, this transformative capability comes with various material, scalability, regulatory, and technical challenges. Addressing these will unlock the full potential of 4D printing in industries such as aerospace, healthcare, and consumer products [71].

4.6.1 Material Limitations and Compatibility Issues in 4D Printing

One of the foremost challenges in 4D printing is the limited availability of materials that can reliably respond to stimuli such as temperature, humidity, light, and magnetic fields. Current smart materials, including shape memory polymers (SMPs), hydrogels, and liquid crystal elastomers (LCEs), offer a narrow range of responsiveness and functionality. While SMPs have gained popularity due to their ability to undergo shape changes with thermal stimuli, their response time is often slow, and their mechanical properties are not robust enough for high-load applications. Moreover, the compatibility of these smart materials with additive manufacturing processes remains a challenge. Many of these materials require specific environmental conditions to be effectively processed, which limits their integration with widely used AM techniques. The lack of materials capable of multi-stimuli responses also limits the broader application of 4D printing. To fully realize 4D printing's potential, there is a growing need for the development of multi-functional, programmable materials that can seamlessly integrate into existing 3D printing frameworks [72].

4.6.2 Scalability and Cost Considerations for 4D Printing Technologies

Scaling 4D printing technologies from the laboratory to industrial manufacturing is fraught with complexities. One of the key barriers is cost. The production of smart materials like hydrogels, LCEs, and shape memory alloys (SMAs) involves complex synthesis processes that are significantly more expensive than conventional thermoplastics used in 3D printing. Furthermore, many 4D printing technologies involve additional post-processing steps, such as thermal treatment or exposure to specific stimuli, which adds time and cost to production cycles. In terms of scalability, current 4D printing methods still face challenges in terms of precision, speed, and repeatability. Most commercial 3D printers are designed for static objects, and adapting them to produce dynamic, stimulus-responsive parts is technically challenging. Advances such as grayscale digital light processing (g-DLP) and multi-material 3D printing techniques show promise in enhancing the scalability of 4D printing by enabling faster production of highly complex, functionally graded materials [73]. Nonetheless, reducing production costs and enhancing scalability will require innovations in both material science and hardware design.

4.6.3 Regulatory Hurdles and Safety Concerns in Implementing 4D Printing

Implementing 4D printing on a commercial scale raises significant regulatory and safety concerns, particularly in industries like healthcare, aerospace, and automotive, where safety standards are stringent. Regulatory bodies will need to develop new guidelines to assess the reliability, durability, and safety of smart materials. For example, 4D-printed medical implants, such as self-adapting stents or tissue scaffolds, will need extensive testing to ensure they remain biocompatible and functional throughout their lifespan [74]. In addition to material safety, environmental concerns arise from the degradation or disposal of smart materials. Many of the current materials used in 4D printing, such as SMPs, are not biodegradable, which could pose environmental risks if widely adopted. Thus, regulatory frameworks will also need to address the environmental implications of 4D-printed products.

4.7 Future Directions and Emerging Trends in 4D Printing

The future of 4D printing is set to be transformative, with exciting developments in new smart materials, advanced design methodologies, and innovative applications across industries. Researchers are actively exploring ways to overcome the current limitations of the technology, focusing on the creation of more sophisticated, multi-functional materials and the integration of artificial intelligence (AI) and machine learning (ML) into the design and manufacturing process.

4.7.1 Development of New Smart Materials

The development of novel smart materials is at the forefront of 4D printing research. Currently, materials like SMPs and hydrogels dominate the field, but their limitations in terms of response time, durability, and environmental sensitivity are driving efforts to discover more advanced alternatives. For example, researchers are exploring magneto-responsive materials and stimuli-sensitive composites that can respond to a broader range of environmental triggers such as magnetic fields and electrical signals.

A promising development is the use of multi-material systems, which combine different properties to create more versatile 4D-printed objects [75]. Researchers are working on materials that can simultaneously respond to multiple stimuli, such as temperature and humidity, allowing for more complex shape transformations. These multi-functional materials could be particularly useful in applications like wearable electronics and responsive architectural elements.

4.7.2 Integration with Artificial Intelligence and Machine Learning

AI and ML are emerging as critical tools in enhancing the precision and functionality of 4D printing. These technologies enable the optimization of material behavior and design parameters, allowing engineers to predict how a 4D-printed object will behave under different stimuli. By using AI to simulate complex transformations, researchers can create more efficient designs and minimize trial-and-error in the prototyping phase [76].

Machine learning algorithms can also help improve the performance of 4D-printed objects by learning from previous experiments and optimizing material formulations for specific applications. This approach is particularly beneficial in biomedical applications, where personalized implants could be optimized for a patient's unique physiological environment [77].

4.7.3 Biomedical Applications

The biomedical sector is poised to be one of the biggest beneficiaries of 4D printing. Researchers are already exploring the potential of 4D-printed smart implants that can adapt to a patient's body over time. For example, 4D-printed stents could expand or contract in response to changes in blood pressure, reducing the risk of complications and improving patient outcomes. Similarly, tissue scaffolds made from shape memory materials could morph into predefined shapes as tissue regenerates, offering new possibilities for regenerative medicine [78].

4.7.4 Sustainable and Regenerative Manufacturing

Sustainability is another key trend in the future of 4D printing. Researchers are developing biodegradable materials and processes that reduce the environmental impact of 4D printing. For example, regenerative living 4D printing involves creating structures that can regenerate or repair themselves over time, minimizing waste and extending product life cycles. This approach is particularly promising for industries like construction, where 4D-printed materials could self-heal or adapt to environmental changes, reducing the need for maintenance and repairs [79].

4.8 Conclusions

This chapter has provided an in-depth exploration of the progression from 3D to 4D printing, highlighting how the addition of time as a functional dimension offers new possibilities for dynamic, adaptive structures across various industries. 4D printing, which builds on the foundation laid by additive manufacturing, introduces transformative capabilities by enabling printed objects to respond to external stimuli such as temperature, humidity, and light. This dynamic behavior has promising applications in sectors like aerospace, healthcare, and architecture, where adaptive and self-assembling materials can significantly enhance functionality and performance. Despite these advancements, several challenges remain in the transition to 4D printing. Key issues include material limitations, such as the narrow range of smart materials available and their limited compatibility with existing additive manufacturing techniques. Scalability and cost also present obstacles to widespread industrial adoption, as do regulatory hurdles surrounding the safety and reliability of 4D-printed products, particularly in critical sectors like medicine and aerospace. However, the future of 4D printing is full of promise. Emerging trends point towards the development of new multi-functional smart materials that can respond to multiple stimuli, the integration of artificial intelligence to optimize design processes, and the expansion of 4D printing into biomedical and regenerative applications. As researchers continue to innovate in both materials' science and additive manufacturing processes, 4D printing is poised to become a transformative technology that redefines the possibilities of manufacturing and design.

Acknowledgments

BBS, and AJ acknowledge the Ministry of Human Resource Development, Government of India, for the graduate fellowship. All authors are grateful to the research scholars and staff of Direct Digital Manufacturing Lab, Department of Mechanical Engineering, & Fiber Optics Systems Lab, Department of Electronics and Electrical Communication Engineering, Indian Institute of Technology, Kharagpur, for their immense help throughout.

References

1. Mahmood, A., Akram, T., Chen, H., Chen, S., On the Evolution of Additive Manufacturing (3D/4D Printing) Technologies: Materials, Applications, and Challenges. *Polymers*, 14, 21, 4698, 2022. doi: 10.3390/polym14214698.
2. Fidan, I., *et al.*, Recent Inventions in Additive Manufacturing: Holistic Review. *Inventions*, 8, 4, 103, 2023. doi: 10.3390/inventions8040103.
3. Tibbits, S., *The emergence of '4D printing,'*, TED Conferences, LLC (Technology, Entertainment, Design), New York, USA, 2013, Accessed: Jan. 24, 2021. [Online]. Available: https://www.ted.com/talks/skylar_tibbits_the_emergence_of_4d_printing.
4. Tibbits, S., 4D printing: Multi-material shape change. *Archit. Des.*, 84, 1, 116–121, Jan. 2014. doi: https://doi.org/10.1002/ad.1710.
5. Samal, B.B., Jena, A., Varshney, S.K., Kumar, C.S., 4D printing: An experimental case study on processing of shape memory polymer by FDM/FFF for nature inspired structures, in: *Advances in Additive Manufacturing Artificial Intelligence, Nature-Inspired, and Biomanufacturing*, A. Kumar, R.K. Mittal,

A. Haleem (Eds.), pp. 361–377, Elsevier, Amsterdam, 2023, doi: 10.1016/B978-0-323-91834-3.00019-3.

6. Singh, S., Mehla, S., Bhargava, S.K., Ramakrishna, S., History and Evolution of Additive Manufacturing, in: *Additive Manufacturing for Chemical Sciences and Engineering*, S.K. Bhargava, S. Ramakrishna, M. Brandt, P.R. Selvakannan (Eds.), pp. 19–51, Springer Nature Singapore, Singapore, 2022, doi: 10.1007/978-981-19-2293-0_2.
7. Beaman, J.J., Bourell, D.L., Seepersad, C.C., Kovar, D., Additive Manufacturing Review: Early Past to Current Practice. *J. Manuf. Sci. Eng.*, 142, 11, 110812, 2020. doi: 10.1115/1.4048193.
8. Kruth, J.-P., Leu, M.C., Nakagawa, T., Progress in Additive Manufacturing and Rapid Prototyping. *CIRP Ann.*, 47, 2, 525–540, 1998. doi: https://doi.org/10.1016/S0007-8506(07)63240-5.
9. Savastano, M., Amendola, C., D'Ascenzo, F., Massaroni, E., 3-D Printing in the Spare Parts Supply Chain: An Explorative Study in the Automotive Industry BT, in: *Digitally Supported Innovation*, pp. 153–170, 2016.
10. Leinster, M., Things pass by, in: *Thrilling Wonder Stories*, vol. 27, pp. 11–35, Beacon Magazines, Inc, New York, 1945.
11. Pei, E., Kabir, I.R., Leutenecker-Twelsiek, B., History of AM, in: *Springer Handbook of Additive Manufacturing*, E. Pei, A. Bernard, D. Gu, C. Klahn, M. Monzón, M. Petersen, T. Sun (Eds.), pp. 3–29, Springer International Publishing, Cham, 2023, doi: 10.1007/978-3-031-20752-5_1.
12. Krishna, R., Manjaiah, M., Mohan, C.B., Chapter 3 - Developments in additive manufacturing, in: *Additive Manufacturing*, M. Manjaiah, K. Raghavendra, N. Balashanmugam, J.P. Davim (Eds.), pp. 37–62, Woodhead Publishing, Duxford, United Kingdom, 2021, doi: https://doi.org/10.1016/B978-0-12-822056-6.00002-3.
13. Rajan, K., Samykano, M., Kadirgama, K., Harun, W.S.W., Rahman, M.M., Fused deposition modeling: process, materials, parameters, properties, and applications. *Int. J. Adv. Manuf. Technol.*, 120, 3, 1531–1570, 2022. doi: 10.1007/s00170-022-08860-7.
14. Savini, A. and Savini, G.G., A short history of 3D printing, a technological revolution just started. *2015 ICOHTEC/IEEE International History of High-Technologies and their Socio-Cultural Contexts Conference (HISTELCON)*, pp. 1–8, 2015, doi: 10.1109/HISTELCON.2015.7307314.
15. Gibson, I., Rosen, D.W., Stucker, B., Development of Additive Manufacturing Technology, in: *Additive Manufacturing Technologies: Rapid Prototyping to Direct Digital Manufacturing*, pp. 36–58, Springer US, Boston, MA, 2010, doi: 10.1007/978-1-4419-1120-9_2.
16. Malone, E. and Lipson, H., Fab@Home: the personal desktop fabricator kit. *Rapid Prototyping J.*, 13, 4, 245–255, Jan. 2007. doi: 10.1108/13552540710776197.
17. Sells, E., Bailard, S., Smith, Z., Bowyer, A., Olliver, V., RepRap: The Replicating Rapid Prototyper: Maximizing Customizability by Breeding the

Means of Production, in: *Handbook of Research in Mass Customization and Personalization*, pp. 568–580, doi: 10.1142/9789814280280_0028.
18. Sufaru, I.-G., *et al.*, 3D Printed and Bioprinted Membranes and Scaffolds for the Periodontal Tissue Regeneration: A Narrative Review. *Membranes*, 12, 9, 902, 2022. doi: 10.3390/membranes12090902.
19. Chua, C.K. and Leong, K.F., *3D Printing and Additive Manufacturing*, 4th, World Scientific, Singapore, 2014, doi: 10.1142/9008.
20. Raviv, D., *et al.*, Active printed materials for complex self-evolving deformations. *Sci. Rep.*, 4, 7422, 2014. doi: 10.1038/srep07422.
21. Ge, Q., Dunn, C.K., Qi, H.J., Dunn, M.L., Active origami by 4D printing. *Smart Mater. Struct.*, 23, 9, 094007, 2014. doi: 10.1088/0964-1726/23/9/094007.
22. Ge, Q., Mao, Y., Yu, K., Dunn, M.L., Qi, H.J., *Active composites and 4D printing*, ICCM International Conferences on Composite Materials, Copenhagen, Denmark, 2015.
23. Gladman, A.S., Matsumoto, E.A., Nuzzo, R.G., Mahadevan, L., Lewis, J.A., Biomimetic 4D printing. *Nat. Mater.*, 15, 4, 413+, Apr. 2016. doi: 10.1038/NMAT4544.
24. Samal, B.B., Jena, A., Mishra, D., Controlled Shape Changing Components by Using 4D Printing Technology, in: *10th International Conference on Precision, Meso, Micro and Nano Engineering (COPEN 10)*, IIT Madras, October, pp. 78–81, 2017, no. October, Accessed: Feb. 05, 2021. [Online]. Available: http://www.copen.ac.in/proceedings/copen10/copen/19.revised_manuscript_262.pdf.
25. Huang, L., *et al.*, Ultrafast Digital Printing toward 4D Shape Changing Materials. *Adv. Mater.*, 29, 7, 1605390, 2017. doi: https://doi.org/10.1002/adma.201605390.
26. Samal, B.B., Varshney, S.K., Kumar, C.S., Four-dimensional (4D) printing through FDM: Effect of infill density and bed temperature on shape memory properties in different thermo-mechanical programming conditions. *J. Mech. Sci. Technol.*, 38, 8, 4313–4319, 2024. doi: 10.1007/s12206-024-0727-3.
27. Bodaghi, M., Damanpack, A.R., Liao, W.H., Adaptive metamaterials by functionally graded 4D printing. *Mater. Des.*, 135, 26–36, Dec. 2017. doi: 10.1016/j.matdes.2017.08.069.
28. Liu, G., Zhao, Y., Wu, G., Lu, J., Origami and 4D printing of elastomer-derived ceramic structures. *Sci. Adv.*, 4, 8, eaat0641, 2018. doi: 10.1126/sciadv.aat0641.
29. Samal, B.B., Jena, A., Mishra, D., Indian Researchers Develop World's Strongest 4D Printed Parts, in: *Manufactur3D magazine*, 2018, Accessed: Jan. 28, 2020. [Online]. Available: https://manufactur3dmag.com/indian-researchers-develop-worlds-strongest-4d-printed-parts/.
30. Zhu, P., Yang, W., Wang, R., Gao, S., Li, B., Li, Q., 4D Printing of Complex Structures with a Fast Response Time to Magnetic Stimulus. *ACS Appl. Mater. Interfaces*, 10, 42, 36435–36442, Oct. 2018. doi: 10.1021/acsami.8b12853.

31. Bodaghi, M., Damanpack, A.R., Liao, W.H., Triple shape memory polymers by 4D printing. *Smart Mater. Struct.*, 27, 6, 065010, Jun. 2018. doi: 10.1088/1361-665X/aabc2a.
32. Wang, Q., Tian, X., Huang, L., Li, D., Malakhov, A.V., Polilov, A.N., Programmable morphing composites with embedded continuous fibers by 4D printing. *Mater. Des.*, 155, 404–413, 2018. doi: https://doi.org/10.1016/j.matdes.2018.06.027.
33. Kuang, X., *et al.*, Grayscale digital light processing 3D printing for highly functionally graded materials. *Sci. Adv.*, 5, 5, eaav5790, 2019. doi: 10.1126/sciadv.aav5790.
34. Schwartz, J.J. and Boydston, A.J., Multimaterial actinic spatial control 3D and 4D printing. *Nat. Commun.*, 10, 1, 791, 2019. doi: 10.1038/s41467-019-08639-7.
35. Li, Z., *et al.*, Bioinspired Simultaneous Changes in Fluorescence Color, Brightness, and Shape of Hydrogels Enabled by AIEgens. *Adv. Mater.*, 32, 11, 1906493, 2020. doi: https://doi.org/10.1002/adma.201906493.
36. Kim, S.H., *et al.*, 4D-bioprinted silk hydrogels for tissue engineering. *Biomaterials*, 260, 120281, 2020. doi: https://doi.org/10.1016/j.biomaterials.2020.120281.
37. Korrayi, R.R., Jena, A., Marupalli, B.C.G., Samal, B.B., Varshney, S.K., Kumar, C.S., Four-Dimensional (4D) Microprinting, in: *Additive Manufacturing with Novel Materials*, pp. 429–457, John Wiley & Sons, Ltd, River Street, Hoboken, USA & Beverly, MA, USA, 2024, doi: https://doi.org/10.1002/9781394198085.ch14.
38. Spiegel, C.A., *et al.*, 4D Printing at the Microscale. *Adv. Funct. Mater.*, 30, 26, 1907615, 2020. doi: https://doi.org/10.1002/adfm.201907615.
39. Hu, Y., *et al.*, Botanical-Inspired 4D Printing of Hydrogel at the Microscale. *Adv. Funct. Mater.*, 30, 4, 1907377, 2020. doi: https://doi.org/10.1002/adfm.201907377.
40. Xin, X., Liu, L., Liu, Y., Leng, J., 4D Printing Auxetic Metamaterials with Tunable, Programmable, and Reconfigurable Mechanical Properties. *Adv. Funct. Mater.*, 30, 43, 2004226, 2020. doi: https://doi.org/10.1002/adfm.202004226.
41. He, C., Zhang, M., Guo, C., 4D printing of mashed potato/purple sweet potato puree with spontaneous color change. *Innov. Food Sci. Emerg. Technol.*, 59, 102250, 2020. doi: https://doi.org/10.1016/j.ifset.2019.102250.
42. Choong, Y.Y.C., Maleksaeedi, S., Eng, H., Yu, S., Wei, J., Su, P.-C., High speed 4D printing of shape memory polymers with nanosilica. *Appl. Mater. Today*, 18, 100515, 2020. doi: https://doi.org/10.1016/j.apmt.2019.100515.
43. Tao, Y., *et al.*, Morphing pasta and beyond. *Sci. Adv.*, 7, 19, eabf4098, 2021. doi: 10.1126/sciadv.abf4098.
44. Kim, K., *et al.*, 4D Printing of Hygroscopic Liquid Crystal Elastomer Actuators. *Small*, 17, 23, 2100910, 2021. doi: https://doi.org/10.1002/smll.202100910.

45. Zhang, C., *et al.*, 4D Printing of shape-memory polymeric scaffolds for adaptive biomedical implantation. *Acta Biomater.*, 122, 101–110, 2021. doi: https://doi.org/10.1016/j.actbio.2020.12.042.
46. Ma, S., *et al.*, 4D printing of PLA/PCL shape memory composites with controllable sequential deformation. *Bio-Des. Manuf.*, 4, 4, 867–878, 2021. doi: 10.1007/s42242-021-00151-6.
47. van Manen, T., Janbaz, S., Jansen, K.M.B., Zadpoor, A.A., 4D printing of reconfigurable metamaterials and devices. *Commun. Mater.*, 2, 1, 56, 2021. doi: 10.1038/s43246-021-00165-8.
48. Li, K., *et al.*, 4D printing of MXene hydrogels for high-efficiency pseudocapacitive energy storage. *Nat. Commun.*, 13, 1, 6884, 2022. doi: 10.1038/s41467-022-34583-0.
49. Sun, X., *et al.*, Machine Learning-Evolutionary Algorithm Enabled Design for 4D-Printed Active Composite Structures. *Adv. Funct. Mater.*, 32, 10, 2109805, 2022. doi: https://doi.org/10.1002/adfm.202109805.
50. Lee, W.H., Alshebly, Y.S., Mohamed Ali, M.S., Nafea, M., Development of a 4D Printed Variable Stiffness Gripper, in: *2022 IEEE 20th Student Conference on Research and Development (SCOReD)*, pp. 151–156, 2022, doi: 10.1109/SCOReD57082.2022.9973965.
51. Cheah, D.S., Alshebly, Y.S., Ali, M.S.M., Nafea, M., Development of 4D-printed shape memory polymer large-stroke XY micropositioning stages. *J. Micromech. Microeng.*, 32, 6, 65006, May 2022. doi: 10.1088/1361-6439/ac68ca.
52. Pandey, H., Mohol, S.S., Kandi, R., 4D printing of tracheal scaffold using shape-memory polymer composite. *Mater. Lett.*, 329, 133238, 2022. doi: https://doi.org/10.1016/j.matlet.2022.133238.
53. Peng, X., *et al.*, 4D Printing of Freestanding Liquid Crystal Elastomers via Hybrid Additive Manufacturing. *Adv. Mater.*, 34, 39, 2204890, 2022. doi: https://doi.org/10.1002/adma.202204890.
54. Liu, H., Wang, F., Wu, W., Dong, X., Sang, L., 4D printing of mechanically robust PLA/TPU/Fe3O4 magneto-responsive shape memory polymers for smart structures. *Composites, Part B*, 248, 110382, 2023. doi: https://doi.org/10.1016/j.compositesb.2022.110382.
55. Samal, B.B., Jena, A., Varshney, S.K., Kumar, C.S., FDM 4D printing: A low-cost approach of shape programming and assessing the shape memory properties using angle measurement methods in hot water actuation testing apparatus. *Mater. Today Proc.*, 101, 7–14, 2024. doi: 10.1016/j.matpr.2023.02.038.
56. Samal, B.B., Jena, A., Varshney, S.K., Kumar, C.S., 4D Printing of Shape Memory Polymers: A Comparative Study of Programming Methodologies on Various Material Properties. *Smart Mater. Struct.*, 32, 7, 074003, May 2023. doi: 10.1088/1361-665X/acda6e.
57. Samal, B.B., Jena, A., Varshney, S.K., Kumar, C.S., 4D Printing Technology: Experimental investigation of thermoresponsive shape-changing honeycomb

and negative honeycomb structures for different proposed programming strategies, in: 대한기계학회 춘추학술대회, p. 10, 2023, [Online]. Available: https://www.dbpia.co.kr/Journal/articleDetail?nodeId=NODE11708468.
58. Li, G., *et al.*, Biomimetic 4D printing of dome-shaped dynamic mechanical metamaterials. *J. Mater. Res. Technol.*, 24, 4047–4059, 2023. doi: https://doi.org/10.1016/j.jmrt.2023.04.039.
59. Wu, W., *et al.*, Metallic 4D Printing of Laser Stimulation. *Adv. Sci.*, 10, 12, 2206486, 2023. doi: https://doi.org/10.1002/advs.202206486.
60. Wang, F., *et al.*, 4D printing of ceramic structures. *Addit. Manuf.*, 63, 103411, 2023. doi: https://doi.org/10.1016/j.addma.2023.103411.
61. Liu, G., *et al.*, 4D Additive–Subtractive Manufacturing of Shape Memory Ceramics. *Adv. Mater.*, 35, 39, 2302108, 2023. doi: https://doi.org/10.1002/adma.202302108.
62. Xu, X., *et al.*, Regenerative Living 4D Printing via Reversible Growth of Polymer Networks. *Adv. Mater.*, 35, 16, 2209824, 2023. doi: https://doi.org/10.1002/adma.202209824.
63. Einstein, A., Die Grundlage der allgemeinen Relativitätstheorie. *Ann. Phys.*, 354, 7, 769–822, 1916. doi: https://doi.org/10.1002/andp.19163540702.
64. Chae, M.P., Hunter-Smith, D.J., De-Silva, I., Tham, S., Spychal, R.T., Rozen, W.M., Four-Dimensional (4D) Printing: A New Evolution in Computed Tomography-Guided Stereolithographic Modeling Principles and Application. *J. Reconstructive Microsurgery*, 31, 6, 458–463, 2015. doi: 10.1055/s-0035-1549006.
65. Bodaghi, M. and Zolfagharian, A., 1 - 4D printing principles and manufacturing, in: *Smart Materials in Additive Manufacturing*, M. Bodaghi and X.X.X.A. Zolfagharian (Eds.), pp. 1–17, Elsevier, Amsterdam, Netherlands, 2022, doi: https://doi.org/10.1016/B978-0-12-824082-3.00014-3.
66. Anas Ansar, S., Singh, A., Aggrawal, S., Soni, N., Chandra Pathak, P., A Systematic Review on 4D Printing Technology. *Mater. Today Proc.*, 2023. doi: https://doi.org/10.1016/j.matpr.2023.02.413.
67. Ahmed, A., Arya, S., Gupta, V., Furukawa, H., Khosla, A., 4D printing: Fundamentals, materials, applications and challenges. *Polymer*, 228, 123926, 2021. doi: https://doi.org/10.1016/j.polymer.2021.123926.
68. Lyu, Z., Wang, J., Chen, Y., 4D printing: interdisciplinary integration of smart materials, structural design, and new functionality. *Int. J. Extreme Manuf.*, 5, 3, 32011, Jul. 2023. doi: 10.1088/2631-7990/ace090.
69. Sahafnejad-Mohammadi, I., Karamimoghadam, M., Zolfagharian, A., Akrami, M., Bodaghi, M., 4D printing technology in medical engineering: a narrative review. *J. Braz. Soc. Mech. Sci. Eng.*, 44, 6, 233, 2022. doi: 10.1007/s40430-022-03514-x.
70. Momeni, F. and Ni, J., Laws of 4D Printing. *Engineering*, 6, 9, 1035–1055, 2020. doi: https://doi.org/10.1016/j.eng.2020.01.015.

71. Demoly, F., Dunn, M.L., Wood, K.L., Qi, H.J., André, J.-C., The status, barriers, challenges, and future in design for 4D printing. *Mater. Des.*, 212, 110193, 2021. doi: https://doi.org/10.1016/j.matdes.2021.110193.
72. Wan, X., *et al.*, Recent Advances in 4D Printing of Advanced Materials and Structures for Functional Applications. *Adv. Mater.*, 36, 34, 2312263, 2024. doi: https://doi.org/10.1002/adma.202312263.
73. Hu, J., Zhu, Y., Huang, H., Lu, J., Recent advances in shape–memory polymers: Structure, mechanism, functionality, modeling and applications. *Prog. Polym. Sci.*, 37, 12, 1720–1763, 2012. doi: https://doi.org/10.1016/j.progpolymsci.2012.06.001.
74. Taylor, S., Mueller, E., Jones, L.R., Makela, A.V., Ashammakhi, N., Translational Aspects of 3D and 4D Printing and Bioprinting. *Adv. Healthcare Mater.*, n/a, n/a, 2400463, 09 July 2024. doi: https://doi.org/10.1002/adhm.202400463.
75. Kuang, X., *et al.*, Advances in 4D Printing: Materials and Applications. *Adv. Funct. Mater.*, 1805290, 2019. doi: 10.1002/adfm.201805290.
76. Pugliese, R. and Regondi, S., Artificial Intelligence-Empowered 3D and 4D Printing Technologies toward Smarter Biomedical Materials and Approaches. *Polymers*, 14, 14, 2794, 2022. doi: 10.3390/polym14142794.
77. Alli, Y.A., *et al.*, Optimization of 4D/3D printing via machine learning: A systematic review. *Hybrid Adv.*, 6, 100242, 2024. doi: https://doi.org/10.1016/j.hybadv.2024.100242.
78. Ramezani, M. and Mohd Ripin, Z., 4D Printing in Biomedical Engineering: Advancements, Challenges, and Future Directions. *J. Funct. Biomater.*, 14, 7, 347, 2023. doi: 10.3390/jfb14070347.
79. Lalegani Dezaki, M. and Bodaghi, M., Sustainable 4D printing of magneto-electroactive shape memory polymer composites. *Int. J. Adv. Manuf. Technol.*, 126, 1, 35–48, 2023. doi: 10.1007/s00170-023-11101-0.

5

4D Printing of Polymers

Sivanagaraju Namathoti[1], Pavan Kumar Gurrala[2], Prakash Chandra[2] and M. R. K. Vakkalagadda[1]*

[1]School of Mechanical Engineering, VIT-AP University, Amaravati, Andhra Pradesh, India
[2]Mechanical Engineering Department, SoT, Pandit Deendayal Energy University - PDEU, Gandhinagar, Gujarat, India

Abstract
4D printing (3D printing of shape memory material (SMM)) has many applications in engineering structures, biomedical scaffolds, electronic devices, and automotive parts. Among all SMM, shape-memory polymers (SMPs) have greater flexibility in manufacturability, higher shape recovery percentage, and capability of responding to various external stimuli (heat, magnetic field, pH value, electric field, etc.). Along with the multiple advantages of using SMPs, 3D printing (3DP) fulfills the requirement of fabricating complex parts, further enhancing the use of SMPs for large applications. SMPs can be 4D printed by various techniques, such as fused deposition modeling (FDM), polyjet printing, digital light printing (DLP), and direct ink-writing (DIW). This chapter presents insights into 4D printing techniques of SMPs, polymer materials, and their behavior. Further, this chapter provides insights about one-way and two-way SMPs along with their applications.

***Keywords*:** Shape memory polymer, 4D printing, external stimuli, FDM, additive manufacturing, SLA, LDM, DLP

**Corresponding author*: ravindra.btw@gmail.com

Bijaya Bikram Samal, Cheruvu Siva Kumar and Shailendra Kumar Varshney (eds.) 4D Printing Technology: Principles, Materials and Applications, (125–164) © 2025 Scrivener Publishing LLC

5.1 Introduction to 4D Printing Techniques of Polymers

The field of additive manufacturing (AM) or 3D printing was founded in 1980 and has greatly developed since then. AM is recognized as a green technology because of its low or no-waste production strategy, distinguishing it from subtractive manufacturing processes [1–3]. 3D printing creates three-dimensional items by layering materials on top of one another, guided by a digital model or design created by computer aided design (CAD). Presently, these technologies, often known as fast prototyping, have emerged as extremely promising instruments for various uses. These include physical prototype manufacturing, product design review, and, most significantly, manufacturing healthcare-related objects and prototypes for living biological structures. Furthermore, 3D printing has applications in manufacturing end-use items, robotics, aircraft, sensing, automation, product component duplication, electronics, and fashion and jewelry products. It is crucial to note that while these objects normally have defined geometry, they provide a great deal of variety and customization possibilities [4–6]. Smart materials with stimuli-responsive behavior are widely used in 4D printing. It is a cutting-edge manufacturing process that expands on existing 3DP technologies. 4D printing distinguishes itself by introducing time as an additional dimension into printed products. This makes creating dynamic structures with adjustable forms, characteristics, or functions possible. Figure 5.1 depicts a schematic illustration of 3D and 4D printing processes. It is now feasible to construct materials that can autonomously morph or adjust according to the external stimuli by using the capabilities of 4D printing [5, 7–9]. The structures created using 4D printing are spatially and temporally dependent, able to change shape and attributes over time in response to diverse environmental stimuli. Temperature [9–13], light [14–17], pH [18–20], magnetic fields [21–23], electricity [24, 25], and moisture [26–29] are among the stimuli. Importantly, these modifications may be pre-designed and carefully controlled, allowing for customized and predictable alterations in printed structures. When subjected to environmental stimuli, 4D printing technology provides significant advantages by enabling the manufacture of smart materials with increased flexibility, expandability, and deformability. Various deformation processes in 4D materials have been widely studied, including phenomena such as the shape memory effect (SME) and shape morphing based on swelling [30–32]. These processes enable the materials

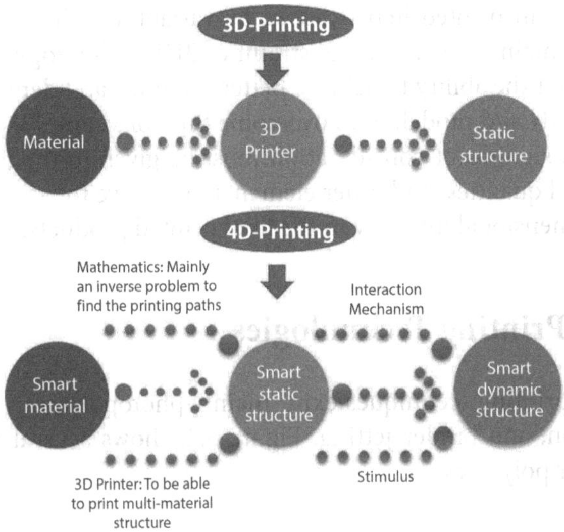

Figure 5.1 Basic difference and schematic representation of 3D and 4D printing.

to respond dynamically and undergo regulated changes in shape or form, broadening their potential uses.

3DP technology allows for the direct realization of three-dimensional items from CAD models, and by adding a time dimension, it is now feasible to make four-dimensional objects. In 3D printing, the CAD model is converted into a printer-readable format before being sliced into precise and microscopic layers using slicer software [33]. This sequential procedure guarantees that the CAD design is accurately reproduced, layer by layer, culminating in the physical manifestation of the intended thing. Subsequently, the formulated model is further processed into a required G & M codes format, which serves as instructions for the computer numerical control (CNC) printer to conduct each layer's sequential deposition [34]. The G-code provides precise instructions regarding the positioning and movement of the printer's components, ensuring the accurate construction of the object layer by layer. There has been a huge increase in the investigation of 3DP technologies during the last few decades. Recent improvements in optimizable 3DP technology have effectively overcome its earlier limits, which included poor production speed [35], reduced end-product resolution [36], and some structural and physical constraints [37]. Growing trends have opened the way for improved capabilities, overcoming existing limitations and creating new opportunities for more efficient and high-quality 3D printing applications. However, dimensional discrepancies in printed items caused by variables such as expansion or shrinkage remain a persistent constraint of 3DP technology. These variations can limit the ability to achieve better accuracy and demand rigorous pre-print study and modeling to overcome this constraint [38]. To address these issues, significant consideration must be given to the printing process, material qualities, and other elements to achieve the necessary precision and dimensional integrity in the final printed products.

5.2 4D Printing Technologies

Several 4D printing techniques exist, mainly photopolymerization, material extrusion, and binder jetting. Figure 5.2. shows several 4D printing processes for polymers.

5.2.1 Stereolithography

The construction of robust structures involves utilizing optical fibers or ultraviolet (UV) light, guided by an X-Y plotter, to selectively expose photopolymer resin. This process enables the resin to solidify and procedure the desired solid structures [38, 39]. By precisely manipulating the UV-light beam, either by following a predetermined path or through sequential scanning across the photosensitive polymer surface, the beam from the UV-light source plays a pivotal role. The interaction between the beam and the photopolymer resin leads to the polymerization process, resulting in the formation of a patterned layer in two dimensions [40]. The SLS technique's schematic design is presented in Figure 5.3. The process involves curing the material layer by layer [41]. This process primarily utilizes epoxy and acrylic resins as the main polymers. This fabrication

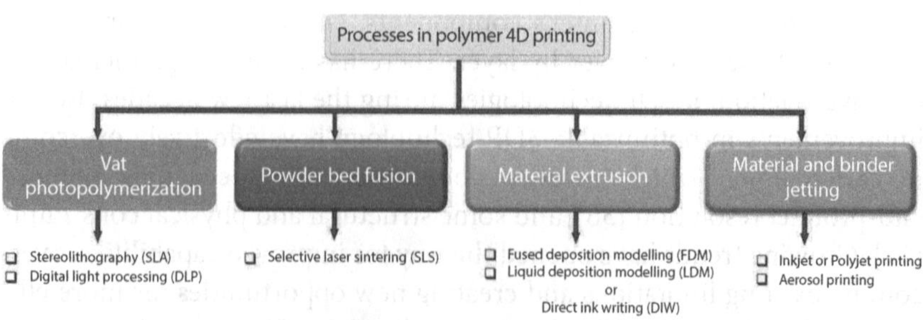

Figure 5.2 Various polymer 4D printing processes.

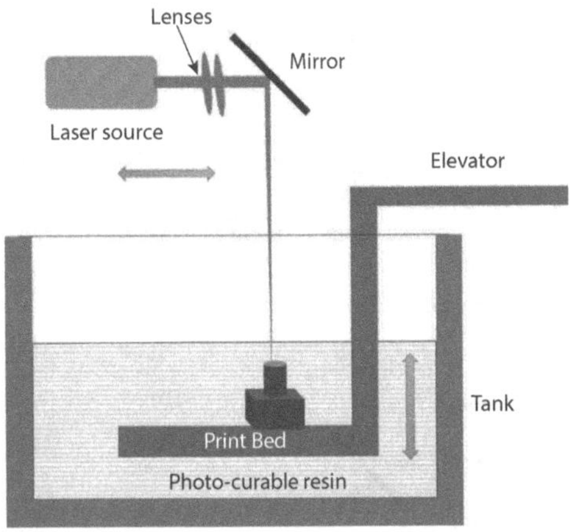

Figure 5.3 The stereolithography process is depicted schematically.

method offers excellent spatial resolution due to a small laser spot diameter. Understanding the polymerization curing processes is crucial for managing the microstructure and properties of the final printed part. Important factors like laser power, scan velocity, and exposure time are critical in influencing both the curing time and printing resolution [42, 43]. A laser beam emitted by the source is directed precisely and focused onto the platform to create three-dimensional objects [44–46]. In the stereolithography setup, a Galvano scanner incorporates a pair of mirrors to regulate the horizontal positioning of the laser beam. SLA offers advantages such excellent resolution, the capacity to 3D prints viscous resins, and the avoidance of nozzle clogging [47]. However, there are important disadvantages to consider. These include significant equipment costs, chemical waste, and the need for post-printing processes.

5.2.2 Digital Light Processing (DLP)

Cross-linking of photopolymer takes place when exposed to light in DLP. DLP requires a photopolymer, UV light and a controlling system to UV light movement. This is usually referred to as a bottom-up arrangement. The production platform moves vertically after each 2D layer is finished. DLP technology can solidify a full layer of monomer using light from a projector. Figure 5.4 shows a schematic illustration of the DLP production method. As a result, compared to SLA (stereolithography), DLP achieves substantially quicker printing and manufacturing speeds. However, it is crucial to remember that DLP often provides lesser printing resolution

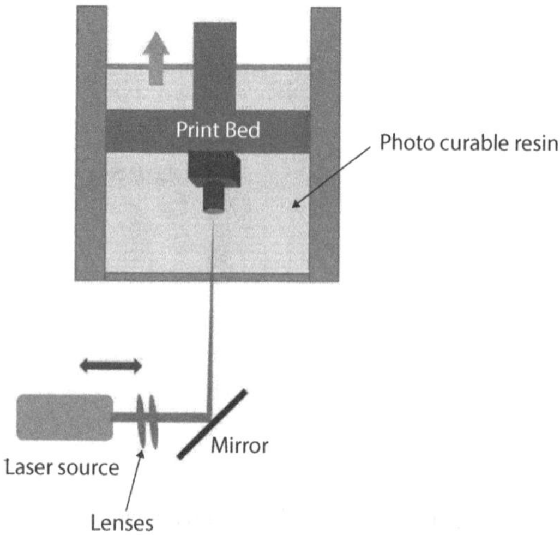

Figure 5.4 The DLP top-down process is represented schematically.

than SLA [48–51]. DLP is frequently compared to traditional lithography, thus the name dynamic mask photolithography. This method encodes information for each structure layer using binary patterns, such as black and white graphics [52, 53].

The projection capacity built into the setup allows for the quick curing of numerous layers simultaneously, resulting in a shorter manufacturing time. DLP generally has lateral resolutions ranging from 10 to 50 μm, governed by the resolution of the digital micromirror device (DMD) [53]. On the other hand, the vertical resolution in DLP can be influenced by depth of light penetration and the degree of curing. Light-absorbing compounds, such as naphthol-containing dyes, can be used to increase vertical resolution. These materials improve accuracy by adjusting curing depth and optimizing total vertical resolution.

5.3 Extrusion 3D Printing

Material extrusion is an AM process that operates by applying mechanical force to extrude printable material through a nozzle [54]. This particular 3D manufacturing technique involves the sequential deposition of melted or semi-melted polymers and polymer solutions layer-by-layer. This process requires polymer in solid form (works for thermo plastic polymers), nozzle and controlling unit for the movement of the nozzle and to maintain nozzle temperatures [55]. There were many techniques involved under extrusion 3D printing, FDM is the most often used technology for printing thermoplastic materials.

5.3.1 Fused Deposition Modeling (FDM)

FDM is an extensively studied melt extrusion method utilized in producing plastic materials. This technique entails transforming the material into a partially molten state by introducing a thermoplastic filament, thereby achieving liquefaction. Once the material is liquefied, it is carefully placed onto the building stage and left to cool and solidify, as shown in Figure 5.5. This procedure repeats for each subsequent layer. To deposit the next layer, the print head or the building stage adjusts its position, guided by a drive wheel, depending on the material used [56]. Most of the thermoplastic materials are frequently employed in FDM [57]. FDM machines offer controlled cooling adjustments to optimize quality, such as minimizing thermal shrinkage of the material and preventing interlayer deterioration. These adjustments enable precise control over the cooling process [58]. The temperature of the platform is regulated through cooling adjustments. Simple configuration of FDM machines offers flexibility and ease of assembly modifications, which helps prevent nozzle clogging during filament softening or when incorporating additional printing materials [59, 60]. FDM is widely acclaimed as one of the most favored 3DP techniques, primarily due to its cost-effectiveness and widespread accessibility [61]. However, this method has certain limitations, including structural defects

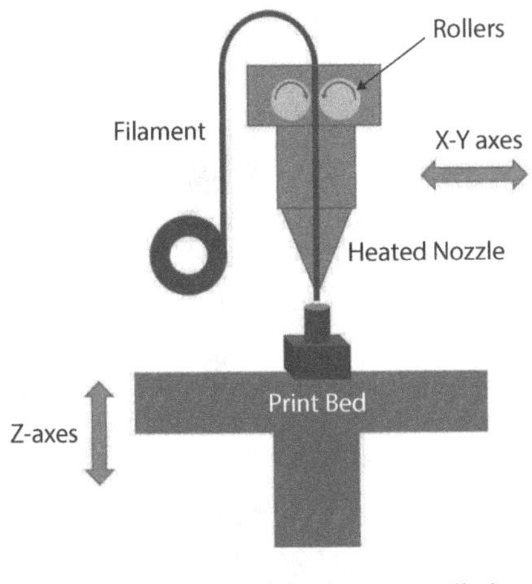

Figure 5.5 FDM printer schematic diagram based on controlled extrusion of thermoplastic filament.

and a decrease in mechanical strength caused by printing imperfections, non-uniformity of filaments, surface roughness, and inadequate interlayer adhesion [62–64]. These defects can be partially addressed by adjusting printing parameters [65]. The resolution in FDM is about 40μm at faster printing rates.

5.4 Liquid Deposition Modeling (LDM) or Direct Ink Writing (DIW)

LDM or DIW is an AM technology that employs extrusion to produce items. This method includes extruding a viscoelastic ink via a syringe. Ink is precisely pushed out via a nozzle using pressure, while a dispenser accurately moves to form shapes by adding layers one after the other. This procedure must use materials that are in a highly viscous liquid condition and can restore their structure after deposition [66]. LDM distinguishes itself as an exceptionally versatile 3D printing process due to its capabilities to deposit different materials, encompassing plastics, ceramics, food, and even live cells, as depicted in the schematic representation in Figure 5.6 [67, 68]. Extruded filaments in LDM must be structurally reliable and self-supporting to prevent ink dispersion or structural collapse. The viscoelastic characteristics of the inks utilized determine the efficiency of extrusion modeling. The printability of these inks is influenced by rheological properties such as viscosity, stress yield, and modulus of elasticity. LDM is another method for

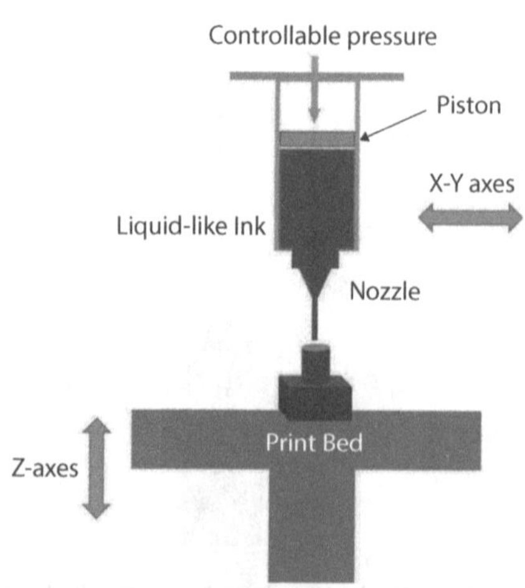

Figure 5.6 Represents a schematic depiction of the LDM process.

3D printing soft materials [69, 70]. DIW usually requires post-processing to harden the extruded inks (typically hard curing).

Furthermore, numerous small compounds and nanofillers can improve response features [71, 72]. Shear thinning behavior is used in LDM to assist deformations, extrusion, and quick filament recovery. The approach, however, confronts difficulties due to the creation of gaps between layers and the fundamentally soft character of the materials being dispensed. To address all various concerns, reinforcements can be added during the post-printing stages to increase mechanical strength and structural fidelity [73]. Due to the vast range of materials accessible and the great adaptability of printing settings, substantial research advances in LDM have been made in recent years.

5.5 Binder Jetting

This AM approach involves carefully applying a binder material to a powder bed to generate cross-sections of the desired structures [74]. Like traditional paper printing, Binder jetting uses a laser beam to partly melt polymer powder particles, resulting in a partial cross-section. The binder, analogous to ink, flows over each layer, producing the final result. Binder jetting is frequently used in 3D printing because of its capacity to build solid layers and its appropriateness for producing large-scale items. The process is represented schematically in Figure 5.7. It is a quick and inexpensive method for processing a wide range of materials, including polymer composites. Binder jetting is further divided into printing technologies that use inkjet and aerosol jet printing.

5.5.1 Inkjet Printing

This method allows liquids to be transformed into both 2D and 3D structures [75]. An inkjet head with multiple nozzles that move along the X-axis is used in the printing arrangement. Each nozzle emits small droplets of photopolymer to build a layer during printing. Instead of dispensing droplets, PolyJet technology ejects layers of curable light-sensitive ink, each approximately 16-μm thick, onto the build plate [76]. A UV flashlight cures the first layer after it is deposited, and additional layers are applied similarly. The ultimate 3D structure is created by repeating the process several times. Using inkjet heads, two distinct materials, called support and construction materials, are deposited in an inkjet configuration [77]. Inkjet printing may handle a wide range of materials, including silver nanoparticles [78],

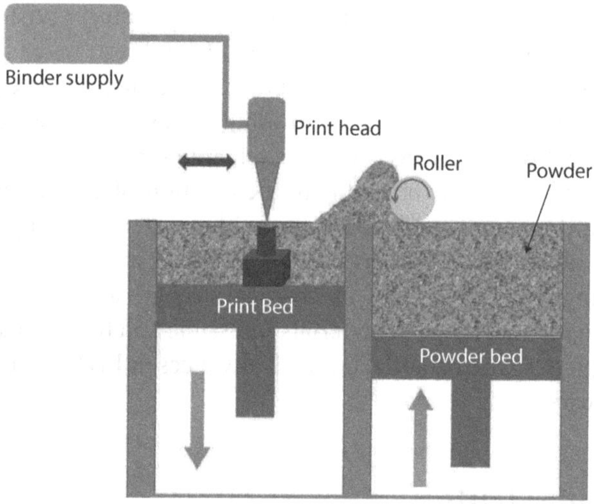

Figure 5.7 Schematic representation of binder jet printing process.

carbon nanotubes [79], and polymers [80], as long as they can be formed into dispersion. Despite not adhering to the final output, the supporting material in 3D inkjet printing plays an important role in building a fully solid 3D structure. It is necessary to support the building materials, especially in places with faults such as voids or overhangs, to construct a structurally stable and dense 3D product [76]. As a result, inkjet printing supplies the building and supporting elements in similar amounts, making it a cost-effective printing technology. Inkjet printing also enables the creation of multimaterial or multicolored structures, enhancing its versatility. Inkjet printing is employed in various disciplines, including electronics [81] and tissue engineering [82]. It is commonly utilized in the electrical area to manufacture devices like sensors [83], antennas [84], and circuit boards [85].

5.5.2 Aerosol Jet Printing (AJP)

AJP employs ultrasonication or pneumatic atomization within an aerosol chamber (see Figure 5.8) to disperse the building material into fine aerosol particles [86–88]. A liquid polymeric material solution and dispersions are employed in aerosol printing. This printing process requires either extremely low viscosity or a viscosity of roughly 2500 MPa and extremely small dispersion particle sizes. Atomization happens throughout the manufacturing process, forming a concentrated aerosol composed of droplets

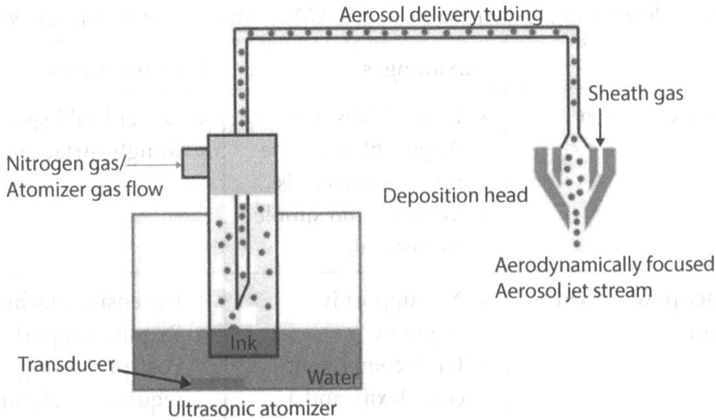

Figure 5.8 Process of aerosol jet printing [91].

ranging from 1–5 μm [89]. An inert gas stream is then used to convey these droplets to the deposition head [90]. The material flow is directed and accelerated by the deposition head through a nozzle with a diameter of 50 to 300m and a length of 20mm. The movement of the construction pieces aids in deposition layer contours in aerosol jet printing [90]. Aerosol jet printing uses a continuous layer-by-layer manufacturing method to ensure an even flow of aerosols throughout the printing process. Following the CAD data, the setup includes a shutter to prevent the stream from reaching the polymer substrate. The aerosol jet technique may treat any soluble substance in a suitable liquid media. Metal [91, 92], polymer [93], carbon nanotubes [94], and composites [95, 96] all are printable using the AJP process. Generally, the lateral resolution of aerosol jet printing is 10μm.

Even though multiple techniques were used for 4D printing, every technique has advantages and limitations, some of which are presented in Table 5.1.

5.6 4D Printing of SMPs and Materials

SMPs are made from a range of materials, including thermoplastic polymers, elastomers, and composites [98–100]. SMPs can "remember" their original shape, and once they are deformed or stretched, they can regain the original shape due to external stimulus [50, 101]. This ability makes SMPs useful in various aerospace, biomedical devices, and soft robotics applications. One of the most common types of SMPs is thermally responsive polymers, which can be triggered by changes in temperature [10, 102].

Table 5.1 Merits and De-merits of various AM techniques for polymers [97].

Process	Advantages	Disadvantages
Polymer extrusion	• Low establishing cost • Applicable for wide range of materials • Versatile and simple to customize	o Lower build speed o Rough surface finish
Polymer powder bed fusion	• No support is required • High component complexity and a diverse material palette • High resolution and precision, with a smooth surface finish	o Expensive machinery o Require support o Post-processing is required to eliminate support
Vat Photopolymerization	• High manufacturing speed • Specific imaging energy is low • A diverse set of materials • High resolution and precision	o Post-curing is required for increased strength o Post-processing has the potential to destroy thin and tiny features
Material Jetting	• A variety of materials with minimal waste • There are several possible materials The look is rough or grainy	o Support materials are not recyclable o The look is rough or grainy o Inadequate strength
Binder Jetting	• Layer fabrication includes support structures • Imaging-specific energy is low • High manufacturing speed • There is no support structure required	o Significant material waste o It is difficult to remove support that has been caught in internal cavities

For example, a thermally responsive SMP may be printed into a particular shape at elevated temperatures and then cooled to a lower temperature to "lock" it into that shape. The SMP will revert to their former shape when the temperature (for thermal responsive SMPs) increases again. Variety of SMPs are available which respond to different stimuli. For example, a light-responsive SMP can be printed into a given shape and then actuated to change its shape by subjecting it to a certain wavelength of light [103, 104]. Overall, SMPs are an attractive category of materials for 4D printing due to their capacity to be printed into a precise 3D complicated shape while retaining their original shape.

5.6.1 Polycaprolactone (PCL) Based Polymers

PCL is a biodegradable polyester that can be used as an SMP since its melting point is low and flexibility to deform to a required shape. When heated beyond its melting point, PCL can be shaped into a new form and cooled to lock in the new shape. Upon reheating, it will revert to its original shape. [105] utilized PCL and epoxidized soybean oil (ESO) to create a biodegradable SMP material. The PCL's mechanical characteristics and biodegradability were enhanced by the ESO. The material's shape memory properties were evaluated, and the impact of the ESO content on these properties was investigated. The required specimens were prepared by melt blending ESO and PCL at different weight ratios and then compression molded into films and they characterized by FTIR, DSC and tensile tests. The tensile characteristics, form recovery, and shape fixity ratios have all greatly enhanced with the presence of ESO, according to the data. The material exhibited the maximum shape fixity and shape recovery ratios with an ESO content of 10 wt% [106].

Shape memory 4D printed polymers possessing two-layer composite structures were prepared by printing the two layers using (FDM) biomaterial printing. The composite structures comprised two well-known thermoplastics, acrylonitrile butadiene styrene (ABS) and PCL, and thermoplastic polyurethane (TPU) filament as an elastomer filament. The thermoplastic layer functioned as a constraint for the elastomeric layer, while the thermoplastic layer's rubber-to-glass transition functioned as a switching phenomenon to stabilize the temporary shape. It also accumulates stress resulting from deformation, allowing for subsequent shape retrieval by reverse-phasing the thermoplastic layer. The results demonstrated that ABS-TPU possessed fixity and a recovery ratio greater than 90%, but the PCL-TPU composite proved broad shape recovering while ensuring fixity at 77.42%. The distinction between the two composites is due to the

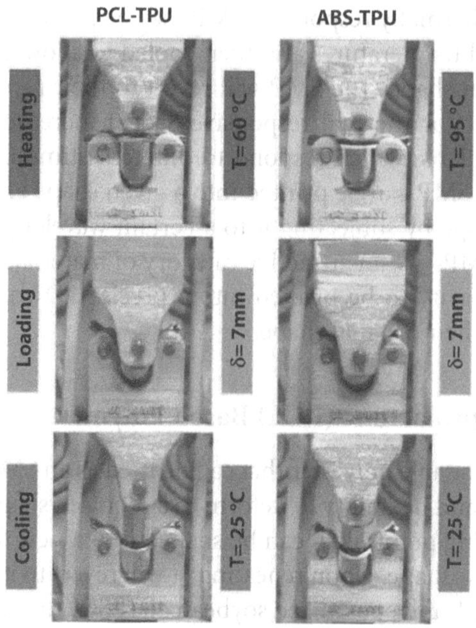

Figure 5.9 Phases and settings of program design for dual ABS–TPU and PCL–TPU composite structures [107].

thermoplastic transition and programming temperature. Scanning electron microscopy (SEM) pictures revealed improved interfacial bonding and part integrity in multi-material 3D printing (Figure 5.9) [107].

5.6.2 Polyurethane (PU)

PU is a common SMP used in various applications. It can be programmed for response to a variety of external stimuli. When subjected to the stimulus, the PU transforms but returns to its prior shape when removed. 4D printing of SMPUs with two-way actuation properties was demonstrated [108] synthesizing a single-network SMPU using a one-step prepolymer mixing method and 4D printing using a DIW technique. The SMPU consists of hard segment (HS) and a soft segment (SS), and the HS and SS molecular weights were tailored to achieve the desired properties. The 4D-printed SMPU exhibited excellent shape recovery (near to 100%). Additionally, the SMPU showed two-way actuation properties, allowing for the recovery of two different permanent shapes. The two-way actuation was achieved by carefully programming the SMPU with two temporary shapes, each corresponding to a different permanent shape. Overall, this study demonstrated

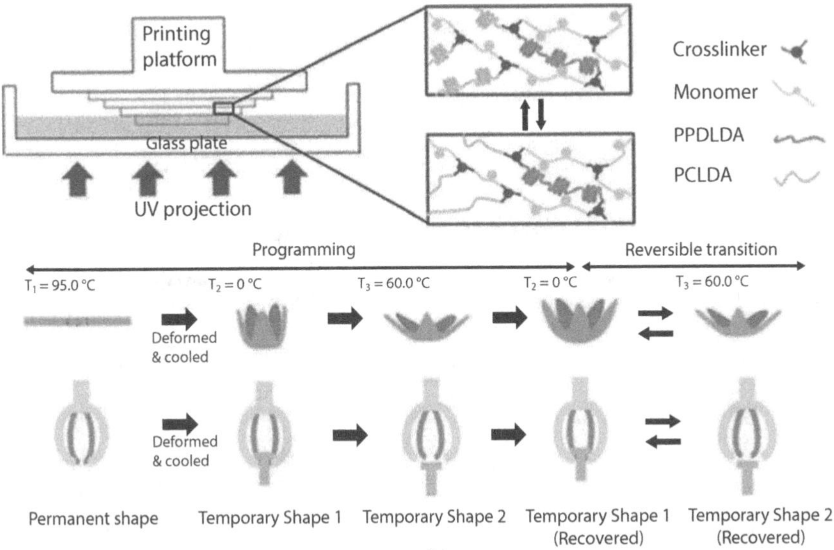

Figure 5.10 4D printed polyurethane-based shape memory polymers [108].

the ability of 4D printing to fabricate SMPUs with superior shape memory and actuation properties, which might be beneficial for biomedical devices and soft robotics (see Figure 5.10) [108].

5.6.3 Polyethylene Glycol (PEG)

PEG is a hydrophilic polymer which can function as an SMP when crosslinked with other materials. PEG is distinctive in that it is able to react to external stimuli, such as temperature, light, and pH. When exposed to the stimuli, the crosslinked PEG chains change shape and then return to their original shape after the stimulus is withdrawn. PEG/poly(ε-caprolactone) (PCL) networks of semicrystalline were developed and investigated for their shape memory behavior. The networks were prepared by crosslinking PEG and PCL precursors using a diisocyanate crosslinker. The resulting networks exhibited a stress-free two-way SME, meaning that the shape recovery occurs without applying external stress. Differential scanning calorimetry (DSC) and tensile tests characterized the shape memory behavior. The DSC analysis revealed that the SME is due to the melting and recrystallization domains of the PCL, which act as physical crosslinks in the network. The tensile tests showed that the network can recover up to 90% of its initial shape after deformation. The shape recovery time has been found to be related to the deformation temperature and the PEG/PCL

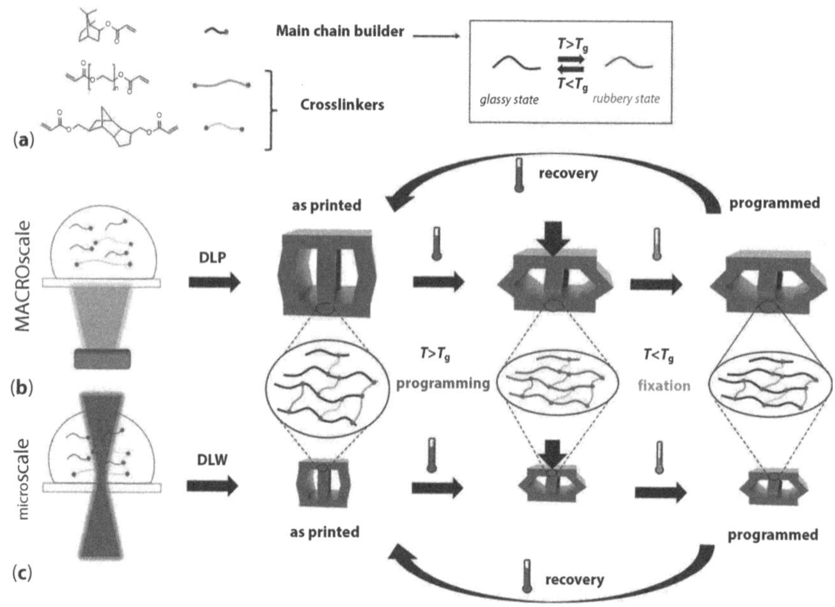

Figure 5.11 Illustration describing 4D printable isobornyl acrylate-based monomer [110].

ratio. These findings suggest that PEG/PCL semicrystalline networks have extensive applications in biomedical devices and soft robotics. Polyethylene glycol-based materials can be investigated for 4D printed SMPs [109].

4D printing at macroscale and microscale can be achieved through novel ink systems using DLP and DLW 3DP technologies. The ink consists of a monofunctional isobornyl acrylate and two soft and hard diacrylate crosslinkers. This technology is compatible with both DLP and DLW printing techniques and may be used for simple frames, double platform structures, infinity rings, and cubic grids, all of which demonstrate excellent shape fixity, recovery, and repeatability. The ink system is also used using DLW at the microscale. The results exhibited excellent shape memory properties and can trap and release micro-objects called microspheres. The flexibility and perspective of the reported system are established by designing smart 4D thermally actuated microstructures (see Figure 5.11) [110].

5.6.4 4D Hydrogels

4D hydrogels, or smart materials, are widely used because they can change shape in response to external stimuli such as temperature, pH, or certain chemicals [111]. Hydrogels are made of crosslinked polymers that absorb and hold water, resulting in a gel-like consistency. The term "4D" in 4D hydrogels refers to the time as fourth dimension, which denotes the materials' potential to change shape or features over time. This contrasts with traditional 3D printing, where objects are created in a fixed shape and size. One common example of 4D hydrogels is stimuli-responsive hydrogels, which are designed to change in response to specific stimuli. For example, a pH-responsive hydrogel could be designed to release a drug when it encounters an acidic environment, such as in the stomach. Shape-changing hydrogels are another type of 4D hydrogel that may change shape in response to external stimuli such as temperature, light, or other environmental conditions. These materials are promising for applications in soft robotics and medication delivery because they may be programmed to release pharmaceuticals or perform tasks in response to specific external stimuli.

These hydrogels are capable of changing their characteristics in response to a variety of physical stimuli. Poly(N-isopropylacrylamide) (PNIPAAm), a widely used temperature-responsive hydrogel, may expand and shrink in response to temperature changes. This property has potential uses in the realm of drug delivery. In this case, the hydrogel is designed to release drugs in reaction to variations in body temperature, falling into a class of hydrogels that vary their properties dependent on pH changes. For example, poly (acrylic acid) (PAA) is a pH-responsive hydrogel that expands or contracts in reaction to pH changes. This property has potential applications in drug delivery, where the hydrogel can be designed to release drugs in response to changes in pH. These are hydrogels that can change their properties in response to light. For example, azobenzene-containing hydrogels change their shape when exposed to light of a specific wavelength. This property has potential applications in soft robotics, where the hydrogel can be designed to respond to light and perform specific tasks. These hydrogels can fold themselves into specific shapes in response to environmental stimuli. For example, a team of researchers at MIT developed a self-folding hydrogel that can fold into a specific shape when exposed to certain chemicals. This has the potential for application in tissue engineering and drugs delivery. In these cases, the hydrogel may be configured to fold into exact shapes, allowing for targeted distribution to certain tissues or organs.

Agarose/PAAm hydrogels created by 4D printing are a hydrogel material that combines the properties of agarose and PAAm hydrogels. This material may change its profile (form) over time in response to a specific stimulus. This material combines agarose, acrylamide (AAm), a cross-linker, a photo initiator, and laponite nano clay, similar to the traditional agarose/PAAm hydrogel material. The process for making 4D printed agarose/PAAm hydrogels are similar to that of traditional agarose/PAAm hydrogels, with the addition of a specific printing technique that enables the creation of complex, three-dimensional structures. The material is deposited layer by layer using a specialized 3D printer that can precisely control the shape and size of each layer. Depending on the application, the stimulus that triggers the shape change in 4D-printed agarose/PAAm hydrogels varies. For example, temperature changes can cause the hydrogel to shrink or expand, while exposure to light can cause the material to undergo a conformational change (see Figure 5.12) [112].

4D printed agarose/PAAm hydrogels have purposes in soft robotics, tissue engineering, and drug delivery. Creating complex, 3D structures which may alter shape when exposed to a certain stimulation, makes these materials particularly attractive for applications requiring precise control over movement and shape. For example, 4D-printed agarose/PAAm hydrogels are used to build soft robots that can move and change profiles in response

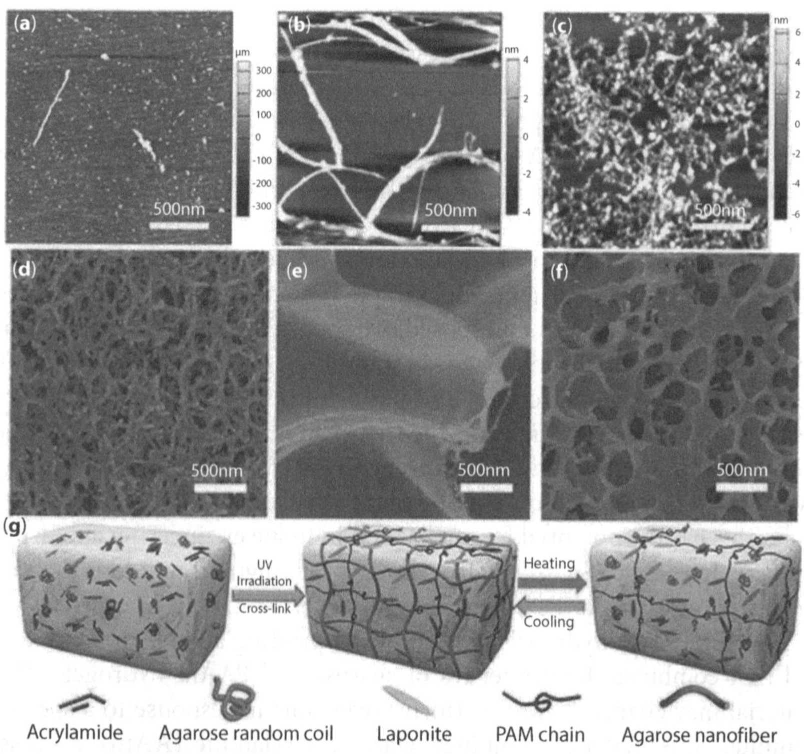

Figure 5.12 AFM, SEM, and schemes describe the structure of 4D-printed agarose/PAAm of hydrogel material. Reproduced with permission [112].

to variations in the external environment. They could also make tissue scaffolds that can change shape over time to support tissue growth and regeneration better. The laponite nano clay reinforcement also improves strength of hydrogel, making it more durable and better able to withstand deformation and damage.

Another example of making tubes is that the tubes are composed of two different types of gels: A gel system is made up of a thermally active swelling gel that reacts to temperature changes and a passive gel that is thermally insensitive. These gels consist of a thermally sensitive swelling component, poly-N-isopropylacrylamide, and a thermally nonresponsive polymeric gel, poly(acrylamide). By carefully aligning the 3D-printed segments of these two gels, the tubes may take on a variety of shapes, deformations and bending. Using finite element simulations and experiments enables the authors to characterize the thermally induced shape changes and demonstrate the tunability and customization of the gel structures. Bioimplants, soft robotics, and biomedical engineering are some of the possible uses for these gel tubes. The article highlights the importance of actualizing and characterizing thermally induced shape changes for developing dynamic gel structures. The approach presented in this article provides a rational and effective way to engineer complex parameters for AM of gel structures with high customization and tunability [113].

One promising approach to creating soft robots is using electroactive hydrogels, which can be engineered to respond to electrical stimuli by changing their shape or size. A 3D-printed electroactive hydrogel is used to achieve soft robotic manipulation and locomotion. The hydrogel is comprised of a polyvinyl alcohol (PVA) matrix with embedded polypyrrole (PPy) nanoparticles, providing electroactive properties. The 3D-printed hydrogel structures are designed to actuate to electrical stimulation. Applying a voltage to the hydrogel can induce bending and twisting motions in the printed structures. They demonstrate that these motions can achieve various tasks, including grasping and releasing objects and propelling the hydrogel forward in a swimming motion. The study highlights the potential of 3D-printed electroactive hydrogels in soft robotics, particularly in applications where the ability to manipulate or move in unstructured environments is critical. The use of 3D printing allows the creation of elaborate and functional designs. Overall, this study demonstrates the promising capabilities of electroactive hydrogels and 3D printing in soft robotics and opens up new possibilities for developing flexible, compliant robots that can perform tasks in various environments (see Figure 5.13) [114].

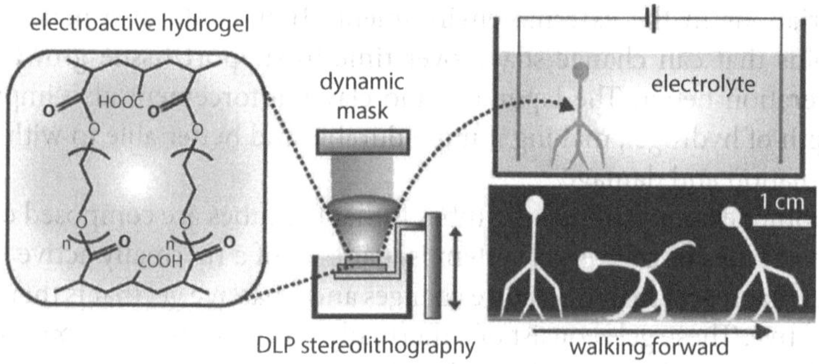

Figure 5.13 4D printed polyvinyl alcohol and polypyrrole-based composite hydrogels. Reproduced with permission [114].

5.7 4D Printing of Two-Way SMPs

Two-way thermo-responsive SMPs exhibit reversible shape change between two permanent configurations when subjected to external stimulus, this the capability of the polymers is known as the two-way SME. The two-way SME is achieved via the application of constant loading and load-free conditions. These two-way SEM are synthesized by combining suitable polymeric macromolecular architecture and a thermomechanical history, known as the shape memory cycle. The two-way SMEs are specifically observed under constant applied stress in two classes of SMPs, also known as semicrystalline networks composed of semicrystalline crosslinked polymers and liquid crystalline elastomers (LCEs). Some notable applications of these SMPs are robotics, textiles, biomedical devices, and delf-folding structures. By understanding the fundamental mechanism and properties of SPMs, it becomes possible to fabricate state-of-the-art advanced materials tailored capabilities in response to external stimuli [115, 116].

The two-way SME can be attained in the semicrystalline network; the desired polymer must be crosslinked and crystallizable. When the polymer is under constant tensile stress, the networks in the polymer matrix get elongated due to entropy elasticity occurring in the rubbery region, followed by crystallization when the temperature is dropped. Conversely, heating above T_m results in melting and contraction of the temporarily oriented crystalline domain. The crystallization is also called crystallization-induced elongation (CIE), whereas its contraction is called melting-induced contraction (MIC). Figure 5.14 exemplifies the shape-memory cycle of the two-way SMPs with precise crystallization and melting temperature (T_c and T_m). Primarily, the material is molded into a permanent form. Subsequently,

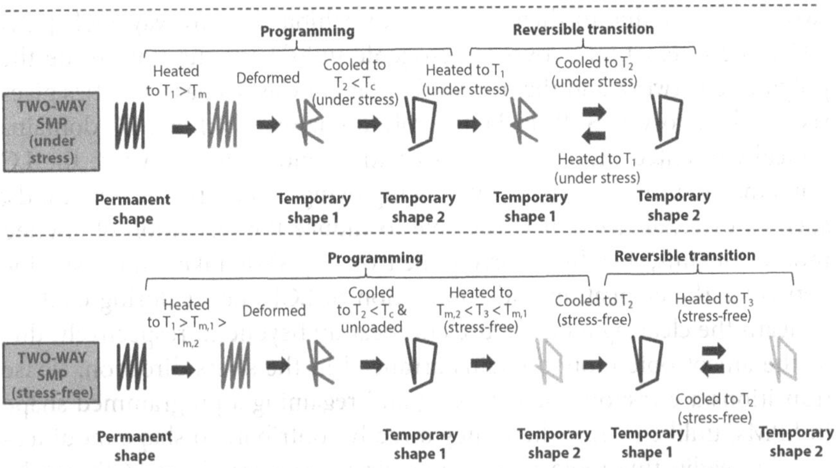

Figure 5.14 Two way shape memory effect and corresponding transition [116].

the material is heated above Tm and distorted under stress to attain the first alterable impermanent shape. After this, the material is cooled below Tc while maintaining stress, inducing crystallization-induced elongation (CIE), and obtaining the second reversible and temporary shape. The cooling process involves two elongations: chain reorientation for entropic elasticity and crystallite formation under stress, enabling further stretching to maintain stress conditions. The structural evolution relaxes stress during crystallite formation [117, 118]. T1 and T2 are usually set at 15-40 °C temperatures above melting point Tm and below crystallization temperature Tc, respectively. After that, the material undergoes material-induced contraction (MIC) and returns to the first reversible temporary shape upon further heating back to T1. The initial reversible impermanent shape's degree of retrieval is characteristically encouraged by various variables, together with materials' physical characteristics and the testing of the environment [119]. Thermal cycling under continual stress, analogous to the one-way SME, permits the controllable transition between the first and second temporary shapes. Similar to the one-way SME, the two-way SME is affected by the change in the entropy of the network and is controlled by the transition temperature (in the present case, T_c and T_m) [120–126]. Furthermore, the concept has been broadened to include triple SMPs in [127–130], which demonstrate a reversible SME across two transition temperatures.

Liquid crystalline elastomers (LCEs), belonging to class (ii), are materials capable of undergoing shape changes and are often categorized as two-way SMPs due to their potential to exhibit the two-way SME [131, 132]. The interaction between mesogenic units' configuration inside the polymeric network and the network's inherent elastic characteristics gives rise to the two-way SME in LCEs. Poly-domains, which are LC domains typically oriented disorderly, coexist with mono-domains, where the LC domains align in a specific way. By employing certain physical methods, such as external stress, it is possible to induce the formation of aligned mono-domains, thereby achieving the two-way SME. Like semicrystalline networks, the elongation and contraction of LCEs occur during cool-off beneath the clearing temperature and heat up beyond it, respectively, due to the anisotropic monodomains arranged in the stress direction. Phase transitions are responsible for fixing and regaining a programmed shape in SPMs, unlike LCEs, where they directly contribute to shape modification by readjusting molecular groups. Previous research on LCEs can be found in sources such as [133–136], and the two-way triple SME has been accomplished [137]. However, the synthesis of two-way shape memory LCEs and modification of their high transition is a challenging task. The two-way SME is not always dependent on the existence of external stress. The two-way SME can be based on internally produced stress with a certain amount of structural and thermos-mechanical history, shaping the memory cycle. Further, self-standing reversible actuation is feasible with the help of reversible bidirectional SME [138].

A diversity of SMPs share the two-way SME, observed in stress-free environments. Semicrystalline employs the application of a two-stage crosslinking technique, chemically crosslinked semicrystalline polymeric networks with either one broad or two melting temperatures and thermoplastic semicrystalline polymeric networks that possess either a broad melting temperature of two temperatures. These polymer materials are further classified into two types: (1) a chemically crosslinked or transition phase that has the maximum transition temperature while keeping its permanent shape; (2) a semicrystalline or elastic amorphous phase, responsible for geometric changes and generating tensile force during compression; and (3) a semicrystalline or liquid crystalline phase, enabling actuation through melting and crystalline temperature. Two-way SME is observed in specific classes of SPMs even without external stress, attributing to the distinct phases within the polymeric network, each contributing to the SME through their unique thermal characteristics [139].

SMPs of class (i) possess semicrystalline networks that are chemically cross-linked and possess either a single broad melting temperature or two different melting temperatures. It takes a programming process and molecular mechanism akin to stress-conditioned two-way SPMs and is responsible for the two-way SME (see Figure 5.14). The process initiates with programming phases like heating to T_1 (greater than both $T_{m,1}$ and $T_{m,2}$), deforming under stress, and cooling to T_2 (T_3 greater than $T_{m,1}$) stress-free after the material has been permanently shaped. At this particular cycle point, the second reversible temporary shape is achieved via partial melting and partial restoration of the shape. By inhibiting the polymer flow above melting temperature (T_m), chemically crosslinked polymer networks have traditionally been thought necessary to achieve the two-way SME. However, recently, some thermoplastic polymers from class (iii) can exhibit the two-way SME even without stress. Thermoplastic polymers have the advantage of reprocessing, which is able and reusable, which negates the necessity of preparing new polymers. However, it is important to highlight that the reversible shape-switching capacity of this kind of SMP is currently comparable to that of chemically crosslinked networks.

Utilizing composites, hybrids, or laminates is another alternative technique to create two-way SMPs. The two-way SMPs are generally comprised of layered polymeric networks composed of shape and non-shape memory polymers or a combination of these polymers. Unbalanced mechanical stresses result from diverse mechanical and thermal properties on heating the structures or if SMPs are used from the form memory effect. This technique also promotes the fabrication of laminated polymers possessing two different layers of SMPs, delivering two-way SMEs even in the lack of stress. The stress-free two-way SME is a new method that overcomes the drawbacks of the traditional method, such as the complex preparation process and limited reversible strain change. The material is first heated to temperatures above its melting point (T1). This causes the material to melt and flow into a new temporary shape. The material is subsequently cooled down to its T_c (T_2). This causes the material to crystallize in a new temporary shape. The reversible transition between the first and second temporary shapes can be achieved through thermal cycling between T_2 and T_3 [140–144].

The two-stage crosslinking method system generates two-way SMPs without stress. Combining a minimum of two polymeric networks leads to the generation of two interpretable networks, resulting in the interlining of the chains inside the matrix. The system is made up of elastomeric components that stretch to extend the shape after cooling, as well as a crystalline phase that causes varied shape shrinkage at high temperatures. The two-stage crosslinking technique is responsible for the system shrinking due to the existence of a crystalline phase while compressing the elastomeric component via heating (above transition temperature). The elastic recovery of the elastomeric component, alternatively, is responsible for crystallization in the direction of force when the system is cooled to lower the switching temperature of the crystalline networks. LCEs (including their composites) have also been incorporated into the scope of the two-stage crosslinking method. Furthermore, employing an approach parallel to the two-stage crosslinking process, new advances have been realized in expanding two-way SMPs comprising dynamic covalent linkages. SMPs crosslinked via dynamic covalent bonding have advantages, outstanding recoverability, re-processibility, and greater mechanical features [139] and [145].

4D printing of multiscale meta materials, which can undergo excessive compression was developed by [179] and [181] and the developed structures were successfully tested for excessive compression in the temperature range of 50 °C to 70 °C. Two novel techniques such as programming during printing and programming after printing were successfully developed and tested using FDM by [180] and the results showed that programming during printing exhibited greater shape recovery. A hot water actuation with complete angle measurement methodology for shape recovery was introduced by [182]. Further, they developed controlled shape change components using f Nitinol sheet [183]. [184] developed, actuators using inkjet 3D printing and tested using joule heating successfully.

5.8 Applications of SMPs

SMPs have unique potential to remember their original form, because of which, they have plenty of applications in various fields. They have applications in the medical field, mechanical engineering, electrically actuated sensors, wearables, and light-emitting diodes [167, 169, 170, 172, 173]. 3D printing of SMPs adds an advantage to printing complex structures. In recent years, many researchers throughout the globe have developed SMPs using 4D printing for various applications in the medical field for drug delivery, biosensors, tissue engineering, biomimetic architectures, regenerative skin, smart stents, bio-medical devices, and smart medical devices. Similarly, on the engineering side, researchers worked on developing thermal actuators, shape-morphing structures, self-rolling objects, smart grippers, wearable sensors, soft electronic devices, and composite structures. Complete details of the type of polymers used along with the 4D printing technique, and the applications are presented in Tables 5.2 and 5.3.

Table 5.2 Various applications (bio-medical) with corresponding techniques for SMPs in 4D printing.

Polymer material	4D printing method	Application	References
Hydrogels	DIW	Smart textiles, robotics, drug delivery	[146, 147]
Hydrogel biomaterials	DIW	Tissue engineering, drug delivery, and biosensors	[148]
Smart hydrogels	DIW	Dynamic and biomimetic tissue structures	[149]
Shape memory hydrogels	DIW	Drug delivery system	[150]
Bio printed scaffolds	DIW	For developing cell membranes	[151, 152]
Collagen, Alginate	DIW	Tissue engineering	[153]
UV Curable Resin	SLA	Biodegradable vascular stents	[154]
Electro spun Polymer Composite Nanofibers	FDM	Bio-devices	[155]
polycaprolactone (PCL) based resin	FDM	For developing smart stents	[156]
SMP gels	FDM	Fabrication of smart medical devices	[157]

5.9 Summary & Future Scope

1. Plenty of techniques exist for 4D printing of polymers. A few economical processes like FDM made 4D printing easier, flexible, and capable of printing complex 3D structures with SMPs for various structures.

Table 5.3 Various applications (mechanical, electrical, thermal and other) with corresponding techniques for SMPs in 4D printing.

Polymer hydrogel composite	DIW	Thermal actuators	[158]
Thermo-responsive PNIPAm	DIW	Structures with shape morphing	[159]
Temperature-responsive PEG hydrogels	DIW	Shape-shifting mechanisms of hydrogels	[158]
Hydrogels	DIW	Structures with shape morphing	[160–162]
Hydrogel composite	DIW	Shape-morphing device	[163]
Functionalized PHEMA, EDA	DIW	Self-rolling objects	[164]
Methylenebisacrylamide, Lepontic	DIW	Actuators	[165]
Laponite nano-clay	DIW	Grippers	[113]
PNIPAAm-Laponite	DIW	Valves for micro fluids	[166]
Sodium carboxymethyl cellulose	DIW	Structures for specific applications	[28]
Light-sensitive liquid resin with magnetically responsive anisotropic stiff particles	DIW	Soft robots	[177]
Methacrylated-polycaprolactone	SLA	Electronic devices	[168]
PCL/UPy-based polymer	DLP	Soft actuators	[174]
PCL	DLP	Electronic devices	[175]
polyethylene glycol diacrylate	DLP	Electronic devices	[176–178]
Poly (styreneblock-methyl acrylate)	FDM	Structures with shape morphing	[179]
Epoxy resin	Inkjet	Structures with shape morphing	[171]

2. A large range of SMPs exists with various material properties, which provides the wider scope of applying the 4D printing technology for a large area of applications.
3. SMPs have huge applications in bio-medical, drug delivery, biosensors, tissue engineering, biomimetic architectures, regenerative skin, smart stents, and medical devices. DIW, FDM, and SLA are widely used techniques for these applications.
4. Besides bio-medical, 4D-printed polymers have applications in actuators, shape-morphing structures, self-rolling objects, smart grippers, wearable sensors, soft electronic devices, and composite structures. DIW, SLA, FDM, inkjet, and PolyJet are the major techniques applicable to these applications.
5. Even though pure SMPs can be used directly for various applications, they need to enhance their properties (like mechanical, shape recovery, thermal, etc.) according to specific application requirements. This could be possible with reinforcements.
6. The use of pure SMPs significantly increases the overall cost. This can be reduced by blending them with conventional polymers.

References

1. O'Donnell, J., Ahmadkhanlou, F., Yoon, H.-S., Washington, G., All-printed smart structures: a viable option?, in: *Proc. SPIE 9057, Active and Passive Smart Structures and Integrated Systems*, p. 905729, 2014, (10 March 2014); https://doi.org/10.1117/12.2045284.
2. Piedade, A.P., 4D Printing: The Shape-Morphing in Additive Manufacturing. *J. Funct. Biomater.*, 10, 1, 9, 2019. https://doi.org/10.3390/jfb10010009.
3. Gibson, I., Rosen, D.W., Stucker, B., *Additive manufacturing technologies*, p. 498, Springer, New York, 2015.
4. Chua, C.K. and Leong, K.F., *3D Printing and Additive Manufacturing: Principles and Applications (With Companion Media Pack) of Rapid Prototyping*, fourth, World Scientific Publishing Company, Singapore, 2014.
5. Pei, E., 4D printing: dawn of an emerging technology cycle. *Assem. Automat.*, 34, 4, 310–314, 2014.
6. Yap, Y.L. and Yeong, W.Y., Shape recovery effect of 3D printed polymeric honeycomb: this paper studies the elastic behaviour of different honeycomb structures produced by polyjet technology. *Virt. Phys. Prototyp.*, 10, 2, 91–99, 2015.

7. Tibbits, S., Mcknelly, C., Olguín, C.J., Dikovsky, D., Hirsch, S., 4D Printing and Universal Transformation, in: *Proceedings of the 34th Annual Conference of the Association for Computer Aided Design in Architecture (ACADIA)*, 2014.
8. Craddock, M., Augustine, E., Konerman, S., Shin, M., Biorobotics: An Overview of Recent Innovations in Artificial Muscles. *Actuators.*, 11, 6, 168, 2022. https://doi.org/10.3390/act11060168.
9. Namathoti, S. and Vakkalagadda, M.R.K., Development of Multiwalled Carbon Nanotubes/Halloysite Nanotubes Reinforced Thermal Responsive Shape Memory Polymer Nanocomposites for Enhanced Mechanical and Shape Recovery Characteristics in 4D Printing Applications. *Polymers*, 15, 1371, 2023.
10. Pandini, S., Inverardi, N., Scalet, G., Battini, D., Bignotti, F., Marconi, S., *et al.*, Shape memory response and hierarchical motion capabilities of 4D printed auxetic structures. *Mech. Res. Commun.*, 103, 103463, 2020.
11. Liu, K., Zhang, Y., Cao, H., Liu, H., Geng, Y., Yuan, W., *et al.*, Programmable reversible shape transformation of hydrogels based on transient structural anisotropy. *Adv. Mater.*, 32, 2001693, 2020.
12. Hua, M., Wu, D., Wu, S., Ma, Y., Alsaid, Y., He, X., 4D printable tough and thermoresponsive hydrogels. *ACS Appl. Mater. Interfaces*, 13, 12689–97, 2021.
13. Meng, H. and Li, G.A., review of stimuli-responsive shape memory polymer composites. *Polymer*, 54, 2199–2221, 2013.
14. Roppolo, I., Chiappone, A., Angelini, A., Stassi, S., Frascella, F., Pirri, C.F., Ricciardi, C., Descrovi, E., 3D printable light-responsive polymers. *Mater. Horiz.*, 4, 3, 396–401, 2017. The Royal Society of Chemistry.
15. Boyle, B.M., French, T.A., Pearson, R.M., McCarthy, B.G., Miyake, G.M., Structural color for additive manufacturing: 3D-printed photonic crystals from block copolymers. *ACS Nano*, 11, 3052–8, 2017.
16. Davidson, E.C., Kotikian, A., Li, S., Aizenberg, J., Lewis, J.A., 3D printable and reconfigurable liquid crystal elastomers with light-induced shape memory via dynamic bond exchange. *Adv. Mater.*, 32, 1905682, 2020.
17. Habault, D., Zhang, H., Zhao, Y., Light-triggered self-healing and shape-memory polymers. *Chem. Soc. Rev.*, 42, 7244–7256, 2013.
18. Hu, Y., Wang, Z., Jin, D., Zhang, C., Sun, R., Li, Z., *et al.*, Botanical-inspired 4D printing of hydrogel at the microscale. *Adv. Funct. Mater.*, 30, 1907377, 2019.
19. Larush, L., Kaner, I., Fluksman, A., Tamsut, A., Pawar, A.A., Lesnovski, P., *et al.*, 3D printing of responsive hydrogels for drug-delivery systems. *J. 3D Print. Med.*, 1, 219–29, 2017.
20. Nadgorny, M., Xiao, Z., Chen, C., Connal, L.A., Three-dimensional printing of pH-responsive and functional polymers on an affordable desktop printer. *ACS Appl. Mater. Interfaces*, 8, 28946–54, 2016.

21. Roh, S., Okello, L.B., Golbasi, N., Hankwitz, J.P., Liu, J.A.C., Tracy, J.B., *et al.*, 3D-printed silicone soft architectures with programmed magneto-capillary reconfiguration. *Adv. Mater. Technol.*, 4, 1800528, 2019.
22. Zhang, Y., Wang, Q., Yi, S., Lin, Z., Wang, C., Chen, Z., *et al.*, 4D printing of magnetoactive soft materials for on-demand magnetic actuation transformation. *ACS Appl. Mater. Interfaces*, 13, 4174–84, 2021.
23. Namathoti, S., Ravindra Kumar, V.M., Rama Sreekanth, P.S.A., review on progress in magnetic, microwave, ultrasonic responsive Shape-memory polymer composites. *Mater. Today Proc.*, 56, 1182–1191, 2022.
24. Ambulo, C.P., Ford, M.J., Searles, K., Majidi, C., Ware, T.H., 4D-printable liquid metal-liquid crystal elastomer composites. *ACS Appl. Mater. Interfaces*, 13, 12805–13, 2021.
25. Liu, Y., Lv, H., Lan, X., Leng, J., Du, S., Review of electro-active shape-memory polymer composite. *Compos. Sci. Technol.*, 69, 2064–2068, 2009.
26. Chen, T., Bakhshi, H., Liu, L., Ji, J., Agarwal, S., Combining 3D printing with electrospinning for rapid response and enhanced designability of hydrogel actuators. *Adv. Funct. Mater.*, 28, 1800514, 2018.
27. Song, Z., Ren, L., Zhao, C., Liu, H., Yu, Z., Liu, Q., *et al.*, Biomimetic nonuniform, dual-stimuli self-morphing enabled by gradient four-dimensional printing. *ACS Appl. Mater. Interfaces*, 12, 6351–61, 2020.
28. Mulakkal, M.C., Trask, R.S., Ting, V.P., Seddon, A.M., Responsive cellulose-hydrogel composite ink for 4D printing. *Mater. Des.*, 160, 108–18, 2018.
29. Huang, L., Jiang, R., Wu, J., Song, J., Bai, H., Li, B., *et al.*, Ultrafast digital printing toward 4D shape changing materials. *Adv. Mater.*, 29, 1605390, 2017.
30. Sundaram, S., Kim, D.S., Baldo, M.A., Hayward, R.C., Matusik, W., 3D-printed self-folding electronics. *ACS Appl. Mater. Interfaces*, 9, 37, 32290–32298, 2017.
31. Ponnamma, D., Chamakh, M.M., Deshmukh, K., Ahamed, M.B., Erturk, A., Sharma, P., Al-Maadeed, M.A.A., Ceramic-based polymer nanocomposites as piezoelectric materials, in: *Smart Polymer Nanocomposites*, pp. 77–93, Springer, Switzerland, 2017.
32. Kumar, S.K., Castro, M., Pillin, I., Feller, J.F., Thomas, S., Grohens, Y., Simple technique for the simultaneous determination of solvent diffusion coefficient in polymer by Quantum Resistive Sensors and FT-IR spectroscopy. *Polym. Adv. Technol.*, 24, 487–494, 2013.
33. Dai, S., Ravi, P., Tam, K.C., pH-responsive polymers: synthesis, properties and applications. *Soft Matter*, 4, 3, 435–449, 2008.
34. Song, X., Pan, Y., Chen, Y., Development of a low-cost parallel kinematic machine for multidirectional additive manufacturing. *J. Manuf. Sci. Eng.*, 137, 2, 021005, 2015.
35. Tumbleston, J.R., Shirvanyants, D., Ermoshkin, N., Janusziewicz, R., Johnson, A.R., Kelly, D., Chen, K., Pinschmidt, R., Rolland, J.P., Ermoshkin,

A., Samulski, E.T., Continuous liquid interface production of 3D objects. *Science*, 347, 1349–1352, 2015.

36. Zheng, X., Deotte, J., Alonso, M.P., Farquar, G.R., Weisgraber, T.H., Gemberling, S., Lee, H., Fang, N., Spadaccini, C.M., Design and optimization of a light-emitting diode projection microstereolithography three-dimensional manufacturing system. *Rev. Sci. Instrum.*, 83, 12, 125001, 2012.
37. Zheng, X., Smith, W., Jackson, J., Moran, B., Cui, H., Chen, D., Ye, J., Fang, N., Rodriguez, N., Weisgraber, T., Spadaccini, C.M., Multiscale metallic metamaterials. *Nat. Mater.*, 15, 10, 1100, 2016.
38. Kang, H.W., Park, J.H., Cho, D.W., A pixel based solidification model for projection based stereolithography technology. *Sens. Actuators A*, 178, 223–229, 2012.
39. Wang, X., Jiang, M., Zhou, Z., Gou, J., Hui, D., 3D printing of polymer matrix composites: a review and prospective. *Comp. Part B Eng.*, 110, 442–458, 2017.
40. Melchels, F.P., Feijen, J., Grijpma, D.W., A review on stereolithography and its applications in biomedical engineering. *Biomaterials*, 31, 24, 6121–6130, 2010.
41. Wei, P., Wei, Z., Chen, Z., He, Y., Du, J., Thermal behavior in single track during selective laser melting of AlSi 10Mg powder. *Appl. Phys. A Mater. Sci. Process.*, 123, 9, 604, 2017.
42. Cho, Y., Lee, I., Cho, D.-W., Laser scanning path generation considering photopolymer solidification in micro-stereolithography. *Microsyst. Technol.*, 11, 2–3, 158–167, 2005.
43. Xu, G.S., Pan, H., Ma, X.M., Luo, S., Qiu, R.H., Investigation of UV light intensity distribution of integral micro-stereolithography system. *Adv. Mat. Res.*, 97, 3985–3988, 2010.
44. Tanzi, M.C., Fare, S., Candiani, G., Chapter 3: Manufacturing Technologies, in: *Foundations of Biomaterials Engineering*, pp. 137–196, 2019.
45. Kim, K., Yeatts, A., Dean, D., Fisher, J.P., Stereolithographic bone scaffold design parameters: osteogenic differentiation and signal expression. *Tiss. Eng. B Rev.*, 16, 5, 523–539, 2010.
46. Heller, C., Schwentenwein, M., Russmueller, G., *et al.*, Vinyl esters: low cytotoxicity monomers for the fabrication of biocompatible 3D scaffolds by lithography based additive manufacturing. *J. Polym. Sci. A Polym. Chem.*, 47, 24, 6941–6954, 2009.
47. Tian, X., Liu, T., Yang, C., Wang, Q., Li, D., Interface and performance of 3D printed continuous carbon fiber reinforced PLA composites. *Compos. A Appl. Sci. Manuf.*, 88, 198–205, 2016.
48. Kim, N., Bhalerao, I., Han, D., Yang, C., Lee, H., Improving surface roughness of additively manufactured parts using a photopolymerization model and multi-objective particle swarm optimization. *Appl. Sci.*, 9, 151, 2019.
49. Han, D., Lu, Z., Chester, S.A., Lee, H., Micro 3D printing of a temperature-responsive hydrogel using projection micro-stereolithography. *Sci. Rep.*, 8, 1963, 2018.

50. Ge, Q., Sakhaei, A.H., Lee, H., Dunn, C.K., Fang, N.X., Dunn, M.L., Multimaterial 4D printing with tailorable shape memory polymers. *Sci. Rep.*, 6, 31110, 2016.
51. Low, Z.-X., Chua, Y.T., Ray, B.M., Mattia, D., Metcalfe, I.S., Patterson, D.A., Perspective on 3D printing of separation membranes and comparison to related unconventional fabrication techniques. *J. Memb. Sci.*, 523, 596–613, 2017.
52. Sun, C., Fang, N., Wu, D.M., Zhang, X., Projection micro-stereolithography using digital micromirror dynamic mask. *Sens. Actuators A Phys.*, 121, 1, 113–120, 2005.
53. Billiet, T., Vandenhaute, M., Schelfhout, J., Van Vlierberghe, S., Dubruel, P., A review of trends and limitations in hydrogel-rapid prototyping for tissue engineering. *Biomaterials*, 33, 26, 6020–6041, 2012.
54. Chiappone, A., Fantino, E., Roppolo, I., Lorusso, M., Manfredi, D., Fino, P., Pirri, C.F., Calignano, F., 3D printed PEG-based hybrid nanocomposites obtained by sol–gel technique. *ACS Appl. Mater. Interfaces*, 8, 8, 5627–5633, 2016.
55. Yardimci, M.A., Guceri, S.I., Danforth, S.C., Agarwala, M., Safari, A., Numerical modeling of fused deposition processing. *Am. Soc. Mech. Eng.*, 69, 1225–1235, 1995.
56. Kazemi, Y., Kakroodi, A.R., Ameli, A., Filleter, T., Park, C.B., Highly stretchable conductive thermoplastic vulcanizate/carbon nanotube nanocomposites with segregated structure, low percolation threshold and improved cyclic electromechanical performance. *J. Mater. Chem. C*, 6, 2, 350–359, 2018.
57. Dudek, P.F.D.M., FDM 3D printing technology in manufacturing composite elements. *Arc. Met. Mater.*, 58, 4, 1415–1418, 2013.
58. Appuhamillage, G.A., Reagan, J.C., Khorsandi, S., Davidson, J.R., Voit, W., Smaldone, R.A., 3D printed remendable polylactic acid blends with uniform mechanical strength enabled by a dynamic Diels–Alder reaction. *Polym. Chem.*, 8, 13, 2087–2092, 2017.
59. Shallan, A.I., Smejkal, P., Corban, M., Guijt, R.M., Breadmore, M.C., Cost-effective three-dimensional printing of visibly transparent microchips within minutes. *Anal. Chem.*, 86, 6, 3124–3130, 2014.
60. Rajan, M., Murugan, M., Ponnamma, D., Sadasivuni, K.K., Munusamy, M.A., Poly-carboxylic acids functionalized chitosan nanocarriers for controlled and targeted anti-cancer drug delivery. *Biomed. Pharmacother.*, 83, 201–211, 2016.
61. Symes, M.D., Kitson, P.J., Yan, J., Richmond, C.J., Cooper, G.J., Bowman, R.W., Vilbrandt, T., Cronin, L., Integrated 3D-printed reactionware for chemical synthesis and analysis. *Nat. Chem.*, 4, 5, 349, 2012.
62. Ventola, C.L., Medical applications for 3D printing: current and projected uses. *P T*, 39, 10, 704, 2014.
63. Ponnamma, D., Sadasivuni, K.K., Cabibihan, J.J., Yoon, W.J., Kumar, B., Reduced graphene oxide filled poly (dimethyl siloxane) based transparent

stretchable, and touch-responsive sensors. *Appl. Phys. Lett.*, 108, 17, 171906, 2016.
64. O'Neill, P.F., Ben Azouz, A., Va´zquez, M., Liu, J., Marczak, S., Slouka, Z., Chang, H.C., Diamond, D., Brabazon, D., Advances in three-dimensional rapid prototyping of microfluidic devices for biological applications. *Biomicrofluidics*, 8, 5, 052112, 2014.
65. Murphy, S.V. and Atala, A., 3D bioprinting of tissues and organs. *Nat. Biotechnol.*, 32, 8, 773, 2014.
66. Wegst, U.G., Bai, H., Saiz, E., Tomsia, A.P., Ritchie, R.O., Bioinspired structural materials. *Nat. Mater.*, 14, 1, 23, 2015.
67. Meyers, M.A., McKittrick, J., Chen, P.Y., Structural biological materials: critical mechanics-materials connections. *Science*, 339, 6121, 773–779, 2013.
68. Ambrosi, A. and Pumera, M., 3D-printing technologies for electrochemical applications. *Chem. Soc. Rev.*, 45, 10, 2740–2755, 2016.
69. Truby, R.L. and Lewis, J.A., Printing soft matter in three dimensions. *Nature*, 540, 7633, 371, 2016.
70. Jiang, Z., Diggle, B., Tan, M.L., Viktorova, J., Bennett, C.W., Connal, L.A., Extrusion 3D Printing of Polymeric Materials with Advanced Properties. *Adv. Sci.*, 7, 2001379, 2020.
71. Priyanka Kumar, A., Multistimulus-responsive supramolecular hydrogels derived by *in situ* coating of Ag nanoparticles on 5-CMP-capped β-FeOOH binary nanohybrids with multifunctional features and applications. *ACS Omega*, 5, 13672–84, 2020.
72. Shi, X., Zheng, Y., Wang, G., Lin, Q., Fan, J., pH- and electro-response characteristics of bacterial cellulose nanofiber/sodium alginate hybrid hydrogels for dual controlled drug delivery. *RSC Adv.*, 4, 47056–65, 2014.
73. Monzón, M.D., Paz, R., Pei, E., Ortega, F., Suárez, L.A., Ortega, Z., Alemán, M.E., Plucinski, T., Clow, N., 4D printing: processability and measurement of recovery force in shape memory polymers. *Int. J. Adv. Manuf. Technol.*, 89, 5-8, 1827–1836, 2017.
74. Miyanaji, H., Zhang, S., Lassell, A., Zandinejad, A., Yang, L., Process development of porcelain ceramic material with binder jetting process for dental applications. *JOM*, 68, 3, 831–841, 2016.
75. Singh, M., Haverinen, H.M., Dhagat, P., *et al.*, Inkjet printing—process and its applications. *Adv.Mater.*, 22, 6, 673–685, 2010.
76. Agarwala, S., Goh, G.L., Yap, Y.L., *et al.*, Development of bendable strain sensor with embedded microchannels using 3D printing. *Sens. Actuators A Phys.*, 263, 593–599, 2017.
77. Lee, H.-H., Chou, K.-S., Huang, K.-C., Inkjet printing of nanosized silver colloids. *Nanotechnology*, 16, 10, 2436, 2005.
78. Kordás, K., Mustonen, T., Tóth, G., *et al.*, Inkjet printing of electrically conductive patterns of carbon nanotubes. *Small*, 2, 8–9, 1021–1025, 2006.
79. Kawase, T., Shimoda, T., Newsome, C., *et al.*, Inkjet printing of polymer thin film transistors. *Thin Solid Films*, 438, 279–287, 2003.

80. Yin, Z., Huang, Y., Bu, N., *et al.*, Inkjet printing for flexible electronics: materials, processes and equipments. *Chin. Sci. Bull.*, 55, 30, 3383–3407, 2010.
81. Boland, T., Xu, T., Damon, B., *et al.*, Application of inkjet printing to tissue engineering. *Biotechnol. J. Healthcare Nutr. Technol.*, 1, 9, 910–917, 2006.
82. Dua, V., Surwade, S.P., Ammu, S., *et al.*, All-organic vapor sensor using inkjet-printed reduced grapheme oxide. *Angew. Chem.*, 122, 12, 2200–2203, 2010.
83. Goh, G.L., Ma, J., Chua, K.L.F., *et al.*, Inkjet-printed patch antenna emitter for wireless communication application. *Virtual Phys. Prototyp.*, 11, 4, 289–294, 2016.
84. Kawahara, Y., Hodges, S., Cook, B.S., *et al.*, Instant inkjet circuits: lab-based inkjet printing to support rapid prototyping of UbiComp devices, in: *Proceedings of the 2013 ACM International Joint Conference on Pervasive and Ubiquitous Computing, ACM*, 2013.
85. Gonzalez, J.A., Mireles, J., Lin, Y., Wicker, R.B., Characterization of ceramic components fabricated using binder jetting additive manufacturing technology. *Ceram. Int.*, 42, 9, 10559–10564, 2016.
86. Secor, E.B., Principles of Aerosol Jet Printing, Flexible and Printed Electronics. *IOP Publishing*, 3, 3, 035002, 2018.
87. Vaezi, M., Chianrabutra, S., Mellor, B., *et al.*, Multiple material additive manufacturing. Part 1. A review: this review paper covers a decade of research on multiple material additive manufacturing technologies which can produce complex geometry parts with different materials. *Virtual Phys. Prototyp.*, 8, 1, 19–50, 2013.
88. Binder, S., Glatthaar, M., Rädlein, E., Analytical investigation of aerosol jet printing. *Aerosol Sci. Technol.*, 48, 9, 924–929, 2014.
89. Ligon, S.C., Liska, R., Stampfl, J., Gurr, M., M€ulhaupt, R., Polymers for 3D printing and customized additive manufacturing. *Chem. Rev.*, 117, 15, 10212–10290, 2017.
90. Hoey, J.M., Lutfurakhmanov, A., Schulz, D.L., Akhatov, I.S., A review on aerosol-based direct-write and its applications for microelectronics. *J. Nanotechnol.*, 2012, 22, 2012. Article ID 324380.
91. Agarwala, S., Goh, G.L., Yeong, W.Y., Optimizing aerosol jet printing process of silver ink for printed electronics. *IOP Conf. Ser.: Mater. Sci. Eng.*, 191, 012027, 2017.
92. Agarwala, S., Goh, G.L., Dinh Le, T.S., An, J., Peh, Z.K., Yeong, W.Y., Kim, Y.J., Wearable Bandage-Based Strain Sensor for Home Healthcare: Combining 3D Aerosol Jet Printing and Laser Sintering. *ACS Sens.*, 4, 1, 218–226, 2019 Jan 25. doi: 10.1021/acssensors.8b01293. Epub 2018 Dec 31. PMID: 30560661.
93. Yang, C., Zhou, E., Miyanishi, S., *et al.*, Preparation of active layers in polymer solar cells by aerosol jet printing. *ACS Appl. Mater. Interfaces*, 3, 10, 4053–4058, 2011.

94. Goh, G.L., Agarwala, S., Tan, Y.J., *et al.*, A low cost and flexible carbon nanotube pH sensor fabricated using aerosol jet technology for live cell applications. *Sens. Actuators B Chem.*, 260, 227–235, 2018.
95. Zhao, D., Liu, T., Park, J.G., *et al.*, Conductivity enhancement of aerosol-jet printed electronics by using silver nanoparticles ink with carbon nanotubes. *Microelectron. Eng.*, 96, 71–75, 2012.
96. Agarwala, S., Goh, G.L., Goh, G.D., Dikshit, V., Yeong, W.Y., Chapter 10 - 3D and 4D printing of polymer/CNTs-based conductive composites, in: *3D and 4D Printing of Polymer Nanocomposite Materials*, K.K. Sadasivuni, K. Deshmukh, M.A. Almaadeed (Eds.), pp. 297–324, Elsevier, Amsterdam, 2020.
97. Lee, J.Y., An, J., Chua, C.K., Fundamentals and applications of 3Dgprinting fornovel materials. *Appl. Mater. Today*, 7, 120–133, 2017.
98. Akbar, I., El Hadrouz, M., El Mansori, M., Lagoudas, D., Toward enabling manufacturing paradigm of 4D printing of shape memory materials: Open literature review. *Eur. Polym. J.*, 168, 111106, 2022.
99. Zhang, B., Li, H., Cheng, J., Ye, H., Sakhaei, A.H., Yuan, C., Rao, P., Zhang, Y.F., Chen, Z., Wang, R., Mechanically robust and UV-curable shape-memory polymers for digital light processing based 4D printing. *Adv. Mater.*, 33, 2101298, 2021.
100. Sosa, R.S.B., Baldos, D.T., Basilia, B.A., 4D Printing of Shape Memory Polymers: A Concise Review of Photopolymerized Acrylate-Based Materials. *Diffus. Found. Mater. Appl.*, 32, 1–12, 2023.
101. Dong, Y., Wang, S., Ke, Y., Ding, L., Zeng, X., Magdassi, S., Long, Y., 4D Printed Hydrogels: Fabrication, Materials, and Applications. *Adv. Mater. Technol.*, 5, 2000034, 2020.
102. Zhang, Y., Huang, L., Song, H., Ni, C., Wu, J., Zhao, Q., Xie, T., 4D printing of a digital shape memory polymer with tunable high performance. *ACS Appl. Mater. Interfaces*, 11, 32408–32413, 2019.
103. Zhang, W., Wang, H., Wang, H., Chan, J.Y.E., Liu, H., Zhang, B., Zhang, Y.-F., Agarwal, K., Yang, X., Ranganath, A.S., Structural multi-colour invisible inks with submicron 4D printing of shape memory polymers. *Nat. Commun.*, 12, 112, 2021.
104. He, Y., Yu, R., Li, X., Zhang, M., Zhang, Y., Yang, X., Zhao, X., Huang, W., Digital light processing 4D printing of transparent, strong, highly conductive hydrogels. *ACS Appl. Mater. Interfaces*, 13, 36286–36294, 2021.
105. Hashemi, M., Rostami, A., Ghasemi, I., Omrani, A., Incorporation of Modified Graphene Nanoplatelets for Development of Bio-based Shape Memory Polymer of Polypropylene Carbonate (PPC)/Polycaprolactone (PCL). *J. Polym. Environ.*, 2715–2726, 2023.
106. Tsujimoto, T., Takayama, T., Uyama, H., Biodegradable Shape Memory Polymeric Material from Epoxidized Soybean Oil and Polycaprolactone. *Polymers*, 7, 2165–2174, 2015.

107. Rahmatabadi, D., Aberoumand, M., Soltanmohammadi, K., Soleyman, E., Ghasemi, I., Baniassadi, M., Abrinia, K., Zolfagharian, A., Bodaghi, M., Baghani, M.A., New Strategy for Achieving Shape Memory Effects in 4D Printed Two-Layer Composite Structures. *Polymers*, 14, 5446, 2022. https://doi.org/10.3390/polym14245446.
108. Jiang, Y., Leng, Q.Y., Yan, Y., Ng, E.L.L., Chee, H.L., Wang, F., Chan, S.Y., Loh, X.J., Wang, J., Chan, B.Q.Y., 4D Printing of Single-Network Shape Memory Polyurethanes with Two-Way Actuation Properties. *ACS Appl. Polym. Mater.*, 4, 8574–8583, 2022.
109. Inverardi, N., Toselli, M., Scalet, G., Messori, M., Auricchio, F., Pandini, S., Stress-Free Two-Way Shape Memory Effect of Poly(ethylene glycol)/Poly(ε-caprolactone) Semicrystalline Networks. *Macromolecules*, 55, 8533–8547, 2022.
110. Spiegel, C.A., Hackner, M., Bothe, V.P., Spatz, J.P., Blasco, E., 4D Printing of Shape Memory Polymers: From Macro to Micro. *Adv. Funct. Mater.*, 32, 2110580, 2022.
111. Champeau, M., Heinze, D.A., Viana, T.N., de Souza, E.R., Chinellato, A.C., Titotto, S., 4D Printing of Hydrogels: A Review. *Adv. Funct. Mater.*, 30, 1910606, 2020.
112. Guo, J., Zhang, R., Zhang, L., Cao, X., 4D Printing of Robust Hydrogels Consisted of Agarose Nanofibers and Polyacrylamide. *ACS Macro Lett.*, 7, 442–446, 2018.
113. Liu, J., Erol, O., Pantula, A., Liu, W., Jiang, Z., Kobayashi, K., Chatterjee, D., Hibino, N., Romer, L.H., Kang, S.H., Nguyen, T.D., Gracias, D.H., Dual-Gel 4D Printing of Bioinspired Tubes. *ACS Appl. Mater. Interfaces*, 11, 8492–8498, 2019.
114. Han, D., Farino, C., Yang, C., Scott, T., Browe, D., Choi, W., Freeman, J.W., Lee, H., Soft Robotic Manipulation and Locomotion with a 3D Printed Electroactive Hydrogel. *ACS Appl. Mater. Interfaces*, 10, 17512–17518, 2018.
115. Chen, S., Hu, J., Zhuo, H., Properties and mechanism of two-way shape memory polyurethane composites. *Compos. Sci. Technol.*, 70, 1437–1443, 2010.
116. Scalet, G., Two-Way and Multiple-Way Shape Memory Polymers for Soft Robotics: An Overview. *Actuators*, 9, 10, 2020.
117. Scalet, G., Pandini, S., Messori, M., Toselli, M., Auricchio, F., A one-dimensional phenomenological model for the two-way shape-memory effect in semi-crystalline networks. *Polymer*, 158, 130–148, 2018.
118. Westbrook, K.K., Parakh, V., Chung, T., Mather, P.T., Wan, L.C., Dunn, M.L., Qi, H.J., Constitutive modeling of shape memory effects in semicrystalline polymers with stretch induced crystallization. *J. Eng. Mater. Technol.*, 132, 041010, 9, 2010.
119. Pandini, S., Baldi, F., Paderni, K., Messori, M., Toselli, M., Pilati, F., Gianoncelli, A., Brisotto, M., Bontempi, E., Riccò, T., One-way and two-way

shape memory behaviour of semi-crystalline networks based on sol–gel cross-linked poly (ε-caprolactone). *Polymer*, 54, 4253–4265, 2013.
120. Kolesov, I., Dolynchuk, O., Jehnichen, D., Reuter, U., Stamm, M., Radusch, H.-J., Changes of crystal structure and morphology during two-way shape-memory cycles in cross-linked linear and short-chain branched polyethylenes. *Macromolecules*, 48, 4438–4450, 2015.
121. Huang, M., Dong, X., Wang, L., Zhao, J., Liu, G., Wang, D., Two-way shape memory property and its structural origin of cross-linked poly (ε-caprolactone). *RSC Adv.*, 4, 55483–55494, 2014.
122. Baker, R.M., Henderson, J.H., Mather, P.T., Shape memory poly (ε-caprolactone)-co-poly (ethylene glycol) foams with body temperature triggering and two-way actuation. *J. Mater. Chem. B*, 1, 4916–4920, 2013.
123. Bothe, M. and Pretsch, T., Two-way shape changes of a shape-memory poly (ester urethane). *Macromol. Chem. Phys.*, 213, 2378–2385, 2012.
124. Raquez, J.M., Vanderstappen, S., Meyer, F., Verge, P., Alexandre, M., Thomassin, J.M., Jérôme, C., Dubois, P., Design of cross-linked semicrystalline poly (ε-caprolactone)-based networks with one-way and two-way shape-memory properties through Diels–Alder reactions. *Chemistry–A Eur. J.*, 17, 10135–10143, 2011.
125. Li, J., Rodgers, W.R., Xie, T., Semi-crystalline two-way shape memory elastomer. *Polymer*, 52, 5320–5325, 2011.
126. Chung, T., Romo-Uribe, A., Mather, P.T., Two-way reversible shape memory in a semicrystalline network. *Macromolecules*, 41, 184–192, 2008.
127. Bai, Y., Zhang, X., Wang, Q., Wang, T., A tough shape memory polymer with triple-shape memory and two-way shape memory properties. *J. Mater. Chem. A*, 2, 4771–4778, 2014.
128. Han, J.L., Lai, S.M., Chiu, Y.T., Two-way multi-shape memory properties of peroxide crosslinked ethylene vinyl-acetate copolymer (EVA)/polycaprolactone (PCL) blends. *Polym. Adv. Technol.*, 29, 2010–2024, 2018.
129. Kolesov, I., Dolynchuk, O., Radusch, H.-J., Shape-memory behavior of cross-linked semi-crystalline polymers and their blends. *eXPRESS Polym. Lett.*, 9, 255–276, 2015.
130. Zotzmann, J., Behl, M., Hofmann, D., Lendlein, A., Reversible triple-shape effect of polymer networks containing polypentadecalactone-and poly (ε-caprolactone)-segments. *Adv. Mater.*, 22, 3424–3429, 2010.
131. Lendlein, A., Progress in actively moving polymers. *J. Mater. Chem.*, 20, 3332–3334, 2010.
132. Zhou, J. and Sheiko, S.S., Reversible shape-shifting in polymeric materials. *J. Polym. Sci., Part B: Polym. Phys.*, 54, 1365–1380, 2016.
133. Burke, K.A. and Mather, P.T., Soft shape memory in main-chain liquid crystalline elastomers. *J. Mater. Chem.*, 20, 3449–3457, 2010.
134. Thomsen, D.L., Keller, P., Naciri, J., Pink, R., Jeon, H., Shenoy, D., Ratna, B.R., Liquid crystal elastomers with mechanical properties of a muscle. *Macromolecules*, 34, 5868–5875, 2001.

135. Burke, K.A., Rousseau, I.A., Mather, P.T., Reversible actuation in main-chain liquid crystalline elastomers with varying crosslink densities. *Polymer*, 55, 5897–5907, 2014.
136. Wang, Y., Huang, X., Zhang, J., Bi, M., Zhang, J., Niu, H., Li, C., Yu, H., Wang, B., Jiang, H., Two-step crosslinked liquid-crystalline elastomer with reversible two-way shape memory characteristics. *Mol. Cryst. Liq. Cryst.*, 650, 13–22, 2017.
137. Wen, Z., McBride, M.K., Zhang, X., Han, X., Martinez, A.M., Shao, R., Zhu, C., Visvanathan, R., Clark, N.A., Wang, Y., Reconfigurable LC elastomers: using a thermally programmable monodomain to access two-way free-standing multiple shape memory polymers. *Macromolecules*, 51, 5812–5819, 2018.
138. Lendlein, A. and Gould, O.E., Reprogrammable recovery and actuation behaviour of shape-memory polymers. *Nat. Rev. Mater.*, 4, 116–133, 2019.
139. Wang, K., Jia, Y.-G., Zhao, C., Zhu, X., Multiple and two-way reversible shape memory polymers: Design strategies and applications. *Prog. Mater Sci.*, 105, 100572, 2019.
140. Gao, Y., Liu, W., Zhu, S., Reversible shape memory polymer from semicrystalline poly (ethylene-co-vinyl acetate) with dynamic covalent polymer networks. *Macromolecules*, 51, 8956–8963, 2018.
141. Dolynchuk, O., Kolesov, I., Jehnichen, D., Reuter, U., Radusch, H.-J., Sommer, J.-U., Reversible shape-memory effect in cross-linked linear poly (ε-caprolactone) under stress and stress-free conditions. *Macromolecules*, 50, 3841–3854, 2017.
142. Li, Q., Zhou, J., Vatankhah-Varnoosfaderani, M., Nykypanchuk, D., Gang, O., Sheiko, S.S., Advancing reversible shape memory by tuning the polymer network architecture. *Macromolecules*, 49, 1383–1391, 2016.
143. Behl, M., Kratz, K., Zotzmann, J., Nöchel, U., Lendlein, A., Reversible bidirectional shape-memory polymers. *Adv. Mater.*, 25, 4466–4469, 2013.
144. Saatchi, M., Behl, M., Nöchel, U., Lendlein, A., Copolymer Networks From Oligo (ε-caprolactone) and n-Butyl Acrylate Enable a Reversible Bidirectional Shape-Memory Effect at Human Body Temperature. *Macromol. Rapid Commun.*, 36, 880–884, 2015.
145. Hanzon, D.W., Traugutt, N.A., McBride, M.K., Bowman, C.N., Yakacki, C.M., Yu, K., Adaptable liquid crystal elastomers with transesterification-based bond exchange reactions. *Soft Matter*, 14, 951–960, 2018.
146. Kuang, X., Roach, D.J., Wu, J., Hamel, C.M., Ding, Z., Wang, T., Dunn, M.L., Qi, H.J., Advances in 4D printing: materials and applications. *Adv. Funct. Mater.*, 29, 2, 1805290, 2019.
147. Gao, B., Yang, Q., Zhao, X., Jin, G., Ma, Y., Xu, F., 4D bioprinting for biomedical applications. *Trends Biotechnol.*, 34, 9, 746–756, 2016.
148. Askari, M., Naniz, M.A., Kouhi, M., Saberi, A., Zolfagharian, A., Bodaghi, M., Recent progress in extrusion 3D bioprinting of hydrogel biomaterials for

tissue regeneration: A comprehensive review with focus on advanced fabrication techniques. *Biomater. Sci.*, 9, 3, 535–573, 2021.
149. An, J., Chua, C.K., Mironov, V., A perspective on 4D bioprinting. *Int. J. Bioprinting*, 2, 1, 3–5, 2016.
150. Dai, W., Guo, H., Gao, B., Ruan, M., Xu, L., Wu, J., Xue, W., Double network shape memory hydrogels activated by near-infrared with high mechanical toughness, nontoxicity, and 3D printability, *Chem. Eng. J.*, 356, 934–949, 2019.
151. Villar, G., *et al.*, Formation of droplet networks that function in aqueous environments. *Nat. Nanotechnol.*, 6, 803–808, 2011.
152. Villar, G., Graham, A.D., Bayley, H., A Tissue-Like Printed Material. *Science*, 340, 48–52, 2013.
153. Yang, X., Lu, Z., Wu, H., Li, W., Zheng, L., Zhao, J., Collagen-alginate as bioink for three-dimensional (3D) cell printing based cartilage tissue engineering. *Mater. Sci. Engineering C*, 83, 195–201, 2018. https://doi.org/10.1016/j.msec.2017.09.002.
154. Fu, M., Liu, F., Hu, L., A novel category of 3D chiral material with negative Poisson's ratio. *Compos. Sci. Technol.*, 160, 111–118, 2018.
155. Deshmukh, K., Sankaran, S., Ahamed, M.B., Pasha, S.K.K., Biomedical applications of polymer composite nanofibers, in: *Polymer Nanocomposites in Biomedical Engineering Applications*, K.K. Sadasivuni, D. Ponnamma, R. Mariappan, M.B. Ahamed, J.J. Cabibihan (Eds.), 111–166, Springer Nature, Switzerland, 2019.
156. Bodaghi, M., Damanpack, A.R., Liao, W.H., Triple shape memory polymers by 4D printing. *Smart Mater. Struct.*, 27, 65010, 2018.
157. Castro, N.J., Meinert, C., Levett, P., Hutmacher, D.W., Current developments in multifunctional smart materials for 3D/4D bioprinting. *Curr. Opin. Biomed. Eng.*, 2, 67–75, 2017.
158. Naficy, S., Gately, R., Gorkin III, R., Xin, H., Spinks, G.M., 4D printing of reversible shape morphing hydrogel structures. *Macromol. Mater. Eng.*, 302, 1600212, 2017.
159. Bakarich, S.E., Gorkin III, R., Panhuis, M.I.H., Spinks, G.M., 4D printing with mechanically robust, thermally actuating hydrogels. *Macromol. Rapid Commun.*, 36, 1211–1217, 2015.
160. Gao, W., Zhang, Y., Ramanujan, D., Ramani, K., Chen, Y., Williams, C.B., Wang, C.C., Shin, Y.C., Zhang, S., Zavattieri, P.D., The status, challenges, and future of additive manufacturing in engineering. *Comp. Aid. Des.*, 69, 65–89, 2015.
161. Bourell, D.L., Beaman, J.J., Leu, M.C., Rosen, D.W., A brief history of additive manufacturing and the 2009 roadmap for additive manufacturing: looking back and looking ahead. *Proc. Rap. Technol.*, 24–25, 2009.
162. Momeni, F., Liu, X., Ni, J., A review of 4D printing. *Mater. Des.*, 122, 42–79, 2017.

163. Gladman, A.S., Matsumoto, E.A., Nuzzo, R.G., Mahadevan, L., Lewis, J.A., Biomimetic 4D printing. *Nat. Mater.*, 15, 4, 413–418, 2016.
164. Nadgorny, M., Xiao, Z., Connal, L.A., 2D and 3D-printing of self-healing gels: design and extrusion of self-rolling objects. *Mol. Syst. Des. Eng.*, 2, 3, 283–292, 2017. The Royal Society of Chemistry, doi ="10.1039/C7ME00023E".
165. Chen, T., Bakhshi, H., Liu, L., Ji, J., Agarwal, S., Combining 3D printing with electrospinning for rapid response and enhanced designability of hydrogel actuators. *Adv. Funct. Mater.*, 28, 19, 1800514, 2018. https://onlinelibrary.wiley.com/doi/ abs/10.1002/adfm.201800514 (accessed Jul. 24, 2019).
166. Jin, Y., Shen, Y., Yin, J., Qian, J., Huang, Y., Nanoclay-Based Self-Supporting Responsive Nanocomposite Hydrogels for Printing Applications. *ACS Appl. Mater. Interfaces*, 10, 12, 10461–10470, 2018. https://doi.org/10.1021/acsami.8b00806.
167. Zheng, S. Y., Shen, Y., Zhu, F., Yin, J., Qian, J., Fu, J., Zheng, Q., Programmed deformations of 3Dprinted tough physical hydrogels with high response speed and large output force. *Adv. Funct. Mater.*, 28, 37, 1803366, 2018. https://onlinelibrary.wiley.com/doi/abs/10.1002/adfm.201803366 (accessed Jul. 24, 2019).
168. Lei, Z., Wang, Q., Wu, P., A multifunctional skin-like sensor based on a 3D printed thermo-responsive hydrogel. *Mater. Horiz.*, 4, 4, 694–700, 2017.
169. Zhu, P., Yang, W., Wang, R., Gao, S., Li, B., Li, Q., 4D Printing of Complex Structures with a Fast Response Time to Magnetic Stimulus. *ACS Appl. Mater. Interfaces*, 10, 42, 36435–36442, 2018. DOI: 10.1021/acsami.8b12853.
170. Kim, Y., Yuk, H., Zhao, R., *et al.*, Printing ferromagnetic domains for untethered fast-transforming soft materials. *Nature*, 558, 274–279, 2018. https://doi.org/10.1038/s41586-018-0185-0.
171. Kokkinis, D., Schaffner, M., Studart, A., Multimaterial magnetically assisted 3D printing of composite materials. *Nat. Commun.*, 6, 8643, 2015. https://doi.org/10.1038/ncomms9643.
172. Bastola, A.K., Paudel, M., Li, L., Development of hybrid magnetorheological elastomers by 3D printing. *Polymer*, 149, 213–228, 2018. doi: https://doi.org/10.1016/j.polymer.2018.06.076.
173. Zarek, M., Mansour, N., Shapira, S., Cohn, D., 4D Printing of Shape Memory-Based Personalized Endoluminal Medical Devices. *Macromol. Rapid Commun.*, 38, 2, 1600628, 2017 Jan. doi: 10.1002/marc.201600628. Epub 2016 Dec 5. PMID: 27918636.
174. Invernizzi, M., Turri, S., Levi, M., Suriano, R., 4D printed thermally activated self-healing and shape memory polycaprolactone-based polymers. *Eur. Polym. J.*, 101, 169–176, 2018. 2018, doi = {https://doi.org/10.1016/j.eurpolymj.2018.02.023.
175. Zarek, M., Layani, M., Cooperstein, I., Sachyani, E., Cohn, D., Magdassi, S., 3D Printing of Shape Memory Polymers for Flexible Electronic Devices. *Adv. Mater.*, 28, 4449–4454, 2016. https://doi.org/10.1002/adma.201503132.

176. Zhao, Z., Wu, J., Mu, X., Chen, H., Qi, H.J., Fang, D., Desolvation induced origami of photocurable polymers by digit light processing. *Macromol. Rapid Commun.*, 38, 13, 1600625, 2017. https://doi.org/10.1002/marc.201600625.
177. Wu, J., Zhao, Z., Kuang, X., Hamel, C.M., Fang, D., Qi, H.J., Reversible shape change structures by grayscale pattern 4D printing. *Multifunct. Mater.*, 1, 1, 015002, IOP Publishing, 2018.
178. Zeang, Z,., Q,.H.J., Fang, D., A finite deformation theory of desolvation and swelling in partially photo-cross-linked polymer networks for 3D/4D printing applications. *Soft Matter*, 15, 5, 1005–1016, 2019. The Royal Society of Chemistry, doi ="10.1039/C8SM02427H".
179. Tao, R., Xi, L., Wu, W., Li, Y., Liao, B., Liu, L., Leng, J., Fang, D., 4D printed multi-stable metamaterials with mechanically tunable performance. *Compos. Struct.*, 252, 112663, 2020. ISSN 0263-8223, https://doi.org/10.1016/j.compstruct.2020.112663.
180. Samal, B.B., Jena, A., Varshney, S.K., Kumar, C.S., 4D Printing of Shape Memory Polymers: A Comparative Study of Programming Methodologies on Various Material Properties. *Smart Mater. Structures*, IOP publishing, 32, 7, 074003. https://doi.org/10.1088/1361-665x/acda6e.
181. Lee, H.M., Han, S.J., Kim, M.-J., Yoon, G.H., Multi-material 4D printing to realize two-phase morphing in self-actuating structures. *Smart Mater. Structures*, IOP publishing, 33, 3, 035007, 2024. https://iopscience.iop.org/article/10.1088/1361-665X/ad21b5.
182. Samal, B.B., Jena, A., Varshney, S.K., Kumar, C.S., FDM 4D printing: A low-cost approach of shape programming and assessing the shape memory properties using angle measurement methods in hot water actuation testing apparatus. *Mater. Today Proceedings*, Elsevier, 101, 7–14, 2024. ISSN 2214-7853. Doi:10.1016/j.matpr.2023.02.038, https://www.sciencedirect.com/science/article/pii/S2214785323005564.
183. Samal, B.B., Jena, A., Mishra, D., Controlled shape changing components by using 4D printing technology, in: *Proceedings of 10th International Conference on Precision, Meso, Micro and Nano Engineering (COPEN 10)*, IIT Madras, pp. 78–81, 2017, Google scholar link: http://www.copen.ac.in/proceedings/copen10/copen/19.revised_manuscript_262.pdf.
184. Akbari, S., Sakhaei, A.H., Panjwani, S., Kowsari, K., Ge, Q., Shape memory alloy based 3D printed composite actuators with variable stiffness and large reversible deformation. *Sens. Actuators A: Phys.*, 321, 112598, 2021. https://www.sciencedirect.com/science/article/pii/S0924424721000595.

6

Polymer Blends and Reinforcements in 4D Printing

Sivanagaraju Namathoti[1], Pavan Kumar Gurrala[2], Prakash Chandra[2], G. Naga Mallikarjun Rao[1] and M. R. K. Vakkalagadda[1*]

[1]*School of Mechanical Engineering, VIT-AP University, Amaravati, Andhra Pradesh, India*

[2]*Mechanical Engineering Department, SoT, Pandit Deendayal Energy University - PDEU, Gandhinagar, Gujarat, India*

Abstract

Shape memory polymers have many applications as they exhibit 100% shape recovery, good shape fixity, and are easy to process. One of the biggest problems with shape memory polymers (SMPs) is the requirement of petroleum-based polluting and non-biodegradable building blocks. To tackle solid waste pollution from petroleum-based plastics, researchers focus on eco-friendly, compostable, and biodegradable polymers. Poly(lactic acid) (PLA) stands out for its biodegradability, physical properties, and cost-effectiveness, including high strength, modulus, transparency, processability, and shape memory. However, PLA's brittleness requires plasticization for flexibility, often achieved with citrate esters, oligomeric lactic acid, poly(ethylene glycol) (PEG), or poly(propylene glycol). Combining blending and reinforcements can customize properties and create multi-stimuli responses in shape memory polymers (SMPs). For example, blending PLA with Fe_3O_4 can enable both thermal and magnetic actuation, enhancing functionality. Furthermore, reinforcements significantly improve their properties and include tunable shape recovery based on the external stimuli. 3D printing of these shape memory blends, polymer composites and understanding their behavior is very important and may further lead to a reduction in their overall cost significantly without sacrificing the required properties. This chapter briefly discusses all the possibilities of SMPs and their blends with the inclusion of reinforcements.

**Corresponding author*: ravindra.btw@gmail.com

Bijaya Bikram Samal, Cheruvu Siva Kumar and Shailendra Kumar Varshney (eds.) 4D Printing Technology: Principles, Materials and Applications, (165–204) © 2025 Scrivener Publishing LLC

Keywords: Polymer blends, 4D printing, reinforcements, shape recovery, CNTs, SME, shape fixity

6.1 Introduction

Polymers are remarkable materials in modern technologies, with outstanding attributes such as excellent elasticity, extraordinary durability, lightweight properties, transparency, affordability, and simplicity of processing. Polymer blends substantially have applications in, healthcare, medical implants & devices, scaffolds, apparel, food, communication, and transportation [1]. A polymer blend contains multiple polymers, resulting in a fresh material with distinct characteristics. Polymer blend research is important in polymer materials because it allows for combining diverse qualities from separate polymer components, resulting in multi-functional capabilities and higher performance. These mixtures have outstanding dual characteristics, making them extremely desirable in research [2–4]. The blending method offers a viable alternative for improving the physicochemical properties of individual polymers by combining different polymer types. This approach proves to be cost and time-effective compared to creating new monomers or developing fresh polymerization techniques. The compatibility of polymer molecules, which is determined by intermolecular interactions, effects the characteristics of these blends [5].

The following are the primary advantages of polymer blending:

- Meet specific demands, new characteristics are developed or existing features are improved.
- Blending SMPs with biodegradable components can enhance biodegradability and biocompatibility, thereby improving their ability to degrade and contribute to pollution control efforts.
- Material processability has improved.
- Exceeding the polymerization step to fulfill the demands of increasing industries [2, 3, 5].

It is now feasible to achieve the desired switch temperature and mechanical characteristics by combining an SMP with a miscible polymer. The features of the SMP may be modified to match specific application needs using this mixing procedure [6]. Blending polymers serve two primary purposes in the realm of SMPs: (i) improving the mechanical, shape recovery, and T_g, etc. according to the requirement and (ii) creating innovative shape memory materials by combining two different polymers with advantages of both materials [7].

Nanocomposites are materials composed of a polymer matrix infused with nanoparticles ranging from 1 to 100 nanometers in size. These nanoparticles can be ceramics, metals or polymers, and they are embedded into the polymer matrix to produce a hybrid material with superior characteristics over typical polymer composites [8, 9]. The choice of nanocomposites that influence the properties of blended polymer nanocomposites in 4D-printed polymers depends on the specific properties and functionalities desired in the final 4D-printed objects. Different types of nanoparticles incorporated within the matrix material to achieve various enhancements are carbon nanotubes, silica nanoparticles, nano clays, metal nanoparticles, thermoresponsive nanoparticles, photocurable nanoparticles, and magnetic nanoparticles.

6.2 Types of Polymer Blends

6.2.1 Miscible Polymer Blends

This blend type develops as a result of uniform molecular-scale mixing of polymer strands. It is well known that these mixtures only show one glass transition temperature. Significant study has been done on PP/PS and PP/PE blends amid miscible blends of polymers [10–13]. PS and PPO are shown as two compatible polymers in Figure 6.1. These components are compatible due to the polymer chains' strictly constrained weight distribution range. Furthermore, the laws of thermodynamics may be used to understand this harmonic relationship [7, 14, 15]. The goal is to reduce the energy needed to mix polymers to make subsequent separation easier. When two polymers have a significant affinity for one another, the resultant mix is more useful, see the below example.

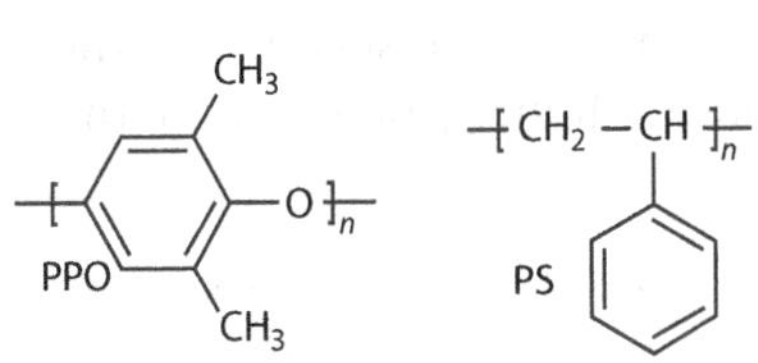

Figure 6.1 Chemical structure of PS and PPO as two compatible polymers.

6.2.2 Immiscible Polymer Blends

Similar to cooking, when mixing polymers, each component must be added precisely at the right time. Contrary to their miscible cousins, immiscible polymer blends do not combine two polymers at the molecular level. Instead, on a micro- or nano-scale, a diversified morphology manifests itself. Immiscible polymer blends, such as PET/PP and polystyrene/polybutadiene (shown in Figure 6.2), serve as examples of this phenomenon. This system develops a sea-island (matrix-disperse) shape when PET and PP chains separate into various stages. This configuration resembles a matrix-disperse pattern. The graphic representation of Figure 6.2 demonstrates how droplets of PP are distributed throughout the matrix of PET.

The features of polymer blends are shaped by the final component arrangement. This final morphology determines a variety of characteristics, including rheology, resistance to environmental stress-rupture, warping resistance, flammability, opacity, pearlescence, economical, and unique film characteristics [13, 16]. These characteristics are possible; thanks to the immiscible blends' specific morphology. From Gibbs free energy of mixing (ΔG), blends could be partially, completely, or completely miscible [13].

Eq. (6.1) provides the polymers' free energy of mixing.

$$\Delta G_m = \Delta H_m - T\Delta S_m \tag{6.1}$$

The two requirements in Eq. (6.2) must be satisfied for miscibility:

$$\Delta G_m < 0 \left(\frac{\partial^2 \Delta G_m}{\partial \varnothing_i^2} \right)_{T,P} > 0 \tag{6.2}$$

6.2.3 Compatibility for Miscible Polymer Blends

A miscible polymer can be distinguished from other polymers or solvents by its ability to produce a homogeneous, smooth mixture. This suggests

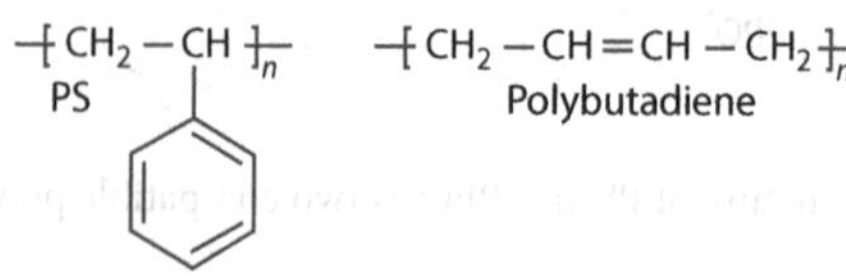

Figure 6.2 Chemical structure of PS and polybutadiene.

that the two components can fully merge at the molecular level, developing a single, cohesive phase. Visual examination, microscopy, thermal analysis, spectroscopy, consideration of processing conditions, and measurement of interfacial tension can all be used to evaluate polymer miscibility.

In the visual inspection approach, the blended polymers are checked visually for any evidence of phase separation or the existence of two dissimilar phases. If the blended polymers appear to be homogenous, they are most likely miscible. Differential scanning calorimetry (DSC) is utilized to evaluate polymer miscibility through a thermal analysis procedure. The thermal behavior of the blended polymers is studied at various temperatures. Polymers are miscible if they have a single glass transition temperature. If the polymers have two unique glass transition temperatures, they are immiscible.

The miscibility of polymers can also be affected by the processing conditions used to combine them. Blending with high shear pressures can enhance mixing and increase the possibility of miscibility. Polymer mixtures should have low mixing energy for easy separation [10–13]. The characteristics of heterogeneous blends are significantly affected by the compatibility of the polymer phases. The force responsible for binding the two polymer surfaces together is interfacial tension, and its magnitude may be used to characterize this compatibility. The blend becomes miscible when the interfacial tension hits zero, suggesting that the phases interact strongly. Thus, the extent of contact between a polymer blend's constituent phases influences its miscibility [7, 14–16].

6.3 Shape Memory Polymer Blends

Polymer blending serves two key purposes in the area of SMPs, (i) Enhancement of current SMP qualities, comprising modifications to elements such as mechanical characteristics, switching temperature, and capabilities of shape memory, and (ii) Novel shape memory materials are being developed, using one polymer as its reversible phase and another one as the fixed phase [7].

6.3.1 Blending of SMP/Conventional Polymers

Blending of SMP along with its suitable polymer allows SMPs to fine-tune both the switching temperature and the mechanical characteristics. Thermoplastic SMPU/phenoxy resin blends have been developed to obtain an adjustable switching temperature that falls between the glass transition temperatures (T_g) of SMPU (soft segment) and phenoxy resin. It can be noted that the SMPU soft segment and phenoxy resin are completely miscible across all compositions within these blends [6]. Jeong and Song conducted blending experiments with thermoplastic SMPU and PVC to adjust its switch temperature and enhance the robustness of the SMPU. It is worth noting that poly (vinyl chloride) is miscible with SMPU's soft segment. In these blends, the switch temperature changed constantly across various component ratios.

Furthermore, biodegradable SMP polylactide has drawbacks, such as higher brittleness while limited durability [17]. Zhang and colleagues used a PA elastomer made from PA-12 and polytetra methylene oxide to improve the toughness of polylactide. Notably, this biodegradable PA elastomer, contributes to the eco-friendly element of the material. The mechanical properties of PA improved significantly due to this change [18].

6.3.2 SMP Blends of Crystalline and Amorphous Polymers

Blending crystalline and amorphous polymers is a simple method for creating innovative shape memory polymers (SMPs) [19]. These blended SMPs have received substantial industry attention compared to covalently crosslinked SMPs. The main benefit is their quick availability, which is obtained by combining two unique polymers. In this method, the crystalline polymer acts as the crosslinking framework, while the amorphous polymer is the switching component. Various polymers have found application in this context, including poly (vinylidene fluoride), polylactide, poly(hydroxybutyrate), poly(ethylene glycol), polyethene, poly(ethylene-co-vinyl acetate), poly(vinyl chloride), poly(vinylidene chloride), as well as copolymers of poly(vinylidene chloride) and poly(vinylidene chloride). Additionally, amorphous polymers like poly (vinyl acetate), poly(methyl acrylate), poly(ethyl acrylate), atactic poly(methyl methacrylate), and numerous other poly(alkyl methacrylate)s and etc. have been utilized in these applications. Mather's and their study group members' contributions to this discipline have been noteworthy [20, 21]. The miscibility of these polymers allows for seamless modifications in the characteristics of the resultant SMPs, making the mix of these polymers desirable.

6.3.3 SMP Blends by Combining Crystalline Polymer with Crystalline Polymer

SMP blends can also be generated by utilizing one crystalline polymer as the stabilizing component and another as the reversible phase. Behl and colleagues created biodegradable blends by combining two polymers, poly(p-dioxanone) (PPDO) and poly(caprolactone), with a poly (alkylene adipate) mediator to function as a compatibilizer. The crystalline PPDO works as the stiff segment in this mix, establishing the fixed shape, while poly(caprolactone) acts as the flexible segment, influencing the switch temperature. It is worth noting that these mixes' mechanical qualities may be modified as needed [22]. Li and colleagues employed an ethylene-butyl acrylate-glycidyl methacrylate terpolymer, abbreviated as EBAGMA, as a reactive compatibilizing agent for the formulation of high-density polyethylene (HDPE) and poly(ethylene terephthalate) (PET) blends. Significantly, the ethylene segment within EBAGMA closely resembles HDPE, whereas the polarity found in the butyl acrylate-glycidyl methacrylate portion of EBAGMA aligns with a segment of PET. In this composition, PET serves the constant phase, while HDPE acts as the reversible phase. It is crucial to note that the utilization of compatibilizers enhances both the blend's compatibility and its shape memory characteristics [23].

6.3.4 SMP Blends by Combining Elastomer with Crystalline or Amorphous Polymer

Polymer blends with the shape memory effect (SME) can be formed by mixing an elastomer and polymer (with low-temperature transition). Weiss and colleagues developed a novel class of shape memory polymers by combining an elastomeric ionomer with low-molecular-weight fatty acids and their salts. The nanophase separation inside the ionomer serves as the foundation for the development of a crosslinking network, while the addition of fatty acids (or their salts) provides a key melting temperature switch. The temperature of the switch may be readily changed by employing fatty acids with varying melting points [24]. Similarly, Zhang *et al.* sought to construct a novel class of SMPs with adjustable characteristics by combining two immiscible polymers, namely styrene-butadiene-styrene tri-block copolymer (SBS) and poly(caprolactone). The SBS component acts as the stabilizing phase in this mix, while poly(caprolactone) acts as the reversible phase. This study also investigates the complicated interplay between shape memory characteristics, poly(caprolactone) loading, and the phase morphology of these blended materials [25].

6.3.5 SMPs Created by Blending and Radiation Crosslinking

SMPs with interpenetrating polymer network (IPN) architectures can be created by combining and crosslinking different elements. In their research, Zhang and their team explored the inclusion of poly (ethylene glycol) dimethacrylate (PEGDMA) into poly(lactide-co-glycolide) (PLGA)/ isophorone diisocyanate (IPDI) systems. This fusion led to the development of IPNs characterized by exceptional shape memory properties and hydrophilic attributes, rendering them highly suitable for biomedical applications [28]. In another study, Zhu and his colleagues enhanced the radiation efficiency of polycaprolactone for biomedical purposes by introducing polymethylvinylsiloxane (PMVS) before subjecting it to radiation crosslinking. The SME of poly(caprolactone) is typically modest, primarily due to its limited effectiveness in radiation crosslinking and subsequent radiation-induced chain breakage [27, 28].

6.4 Reinforcements in 4D Printing

4D printing is a promising and recently established technology with various latent applications. It is in the primary phases of expansion, but it can transform the manufacturing of products. In the current segment, we have discussed incorporating a diversity of reinforcements into the polymer matrix via 4D printing. Reinforcements are crucial in triggering required properties, shape recovery, and incorporating desired properties. These reinforcements could be fiber reinforcements or nanoparticles.

6.4.1 Fiber Reinforcements

When increasing the mechanical strength of SMPs, chopped microfibers and textiles surpass micro and nano-sized particles. When exploring the material in the fiber orientation of the composite, the significant elastic modulus of fibers or textiles sometimes covers the SME. The standard method for evaluating SME includes the composite bending in the direction perpendicular to the fibers rather than doing tensile tests along the fiber direction. The directions in reference [29], provide specific protocols for conducting SME testing using bending.

Fiber/fabric-reinforced SMPs are known as elastic memory composites (EMCs) in self-deployable space constructions [30–33]. The use of chopped glass fiber, unidirectional Kevlar fiber, and woven glass fiber as reinforcements for SMPU has been shown to improve SMPs strength and stiffness [34–36]. Similarly, research using epoxy laminates reinforced with carbon fiber, glass fiber, and Kevlar fiber shows that microfibers and textiles could be used to increase the stiffness of epoxy SMP resins [37, 38]. SMP composites prove highly effective in mitigating vibrations, as they excel at absorbing vibrational energy by undergoing shape deformations in proximity to their glass T_g. An innovative vibration control material was developed by constructing a sandwich-structured composite sheet using SMPU in conjunction with an epoxy beam [39]. Surprisingly, when used alone, this composite laminate had an impact strength up to four times that of the epoxy beam.

The Joule heating phenomena observed in 3D printed structures, reinforced with carbon fibers (CF) integrated within either PEEL or PLA, were achieved through the utilization of a specially designed fused filament fabrication (FFF) device. They crafted bar-shaped samples with a pristine polymer layer and a subsequent layer fortified with continuous CF. The FFF specimens directly correlated with the electrical power input and heat elevation. The data disclosed two distinctive gradients: one abrupt slope encompassing room temperature to the T_g, and a more gradual slope from T_g to the melting point (Tm). This indicated that the heat conductivity of the various polymer matrices had decreased. Samples containing PLA (as a matrix material) required 0.45 W of electricity to achieve PLA's T_g which is 57 °C and 2.4 W to approach melting initiation (155 °C).

Bilayer constructs, functioning as cantilever beams, displayed upward bending with temperature rise and downward bending during cooling, attributed to distinct layer stiffness and thermal characteristics. The degree of bending corresponded to applied power and mirrored the power-temperature trend. However, some specimens could not entirely restore their shape post-heating and cooling, accredited to polymer softening and amplified shear stress at the fiber-matrix interface beyond T_g. This study underscored the potential to govern deformation in such printed bilayer structures by regulating the applied power magnitude [40].

Rodriguez *et al.* [41], introduced a significant advancement in 3D printing through their innovative work involving bio-based thermoset composites. They employed environmentally benign epoxidized soybean oil as the foundational background medium, aligning their approach with sustainable principles. They fine-tuned mechanical attributes by skillfully introducing varying proportions of bisphenol F diglycidyl ether in the background. Infusing carbon nanofibers (CNF) as fillers notably augmented the printed structures' electroactivity. Through adept manipulation of the composition ratios of the thermoset resin constituents, they achieved precise modulation of key characteristics such as mechanical behavior, T_g, and transport properties. This versatile methodology granted them exceptional control over the resultant material properties. Leveraging advanced 3D printing techniques, the team ingeniously fabricated intricate, origami-inspired multi-material structures. These complex constructs demonstrated a remarkable ability to undergo controlled alterations in shape, triggered either by direct thermal modulation or the application of electric (Joule) heating. This visual showcases the strategic programming of its configuration and the subsequent retrieval of its intended shape. In essence, the research conducted by Rodriguez and coworkers emphasizes the considerable potential inherent in bio-based thermoset composites, harnessed through 3D printing, to engender multifunctional structures capable of intricate and precisely controllable shape transformations, instigated by a diverse array of stimuli.

There's a synergistic interaction between Joule heating and hygroscopic actuation in 3D-printed polyamide composites. They used two different polyamides, PA6 and PA6-I, as the base for shape memory polymers (SPMs). To create multifunctional behavior, they incorporated PA6-I-CF filaments into the printed constructs to create Joule heating elements. The 3D-printed samples had a strategic configuration consisting of the hygroscopic stratum of PA6-based material next to an electrically active PA6-I-CF-based stratum material. The arrangement formed a U-shape at the stratum's border. This combination permits the assembly to take up upright morphology when completely drenched with moistness and bends upon dehydrated conditions. A controlled application of Joule heating, a slow intrinsic characteristic of PA6, was used for the desiccation process. The width of the distinct strata was cautiously adjusted with the help of bimetallic strip model estimations to accomplish optimal curvature-based actuation. The accumulation arrangement of the CF-reinforcement to the aforementioned polymeric materials served a dual purpose: it enhanced mechanical properties such as difficulty and molded the shape of the electrically conductive pathway. Importantly, quantified optimum temperatures produced by Joule heating remained within the T_g range of PA6 and PA6-I. Consequently, thermal stress was minimal, and hygroscopic stress overshadowed the propulsion mechanism within the printed structures. However, It is worth noting that this research did not quantify the T_g reduction due to moisture absorption [42].

6.4.2 Carbon Nanotubes (CNTs) or Carbon Nanofibers (CNFs)

CNTs and CNFs are most successful in enhancing characteristics of SMP composites [43–53]. Furthermore, CNTs and CNFs can impart electrical and infrared-light-active shape memory properties on the resulting SMP composites. Approaches designed to enhance the dispersion of CNTs/CNFs within the polymer (as matrix) include employing twin-screw extruders, introducing surfactants, implementing surface oxidation, conducting functionalization (chemically) of CNT/CNF surfaces, and utilizing *in-situ* polymerization methods [46, 54–58]. In 4D printing, CNTs can help create objects with enhanced structural integrity and responsiveness to electrical stimuli.

Utilizing 3D and 4D printing techniques to create electro-responsive poly (D, L-lactide-co-trimethylene carbonate) (PLMC)/CNT nanocomposites capable of shape transformation, obtained it using solution-based DIW techniques. A copolymerization strategy was implemented, combining d, l-lactide and trimethylene carbonate, to improve the suitability of PLA for biomedical applications. This effectively tuned the T_g of the resultant PLMC to 49°C. The constructions generated through four-dimensional printing exhibited notable electrical conductivity, even in the case of porous scaffolds. This is attributed to well-controlled filler dispersion levels. Remarkably, these composite-fabricated scaffolds demonstrated rapid shape-changing behavior within a mere 16 seconds upon applying a 25 V electric stimulus. This study has shown the necessity of using 4D printing to create smart materials for altering the morphology while reacting to external stimuli. These materials find diverse applications in biomedical engineering, robotics, and sensor technology [59].

A new type of SMPC was prepared the 3D printing technique. The SMPC is made of a poly (ethylene glycol) diacrylate/poly (hydroxyethyl methacrylate) matrix with embedded MWCNTs. The photocurable material was composed of (meth) acrylate-containing polymers altered to modulate the thermomechanical characteristics of the surrounding polymeric environment, and the effect of CNTs was examined. Electrical measurements showed that the CNTs made the SMPC electrically conductive, which enabled it to be remotely heated using Joule heating. The SMPCs presented decent fixity and regaining of shape and demonstrated the technique for regulating carefully electro-activated regain of shape. This work covers the way for the 4D printing of smart devices [60]. The cellulose nanofibers (CNF) were incorporated into the PLA/PBAT polymer composites for FFF printing. The CNF improvised the PBAT droplet's transformation into PBAT fibrils upon extrusion, and the attenuated PBAT fibrils intermingled robustly with PLA and CNF. This enhanced the interfacial compatibility amongst miscellaneous components, improving the strain, fixity regain, and retrieval rates.

Furthermore, CNF insertion increased PBAT crystallization temperature by 30 K and PBAT transition temperature from PLA T_g. This permitted a triple-shape effect, which was impractical in the absence of CNF. In other words, the CNF made the PBAT fibrils stronger and more interconnected, which made the composites more ductile and elastic. The higher crystallization temperature of PBAT also made it more stable and less likely to deform. These changes allowed the composites to recover their shape after being deformed, even after being exposed to high temperatures. The triple-shape effect is a property that can vary the shape three times. In the case of the CNF-reinforced PLA/PBAT composites, the first shape change occurs when the composites are heated above the T_g of PLA. The second shape change occurs when the composites are cooled below the crystallization temperature of PBAT. The third shape change occurs when the composites are heated above the crystallization temperature of PBAT again. The triple-shape effect is a promising property for SMPC actuators and biomedical device applications.

3D printing with digital light processing (DLP) technique was used for SMPC and was synthesized by mixing thermosetting poly(ethylene glycol) diacrylate/poly(hydroxyethyl methacrylate) integrated with MWCNTs. MWCNTs influenced the photo reactivity in the composite system, and rheological characteristics were scrutinized. The consequence of MWCNTs on the photo reactivity and rheological characteristics were scrutinized, and the effect of the photocurable (meth)acrylate system ingredients was varied to control the thermos-mechanical characteristics. The increase in CNT concentration from 0.1 to 0.5 wt.% was responsible for delaying the start of photopolymerization. These results demonstrate the MWCNTs' UV protection function, which absorbs light and reduces the photo initiator's efficacy. The shape fixity and shape retrieval ratios of the 3D-printed SMPCs were good (Rf = 100%, Rr > 95%). Modifying the location of Cu electrodes, consecutive regaining of diverse sections of a printed zig-zag-shaped displayer was attained. They demonstrated the potential of DLP of SMPCs with tunable properties and various applications [60].

3D printing of composites synthesized via incorporating PLA (polylactic acid) and MWCNTs at variable composition (0.1 to 0.6 wt.%). Subsequently, processing and extradition of the composite system were performed to generate filament. Finally, the 3D printing of these filaments was performed via the FFF with the help of a 275 W near infrared lamp as a light source, and the thermal response of the resulting material was investigated. Thermal treatment via IR spectroscopy revealed that pure PLA reached a temperature of around 90°C, whereas 0.6 weight percent MWCNT doped PLA obtained a higher temperature of around 170°C.

These results confirm the photo-responsive nature of the composite material, where MWCNT contributes to higher temperature changes upon light exposure. Rectangular actuators (60 mm × 6 mm) were made from a composite with MWCNTs (of 0.5 wt.%) and a photo-responsive matrix material. These actuators were fabricated using a 3D printing technique, which involved depositing two layers of the composite material, each of 200 μm, on a substrate consisting of a 300 μm-thick piece of paper. Different infill densities were used to create the actuators, which correspond to the interior structure of the 3D-printed device. Infill densities tested were 100%, 66%, 60%, and 57%. The study discovered that the density of infill affects the execution of photo-responsive SMPs. A 100% infill density resulted in extended response and retrieval intervals. Infill densities ranging from 57% to 66% exhibited comparable response times. However, when the infill density was set at 66%, the material displayed the fastest recovery in regaining its shape. This underscores the influence of infill density on the material's capacity to respond to and bounce back from temperature fluctuations. The 3D printed a flower-like structure that exhibited opening and closing behavior in response to turning the infrared light on and off. This study highlights the potential of using composite materials, such as PLA and MWCNT, for creating 3D-printed objects with photo-responsive properties, enabling them to change shape based on external stimuli like light [61].

Lightweight and highly conductive nanocomposites were created by combining Ag-coated carbon nanofibers (Ag@CNF) as mixed fillers in a PLA-DCM (dichloromethane) system. The nanocomposite system was prepared by incorporating Ag@CNF into a mixture of PLA and DCM. The use of Ag@CNF was intended to improve the conductivity (electrical) of the composite. The aforementioned composite polymeric material can be used to create smart grippers that open and close in response to an electric current. Moreover, the nanocomposites were printed under ambient conditions. Isolated coils could be 3D printed using these nanocomposites. The nanocomposites exhibited a low percolation threshold, meaning that a small volume fraction (<6%) of the conductive filler was enough to establish a continuous conductive network throughout the material. The resulting nanocomposites demonstrated high electrical conductivity (>2.1 × 105 S/m) without requiring post-treatment. The qualitative investigation of electrical conductivity of the printed nanocomposites by integrating an as-printed cubic framework and a freeform curved into an electrical circuit. The printed composite material was used to power a red light-emitting diode. The LED was powered at 2.5 V and 100 mA current, demonstrating the immediate utility of printed nanocomposite material. These Ag@CNF in a PLA-DCM mixture can find potential applications such as smart grippers [59].

Gunes and colleagues used a melt-mixing process following the chain extension of SMPU to create composites of CNFs and SMPU. The composite was made with CNFs with 60-200 nm of diameter and 30 - 100 micro meters of length [62]. Also investigated and used was an *in-situ* polymerization approach to synthesize composites of MWNTs and SMPU after chemically functionalizing the MWCNTs. We also discovered that carbon nanotubes (CNTs) helped to improve the shape recovery stress of SMPUs. CNTs have enhanced form memory polyvinyl alcohol (PVA) and the SMPU system [63]. Namathoti, S., Vakkalagadda, M.R.K. also prepared MWCNTs/SMPU composites using a twin-screw extruder to achieve better dispersion. The effort resulted in notable improvements in the SMPs' mechanical strength and shape retrieval capabilities [64]. They used 3D printing to construct specimens. The strength and shape retrieval capabilities of the SMPs are improved by this approach [64]. However, a large number of CNTs/CNFs may be added before the SMPs' properties are appreciably altered. Furthermore, the crystallization of the soft segments inside SMPUs may be affected differently by varied CNT and CNF diameters [65]. The outcomes of experiments to understand the effect of nano fillers on the SME of diverse SMPs must be constant. This lack of homogeneity may be exacerbated by the difficult process of producing composites.

6.4.3 Graphene Reinforcements

Graphene is a promising nanomaterial with a diversity of potential uses. Combined with polymers, it can create polymer-graphene (nanocomposites) with improved properties. This makes it a promising technology for creating smart materials that adapt to their environment. The combination of 4D printing and polymer-graphene nanocomposites possesses the potential to generate novel materials with exclusive and tunable properties. These materials may find application in transducers, actuators, and biomedical devices. Some of the probable applications of 4D-printed polymer-graphene nanocomposites comprise the growth of sensors that can detect changes in temperature, pressure, or other environmental conditions. Figure 6.3 shows the graphical representation overall process, like 4D-printing of actuators that can alter the structure or size counter to an external stimulant; biomedical devices that can deliver drugs or other therapies to specific areas of the body; structural materials that can self-repair or adapt to changing loads; textiles that change color or texture in reaction to their surroundings, and more. Very few research studied were carried

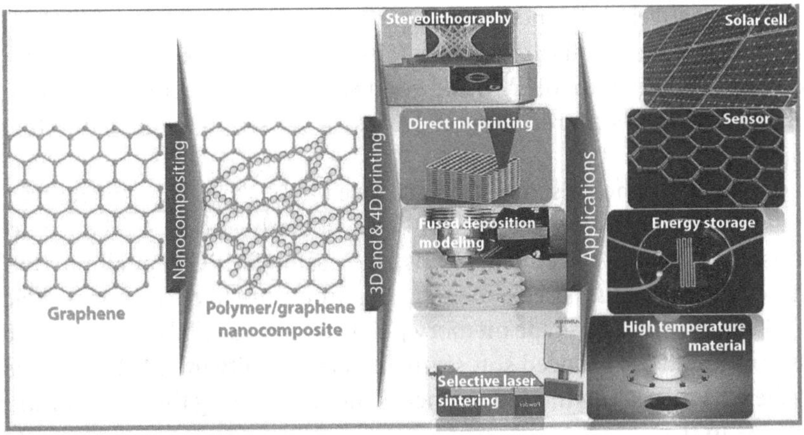

Figure 6.3 4D-printed graphene/polymer nanocomposites for miscellaneous applications in sensing, energy storage, solar cells and high-temperature materials [66].

on 4D-printed polymer-graphene nanocomposites but they have promising applications. As technology continues to develop, we can expect to see these materials being used in many products and devices [66].

PLA/graphene-based composite filaments were fabricated with single-layer structures with electrical heat responsiveness. These structures were set at 70 °C, and by applying an electric current, the aforementioned complex system regained its original shape. Cu tape electrodes were affixed to the structure's ends to enable electric heating, with wires soldered to the tape for current conduction. The relationship between voltage and performance was evident; a 50 V voltage induced a temperature of roughly 50 °C, while elevating it to 100 V raised the temperature close to 110 °C. Moreover, the deformation angle and composite temperature were positively correlated with the applied voltage. Notably, with increasing heating cycles, the retrieval angle exhibited a declining trend. This research showcases the potential of electrically triggered SME in PLA/graphene-based structures and underscores the influence of voltage on their thermal and morphological responses [67].

6.4.4 Carbon Black Reinforcements

Polymer composites containing conductive carbon black (CB) have a wide range of uses in the electrical and electronics sectors, serving as polymer conductors, semiconductors, and heat transfer mediums. CBs are less effective than CNTs and CNFs in improving properties. Moreover, the presence of CBs negatively influences the shape retrieval performance of SMPs. Li and colleagues created carbon black (CB) and polyurethane composites to achieve a combination of SME and good electrical conductivity. A solution mixing and solution precipitation process was used to manufacture these composites.

Surprisingly, the presence of CBs reduced both shape recovery ratio and the speed of form recovery in SMPU, particularly at high CB concentrations. This drop was ascribed to CBs, decreasing SMPU crystallinity [68]. Gunes and colleagues used a melt mixing procedure to create carbon black (CB) composites and shape memory polyurethane (SMPU). Their research also indicated the reduced soft segment's crystallinity in SMPUs, ascribed to the CBs restricting the soft segment's mobility during the crystallization process [62].

TPU (Thermoplastic Polyurethane) and CB (Carbon Black) based composites were designed and fabricated using a commercial FFF 3D printer. The CB component possesses a high photothermal conversion efficiency of 92% in the composite system. The photo-responsive was generated by incorporating CB into shape memory TPU, resulting in a shape change in response to light exposure. The shape retrieval for these photo-responsive materials is triggered by exposing them to light. In the present scenario, 76 mW/cm^2 of solar light for 160 seconds was adequate to restore the shape. The TPU-CB composite has a tunable CB content (0.1 to 15% by weight).

Interestingly, ameliorated CB content led to elevated achieved temperatures. For instance, when heated with a xenon light source, TPU stuffed with 15 wt.% CB acquired higher temperatures than unfilled TPU. The variation in CB content offers a way to customize the response of the 3D-printed objects produced using FFF technology. Different CB content levels result in different temperature changes upon light exposure, allowing for control over the shape-changing behavior. For the 3D printing process (3DP), a CB content of 10 wt.% was chosen. This specific CB content yielded filaments with a diameter that was wanted for the printing process [69].

6.4.5 Nano Clay Reinforcements

Organic-exfoliated nano clay can be used to make functional composites with better characteristics in some circumstances. It is also an effective method to improve shape recovery and other mechanical characteristics in SMPs. Cao and Jana used a Brabender Plastic order to create composites of organoclay and shape memory polyurethane (SMPU). They maintained a molar ratio of 1:1 between isocyanine -NCO and alcoholic -OH groups, with the -CH_2CH_2OH groups derived from the quaternary ammonium ions in the organoclay contributing to a portion of the overall -OH functionality. The tethered reactive clay inserted into the polyurethane chains formed a novel network structure, placing additional constraints on SMPU chain mobility, particularly inside the hard regions. Surprisingly, the nanocomposite containing 1% organoclay demonstrated roughly 20% greater recovery stress than pure SMPU [70, 71].

Nano clay particles with high aspect ratio can enhance the mechanical and barrier aspects of the composite (e.g., gas and moisture resistance) of the polymer matrix. They are often used in polymer nanocomposites to enhance stiffness and dimensional stability. Nanoscale layered silicates (nano clays) have attracted researchers in the field of 3DP resin because they can improve the durability and stiffness, resistance to flammability, and shape memory features of the resin. The expanded surface-to-volume ratio and platelet shape, which promote strong interfacial adhesion and allow the nano clay platelets to orient freely inside the polymer matrix, were most likely responsible for the earlier features [72–75].

Using stereolithography, the dispersal, arrangement, dimensions, and composition of nano clay-modified commercial acrylate-based monomer [73]. The nano clay platelets composite with acrylate-containing polymers were well dispersed in the monomer and aligned along the build direction. As a result, the mechanical qualities of the printed components improved significantly. Nadim *et al.* [74], used nano clay and fumed silica to improve the properties of epoxy resin ink. They found that adding nano clay and fumed silica increased the ink's rheology, printability, strength, and T_g. Li *et al.* [76], used nano clay to improve the properties of a bilayer structure hydrogel mixture. They found that adding nano clay increased the hydrogel's swelling rate and pore size, improving its mechanical properties. These studies highlight the future of nano clays to enhance the properties of 3DP resins. Further research is needed to optimize the dispersion, alignment, and content ratio of nano clays in resins to achieve the best possible results [8, 77].

Li *et al.* [78], fabricated lightweight nanocomposites possessing excellent SME via incorporating clay nanoparticles (25 wt%) into the polyimide/methacrylate resin that finds applications in the UV-assisted DIW 4D printing. The silica incorporation enhanced the resulting composite material's shear thinning and storage modulus. The DLP printing techniques were also investigated for the photo resin to improve the shape fixity and retrieval to 99.8% and 98.3%, correspondingly with a glass T_g of 160 °C. Furthermore, DIW extrusion molding was used to improve shape memory polyimide-based components, allowing them to restore their original shape in a matter of minutes. Design variety and spatial shape memory of

4D printing are demonstrated by the creation of a self-folding box and a stimuli-response gripper.

6.4.6 Metal/Metal Oxide Nanoparticles

The polymer composite with Cu, ZnO nanoparticles, Ag nanoparticles, gold, or Cu can introduce electrical conductivity, antimicrobial properties, or optical effects to the polymer matrix. Until now, a few more nanofillers have been hooked into 3D-printed polymeric resins, vital for understanding miscellaneous 4DP usages. The incorporation of the CNC particles as the reinforcements in the matrix of the stereolithographic resins demonstrates that the storage modulus values increase consistently on supplementing the CNC content for lower and above T_g. The modulus of the rubbery area was enhanced by 166, 233, and 587%, respectively, for CNC/SLR nanocomposites with CNC concentrations ranging from 0.5, 1.0, and 5.0 wt% [79].

Barium titanate nanoparticles (BTNs) possessing piezoelectric properties with variable loadings (0–30 wt%) were incorporated into the mixture ink composed of PLA, polyesteramide (PEA), and short carbon fiber solubilized in DCM using DIW printing. The composite components had up to 95% shape memory and sensing capabilities, and the long-lasting sensor could withstand temperatures ranging from 23 to 100 degrees Celsius and more than 5000 operation cycles. In 4D printing, these nanoparticles can create responsive, conductive, or observable attractive objects [80]. Choong *et al.* established SPM/nano-silica composite resins for DLP 4D printing. They observed that adding nano-silica particles to the SMP resin resulted in quicker curing and better mechanical and shape memory capabilities. The shape memory performance of the SiO2-SMP composites was outstanding, with almost complete shape fixity and shape retention [81].

Siang *et al.* [82] investigated composites of ZnO nanoparticles with acrylate-based photo resin for their mechanical properties, printability, and precision of the dimensions. Experimental investigations reveal that the ZnO acted as the photo initiator during the polymerization process, facilitating layered curing in DLP 3D printing under the 405 nm UV light wavelength.

Gold nanoparticles (AuNPs) nanocomposite with acrylates were created, with the AuNP serving as the photothermal initiator to create programmable light-activated 4D SMPCs. The outcomes indicated that cost-effective LED light was turned into heat, allowing the shape shift at the polymer's glass T_g. These polymers could recover more than 95% of the original shape. Numerical simulations further corroborate the experimental findings. Also, the shape memory retrieval outcomes were speculated using mathematical simulations. The authors summarized that this method could remotely control morphing, specifically in actuators and soft robots [8]. The radio frequency identification (RFID) package is easily synthesized and connected with a surface acoustic wave transponder (SAW) using a multiple-jet 3D printer. The polymeric composite material used for the application above was composted with acrylate and 50 wt% Ag-nanoparticles ink. The role of the Ag nanoparticles ink in the composite system was to fabricate the antenna for the wireless readout. The SAW package prototype in a flip-chip design achieves an impressive signal-to-noise ratio (SNR) of around 30 dB over an antenna-bridged radio frequency connection with a tolerable expanse of 3 cm [83]. Ikram *et al.* investigated [84] the effect of a metal-matrix composite system including Ag nanoparticles on the physical and thermal characteristics of poly (methyl methacrylate) photopolymer resin synthesized using SLA 3D printing [84]. This study reveals that incorporating two wt.% Ag in the polymer matrix enhanced the durability towards high temperature by 6.6% and overcame the loss in weight by 8%, as compared to the unmodified resin.

Epoxy-based shape memory polymer interpenetrating network (SMP-IPN) composites with the capability of 4D printing, encompassing both printing and subsequent shape transformation over time, have been successfully engineered. These display excellent strength and thermal stability. The formulation includes a blend of commercial photocurable acrylates, thermally curable epoxy, and silica nanoparticles in varying weight ratios. The incorporation of silica nanoparticles in the composite system plays a crucial role in modulating viscosity and inducing shear-thinning behavior. The multistep synthesis is initiated with the UV-assisted printing and the acrylate network. The first step, establishing the initial shape and structure of the printed object. In the subsequent second step, the epoxy oligomer underwent thermal curing, leading to the formation of an IPN composite structure. The UV-assisted DIW technique, the successful printing of intricate designs, including lattices, gear wheels, swirl bows, and locking mechanisms, became achievable. For instance, a lattice structure is cited as an exemplar of a printed design.

6.4.7 Thermo-Responsive Nanoparticles

Thermoresponsive nanoparticles enhance the shape-changing behavior of 4D-printed objects by triggering transformations at specific temperatures. The printed IPN composites displayed clear SMEs, meaning they could modify shape by reacting to external stimuli. A specific example is given where the material could undergo complete shape change within 10 seconds at a temperature of 104°C, ameliorating its T_g of approximately 76°C. The printed IPN composites exhibited enhanced mechanical toughness, which indicates their ability to withstand external forces without breaking [85].

Wu *et al.* investigated using gas-filled micro balloon pore formers to fabricate 3D-printed porous structures with shape memory capabilities. The T_g of the micro balloons significantly contributed to the mechanical characteristics of the subsequent printed polymer. Lower T_g micro balloons resulted in more compliant structures and a higher retrieval ratio after compression. This is because lower T_g micro balloons soften at lower temperatures, which allows the printed structures to deform more easily. The findings suggest that gas-filled micro balloons can be used to create wearable protective padding and cushions that are comfortable and conform to the body's shape. By optimizing the T_g of the micro balloons to match human body temperature, it may be possible to create materials that provide the desired level of protection without sacrificing comfort. In addition to wearable protective padding and cushions, gas-filled micro balloons could be used to create various other materials with shape memory properties, such as biomedical implants, soft robotics, energy absorption materials, soundproofing materials, and filtration materials. The research in this area is still ongoing, but the potential applications are vast. 3D-printed porous structures with shape memory properties have the capability to revolutionize industries [86].

6.4.8 Photo-Curable Nanoparticles

Nanoparticles can undergo polymerization when exposed to light and fit in the polymer matrix. This allows for precise control over the material's response to light, particularly useful for 4D printing applications. Zhang *et al.* used DLP to 3D print a photocurable self-heal resin. The resin is made of benzyl methacrylate (BMA), poly(ethylene glycol)-dimethacrylate (PEGDMA), and linear polycaprolactone (PCL). The PCL content was varied to study its effect on the thermos-mechanical behavior, viscosity, and demonstrate the auto-curable property of the resin. The results showed that the mechanical aspects of an impaired structure demonstrate 100% repositioning when the composite system contains 20 wt.% of PCL [87].

An advanced 4D printing method, UV-assisted DIW, was employed to fabricate epoxy-based SMP- IPN composites characterized by remarkable mechanical properties and thermal stability. The composite polymer on curing generated a complex acrylate network via the UV-facilitated printing technique. The epoxy oligomer was then thermally cured to form the IPN composite. The aforementioned printed IPN composites exhibit indistinct SMEs that include an absolute change in the shape of the polymer composite within ten seconds at 104 °C (greater than the T_g of 76 °C) and improved mechanical toughness [85].

6.4.9 Magnetic Nanoparticles

Magnetic nanoparticles can be used to create materials that respond to magnetic fields. This can be leveraged for actuation and shape changes in 4D printing. Lin *et al.* developed a magnetically responsive PLA composite by adding Fe_3O_4 particles. The composite was processed via the well-established FFF tool. The composite had a shape recovery rate (Rr) of over 94%, even with up to 20 wt.% of magnetic particles. The AMF frequency of 27.5 kHz and field strength of 4 kA/m assisted in evaluating the composite system's magnetically induced shape memory feature. They found that the PLA composite with 10 wt.% of Fe_3O_4 experienced sufficient inductive heating to initiate shape retrieval. The aforementioned composite was used to manufacture occludes for atrial septal defect (ASD), a congenital heart condition. The printed occludes recovered the original shape in 16 seconds in the AMF [88]. Zhao *et al.* created a PLA-15 wt.% Fe_3O_4 filament using a twin-screw extruder in a similar investigation. The composite filament was used to print tracheal scaffolds with a retrieval time of 35 seconds. *In vitro* testing of the printed scaffolds showed potential for artificial tracheal scaffold repair. The findings of these studies suggest that magnetically responsive PLA composites are a promising material for biomedical applications. The composites can be used to print devices remotely controlled using an AMF [69]. The interactions of nanoparticles with the polymer matrix, as well as nanoparticle dispersion, are essential elements that must be carefully studied throughout the formulation and processing of blended polymer nanocomposites for 4D printing.

6.5 Applications of Blended and Reinforced SMPs in 4D Printing

Blended polymer nanocomposites are a type of nanocomposite where two or more different polymers are blended with nanoparticles. Incorporating nanoparticles into blended polymer nanocomposites can significantly influence their properties, especially in the context of 4D printing. Nanoparticles can reinforce the polymer matrix, improving their desired properties. This is crucial in 4D printing, as the printed objects need to maintain their structural integrity while undergoing shape changes. This can be useful in creating 4D-printed objects with built-in sensors or actuators that respond to electrical stimuli. Some nanoparticles, such as SPMs or those with photo responsive properties, can enable the 4D-printed objects to respond to specific trigger. Nanoparticles can also influence the printing process itself by improving the flow properties of the blended polymer nanocomposite material. This can lead to more precise and accurate 4D printing outcomes. Nanoparticles can modify the surface properties of printed objects, affecting factors like adhesion, friction, and wettability. These properties can be crucial for the objects' interaction with their environment. In some applications, such as in the medical field, blended polymer nanocomposites with nanoparticles can be designed to be biocompatible, allowing for 4D-printed objects that interact safely with biological systems. However, it is important to note that incorporating nanoparticles into polymer blends for 4D printing also presents challenges. Obtaining consistent distribution of reinforcements within the polymer matrix, controlling their alignment for desired shape changes, and optimizing printing parameters are some of the technical hurdles that must be addressed. In summary, the influence of nanocomposites in blended polymer nanocomposites for 4D printing is substantial. By tailoring the type and concentration of nanoparticles, engineers and researchers can create materials that exhibit unique and desirable properties for creating complex, responsive, and functional 4D-printed objects. Shown in Figure 6.4, is the 4D printing in different engineering fields with smart materials for wide-ranging applications [8].

6.6 Characteristics of 4D Printed Polymer Blends and Nanocomposites

It is critical to understand the changes in the properties of SMPC, whether they are formed by blending or reinforcing. Various researches have investigated the characteristics of SMP composites using traditional processing methods; however, there has been a limited focus on the unique arena of 4D printing. This section looks primarily into the properties of SMP composites and mixes made entirely by 4D printing. Minzimo Song and colleagues [89], used 4D printing using a PLA/PCL-based bio-polyurethane

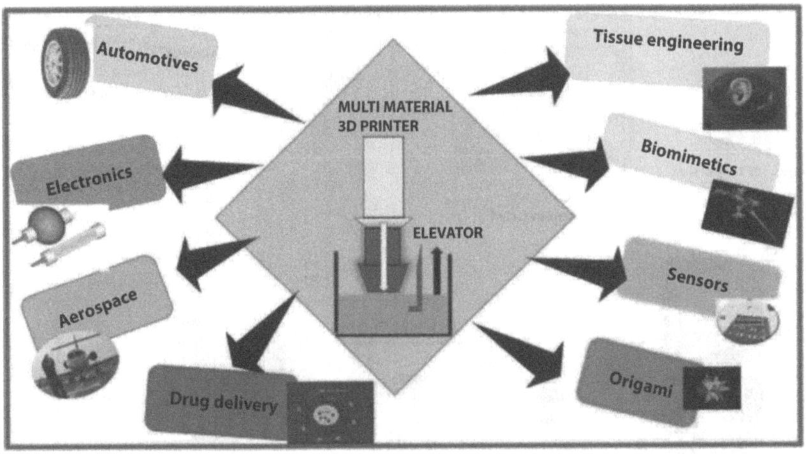

Figure 6.4 Depicts multiple applications of multi-material 4D printing [8].

(CBPU) combination. This bio-polyurethane has three notable properties: To begin, the printed material has exceptional mechanical qualities, with a break elongation of 584%. Second, when subjected to heat stressors, the printed models spontaneously deformed between their initial and temporary forms, attaining an amazing 98% shape recovery ratio. Finally, the printed models' thermal reaction temperature was quite similar to that of living organisms. This study's 4D-printed bio-polyurethane material has tremendous promise for various applications. Figure 6.5 illustrates the 4D printing process of CBPU and bio-polyurethane (BPU) dispersions.

Cheng Lin *et al.* [90], investigated the characterization of PLA and PLA/PBS shape memory blends (fabrication is shown in Figure 6.6). Their research revealed that PBS/PLA blend gave an improvement in toughness compared to pure PLA, all the while maintaining its strength and modulus. Furthermore, the excellent printability of this blend was demonstrated through the creation of various configurations, such as starfish and endoluminal stents as shown in Figure 6.7.

Comprehensive morphological and thermo-mechanical investigations were performed on the shape memory thermoplastic polymer blend of PCL/SEBS, which was created by fused filament manufacturing. This investigation aimed to systematically explore characteristics, which could be adjusted by modifying the blend's composition. The material has a dual-phase structure with SEBS and PCL coexisting. At differing composition ratios, its immiscibility causes the formation of either continuous or scattered morphologies within the material. These morphologies, in turn, are critical in determining form fixity and recovery ratios [91]. 4D printed

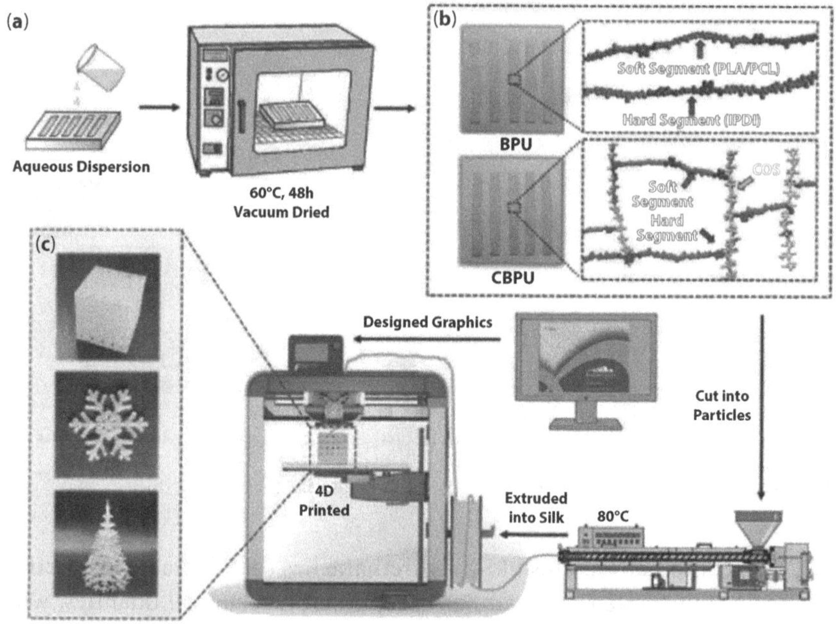

Figure 6.5 BPU and CBPU 4D printing process: (a) molding and printing route; (b) BPU and CBPU microstructure schematics; (c) cube, snowflake, and tree models after printing [89].

PLA/PCL mixes with varying infill levels were created. Remarkably, the infill density measurements had minimal impact on the excellent properties of the materials, maintaining a commendable shape fixation rate (R 95%) along with an acceptable recovery rate of approximately 80%. However, it was noted that increasing infill densities tended to slow down the recovery speed, while all 4D printed samples displayed form memory recovery durations of less than 22 seconds [92]. The PCL/POE blends shape memory blends were 4D printed, and the compatibility of the blends increased as POE content increased, and at a POE content of 60 weight percent, a co-continuous structure gradually formed, according to the SEM images and all blends showed excellent toughness, even though the other mechanical properties of the blends decreased as the POE content increased.

Additionally, the study discovered that the printing configurations and filling rate had a negligible impact on how well the PCL/POE blends performed in shape memory [93]. 4D printed PLA/TPU blends of different compositions were fabricated using fused deposition molding. The SME, DMTA analysis, and SEM techniques were carried out for different combinations. TPU's role as a recovery enhancer is critical since it boosts the

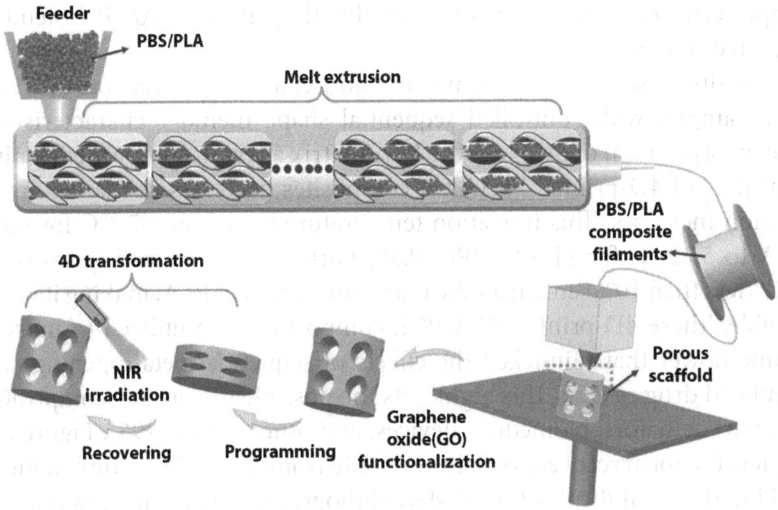

Figure 6.6 Fabrication of 4D printed PBS/PLA shape memory blend [90].

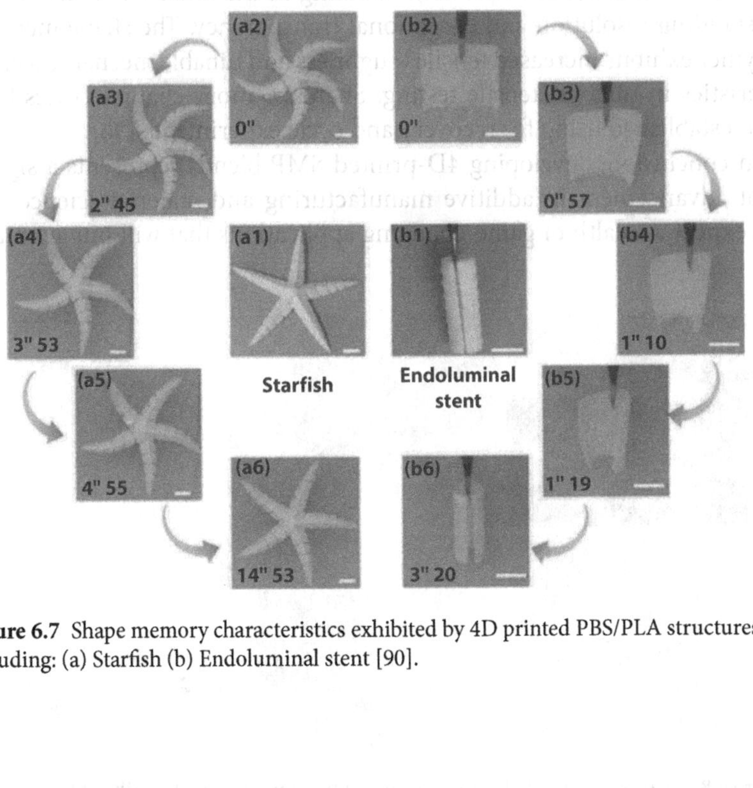

Figure 6.7 Shape memory characteristics exhibited by 4D printed PBS/PLA structures, including: (a) Starfish (b) Endoluminal stent [90].

shape recovery ratio. The recovery of the 4D-printed PLA/TPU blend is depicted in Figure 6.8 [94].

The direct-write (4D printing) method enables the creation of PLA/PCL samples with controlled sequential shape memory characteristics. The incorporation of PCL into the PLA matrix alters the properties significantly. T_g of 4D-printed PLA/PCL composites decreases as the PCL proportion increases. This transition temperature goes from 67.2°C for pure PLA to 55.2 °C for PLA0.50/PCL0.50. Furthermore, a small quantity of PCL (less than 10%) enhances the maximum strain of PLA, making it more flexible. These 4D-printed PLA/PCL composites were utilized to make a bionic flower that mimicked the effects of sequential petal opening and regulated drug release. This highlights the possible use of such composites in smart actuators, biomedical devices, and other devices [95]. Figure 6.8 depicts the form recovery of a flower made from a PLA/PCL combination.

Liquid crystal display (LCD) stereolithographic 3D printing is utilized for the 4D printing of SMP. This process involves the blending of epoxy acrylate (EA) with varying ratios of isobornyl acrylate (IBOA) and trimethylolpropane triacrylate (TMPTA), resulting in materials characterized by outstanding resolution and exceptional transparency. The shape memory polymer exhibits increased tensile toughness and tunable mechanical characteristics in uniaxial tensile testing. Shape memory characteristics have been established using fix-recovery and cycle experiments [96].

In conclusion, developing 4D-printed SMP blends represents a significant advancement in additive manufacturing and material science. We can expect a wealth of game-changing applications that will impact many

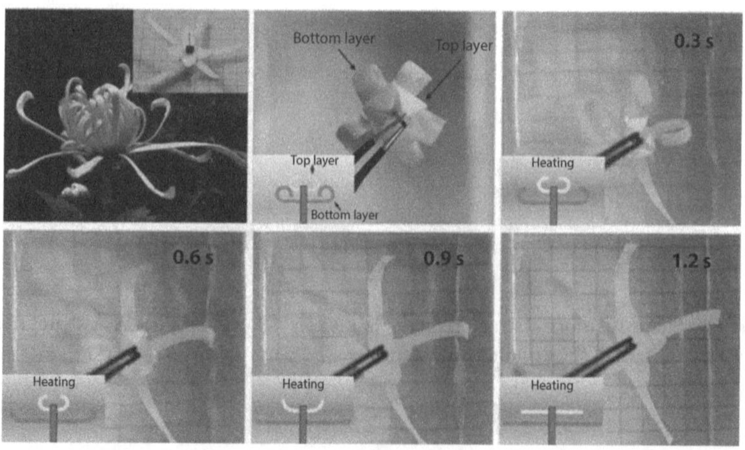

Figure 6.8 Shape recovery of 4D-printed chrysanthemum built by PLA/PCL blend [96].

industries as researchers continue to investigate and develop these novel materials. These applications will offer solutions to challenging issues and open new doors in materials engineering. Xinyu Dong *et al.* [97], created an electroactive shape memory structure made of PLA/CNT using fused deposition molding and altering the weight percentages of CNT. When the CNT concentration rose within a certain range, the SMPC's electrical conductivity, thermal conductivity, and shape recovery ratio all increased. At room temperature, the shape fixation ratio was 100%, whereas the form recovery ratio was 90% at a certain voltage. Figure 6.9 demonstrates the process of making PLA/CNT composite filaments.

Han Liu *et al.* [98], created magneto-responsive 4D printed SMP nanocomposites by incorporating Fe_3O_4 nanoparticles into PLA/TPU by direct melt mixing. The addition of magnetic Fe_3O_4 particles facilitated shape memory behavior, allowing for distant and contactless activation. According to the findings, all PLA/TPU/Fe_3O_4 ratios obtained exceptional form fixation rates of up to 99% and excellent shape recovery ratios of up to 96.4%. The form memory response accelerated dramatically as the Fe_3O_4 level rose, with reaction times as short as 40 seconds. Figure 6.10 depicts sample studies in which a bionic flower-like model was distorted into a temporary shape using external force. The bionic flower with variable Fe_3O_4 levels demonstrated rapid reactions and good form recovery, providing important insights into magneto-responsive shape memory behavior.

By direct ink writing, Xue Wan *et al.* [99], constructed a feasible paradigm 4D printed PLMC/PTMC/Fe_3O_4 shape memory polymer nanocomposites. With various PLMC/PTMC compositions, they optimize triple SMP properties best, and the material design exhibits exceptional repeatability and dependability over numerous shape memory cycles. Fe_3O_4

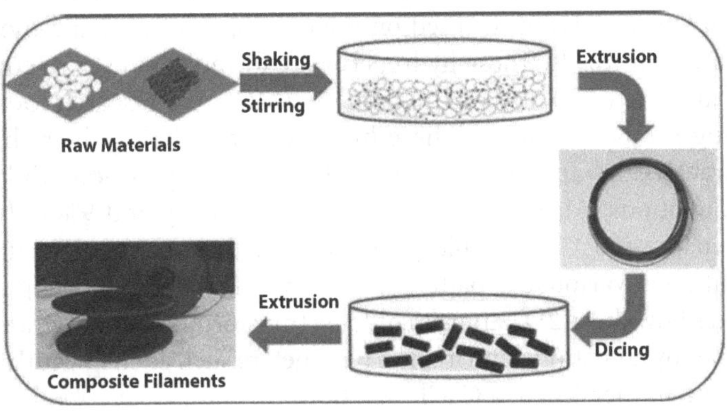

Figure 6.9 PLA/CNT shape memory polymer nanocomposites [97].

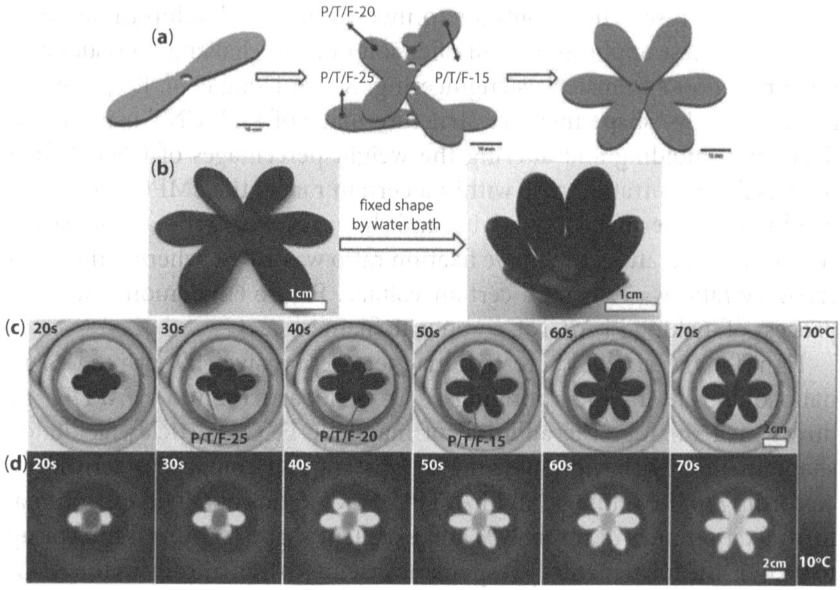

Figure 6.10 (a) Schematic design of the bionic flower modeling method, (b) bionic flower shape programming, (c, d) the restoration of shape memory in a biomimetic floral structure is tracked and analyzed using digital pictures and infrared thermal imaging [98].

nanoparticles were added to the triple SMP for them to have more degrees of freedom and remote magnetic responsiveness. This developed method enables the development of 4D printed structures with region-specific, sequential shape changes that are accurately driven by magnetic fields and thermal stimuli. This accomplishment is made feasible by the incorporation of a multi-material design and the use of a triple-shape memory response. Furthermore, these structures are extremely biocompatible.

Jiahui Lai *et al.* [100], worked on the development of shape-morphing 4D printed hydrogels. These hydrogels were formulated using a combination of alginate (Alg) and methylcellulose (MC). The research focuses on analyzing the printability of these hydrogels considering their rheological characteristics and ability to retain form integrity. Among the hydrogel formulations, Alg3/MC9, including 3% (w/v) Alg and 9% (w/v) MC, displayed the most acceptable printability. It demonstrated accuracy in the printing of various 2D patterns, both simple and complex. Figure 6.11 illustrates how these 2D patterns could be transformed into programmable 3D shapes by adjusting deformation parameters such as strip width, orientation, and immersion in a CaCl2 solution.

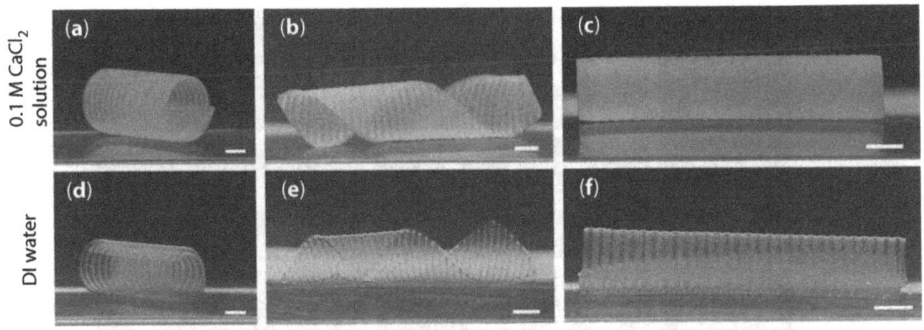

Figure 6.11 3D shape deformations of hydrogels featuring tube-curling, helix, and rolling structures, created through 4D printing, were observed when immersed in a 0.1 M CaCl2 solution and DI water. These hydrogels incorporated well-aligned strips [100].

Jheng-Wun Su *et al.* [101], created 4D printed composites employing water-borne polyurethane paint as a foundation material. They did this by integrating carboxymethyl cellulose (CMC) and silicon oxide (SiO2) nanoparticles into the paint. CMC and SiO2 had a vital role in increasing the rheological characteristics of the polyurethane paint, resulting in a printable precursor. This advancement improved additive manufacturability and printing resolution. Moreover, the composite precursors sped up the drying process of the polyurethane paint while keeping its natural shape memory features. The 3D printed composite parts outperformed components made entirely of pure polyurethane in terms of mechanical strength. Figure 6.12 illustrates components produced through polyurethane-CMC 4D printing and visually demonstrates the time-dependent shape changes triggered by heat.

Yang Li *et al.* [102], investigated the behavior of 4D printed PLA reinforced with CNT under the influence of a DC voltage in their study. They investigated the temperature-dependent volume resistivity, temperature distribution caused by Joule heating, and electro-activated shape memory properties of this composite. The temperature-dependent changes in volume resistivity in CNT/PLA composites resulted in shrinkage and changes in CNT contact resistance.

Yu Ying Clarrisa Choong *et al.* [81], used digital light processing to 4D print shape memory polymers using nano silica as a filler, evaluating the effect of nano silica on the 3D printing and the SMPs characteristics. The SiO_2-SMPs displayed amazing shape memory performance, exhibiting properties such as 100% shape fixity, 90–97% shape recovery, and, significantly, a doubling of the life cycle compared to plain SMPs. This holds significant promise for the production of advanced reinforced composites;

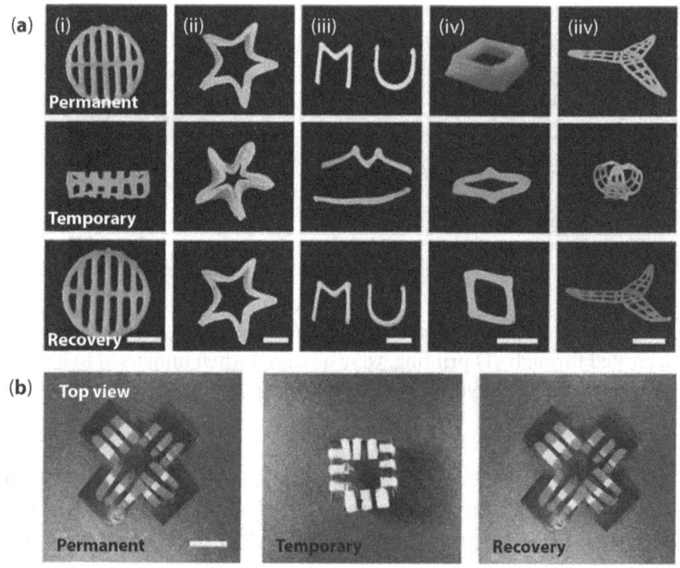

Figure 6.12 4D-printed designs created of PU/CMC composites [101].

thanks to the excellent curing characteristics, effective dispersion, and enhanced mechanical and shape memory properties. The inclusion of nano silica particles in 4D-printed SMPs sheds light on how these particles influence the formation of SiO2-SMP resin and the processes reliant on photopolymerization.

Yang Liu and colleagues [103], investigated 4D-printed angle-ply laminated and rectangular braided preforms using two types of filaments as the basic material for SMP: PLA and carbon nanotube-reinforced PLA (CNT/PLA). Angle-ply laminated preforms and composites outperformed rectangular braided preforms and composites in terms of recovery force and flexural characteristics. They did, however, lag behind the latter in terms of shape recovery rate. Both the preforms and composites maintained identical yarn orientation (45°) and volume fraction (38%). The incorporation of a silicone elastomer matrix into angle-ply laminated and braided CNT/PLA preforms resulted in a significant improvement in the final shape recovery ratio of the resultant composites. The addition of CNT into the PLA printing filament aided in the specimens' faster form recovery. Furthermore, the inclusion of CNT and the silicone matrix to the preforms significantly improved the 4D printed composites' recovery force, flexural load, and flexural modulus.

Jayantha *et al.* [104], conducted an investigation into the structural efficiency, shape memory behavior, and photothermal effects of two materials:

pristine SMP and a composite of shape memory polymer enhanced with MWCNT. The inclusion of MWCNT had an impact on the quality of the fused deposition molding process. However, the MWCNT-enhanced shape memory polymer composite (SMPC) exhibited enhanced impact energy absorption and compression strength. It also enabled the actuation of SMPCs using light, a crucial feature for applications in space engineering.

Jaydeep Chowdhury *et al.* [105], investigated the micro-stereolithography 4D printed integrating Graphene and acrylate-based SMP nanocomposite mechanical characteristics. Graphene nanoparticles in the SMP matrix improve all mechanical and surface qualities. It has been found that adding 0.3 weight percent graphene nanoparticles yields the best results. Hongqiu Wei *et al.* [106], constructed PLA-Fe_3O_4 4D active changing structures using direct-wire printing. 4D shape-changing objects with quick remote control and magnetic guidance capabilities are successfully integrated with the addition of iron oxide. This research shows how to print shape memory nanoparticles and polymers, which provides a feasible way for constructing 4D active shape-changing objects with a variety of beneficial features.

6.7 Summary and Future Scope

To summarize, polymer blending plays a crucial role in shape memory polymers (SMPs) because it enhances existing SMP qualities by modifying factors such as mechanical properties, transition temperatures, and shape memory abilities. Moreover, it enables the development of new SMPs, using one polymer for its reversible phase and another for its fixed phase. Nevertheless, it's important to acknowledge that while there are many potential applications, some remain largely unexplored, requiring further research into the underlying processes. The upcoming section will discuss both challenges and prospects in polymer blending and reinforcing in 4D printing.

6.8 Challenges

Challenges in 4D printing include limited accessibility, particularly in achieving biocompatibility for medical applications. While 4D printed materials can respond to various stimuli, understanding their behavior under multiple stimuli and achieving complete recovery after stimulus removal are ongoing challenges. The anisotropic nature of 3D-printed materials complicates comparisons with traditional isotropic materials, especially regarding how properties change over time in 4D printed materials. Comparing 3D and 4D printed materials is difficult, as 4D materials, primarily polymer-based, have limited high-temperature applications compared to the wide range of materials available for 3D printing. Ideal printable materials for 4D printing have low glass transition temperatures and suitable viscosities. Finding suitable replacements for shape memory polymers (SMPs) as 4D printing materials requires meeting these criteria. Improving mechanical properties in conventional 3D formations can trigger smart effects conducive to 4D printing. However, 4D printed structures face challenges with gradual degradation after multiple cycles, requiring advancements in materials and fabrication processes. Integrating machine learning into 4D printing algorithms could enhance the process's intelligence and autonomy.

6.9 Future Prospects

Although relatively new, 4D printing has recently gained interest for its ability to create structures that can change function or shape over time when exposed to external stimuli. Currently, 3D printing is mainly used for prototyping, but advancements in 4D printing are shifting its focus towards producing real, usable products for consumers. Researchers over the past 8 to 10 years have made significant contributions to this progress. Looking ahead, 4D printing technology is expected to be revolutionary, requiring collaboration among engineers, chemists, material scientists, and other disciplines. For example, in medicine, more research is needed to develop materials compatible with body tissues and understand the life cycle of implanted materials. The biodegradability of 4D printed materials also requires attention. Harnessing the properties of 3D and 4D printed materials remains a challenge, necessitating further research to understand how these properties change under different stimuli. The constantly evolving technology opens up nearly limitless potential applications and benefits for 4D printing, promising significant advancements in manufacturing and everyday life in the near future.

References

1. White, J.L., First in a Series: Pioneers of Polymer Processing: Thomas Hancock - The First Polymer Processing Innovator. *Int. Polym. Proc.*, 11, 4, 298–298, 1996.

2. Markovic, G. and Visakh, P.M., *1 - Polymer blends: State of art, Recent Developments in Polymer Macro, Micro and Nano Blends*, pp. 1–15, Woodhead Publishing, Amsterdam, The Netherlands, 2017.
3. Thomas, S., Shanks, R., Chandrasekhara kurup, S., *Nanostructured Polymer Blends*, pp. xv–xvi, William Andrew Publishing, Norwich, NY, 2014.
4. Mark, J.E., Physical Properties of Polymers Handbook. *J. Am. Chem. Soc.*, 130, 3, 1111–1111, 2008.
5. Chen, Z., Mo, X., long He, C., Wang, H., Intermolecular interactions in electrospun collagen–chitosan complex nanofibers. *Carbohydr. Polym.*, 72, 3, 410–418, 2008.
6. Jeong, H.M., Ahn, B., Kim, B., Miscibility and shape memory effect of thermoplastic polyurethane blends with phenoxy resin. *Eur. Polym. J.*, 37, 2245–2252, 2001.
7. Meng, Q. and Hu, J., A review of shape memory polymer composites and blends. *Composites, Part A*, 40, 11, 1661–1672, 2009.
8. Alsaadi, M., Hinchy, E.P., McCarthy, C.T., Moritz, V.F., Zhuo, S., Fuenmayor, E., Devine, D.M., Liquid-Based 4D Printing of Shape Memory Nanocomposites: A Review. *J. Manuf. Mater. Process.*, 7, 35, 2023.
9. Naga Mallikarjun Rao, G. and Vakkalagadda, M.R.K., A review on synthesis, characterization and applications of nanoparticles in polymer nanocomposites. *Mater. Today Proc.*, 9, 68-80, 2023.
10. Paul, D.R., Control of phase structure in polymer blends, in: *Functional polymers*, pp. 1–18, Plenum Press, New York, 1989.
11. Scobbo, J.J. and Goettler Jr., L.A., Applications of polymer alloys and blends, in: *Polymer blends handbook*, pp. 951–76, Springer, Netherlands, 2003.
12. White, J.L. and Bumm, S.H., Polymer blend compounding and processing, in: *Encyclopedia of polymer blends*, pp. 1–26, Wiley-VCH, Weinheim, 2011.
13. Mishra, J., Tiwari, S.K., Abolhasani, M.M., Azimi, S., Nayak, G.C., 2 - Fundamental of polymer blends and its thermodynamics, in: *Micro and Nano Fibrillar Composites (MFCs and NFCs) from Polymer Blends*, pp. 27–55, Woodhead Publishing, Composites Science and Engineering, 2017.
14. George, S.M., Puglia, D., Kenny, J.M., Causin, V., Parameswaranpillai, J., Thomas, S., Morphological and mechanical characterization of nanostructured thermosets from epoxy and styrene-block-butadiene-block-styrene triblock copolymer. *Ind. Eng. Chem. Res.*, 52, 91219, 2013.
15. Jyotishkumar, P., Ozdilek, C., Moldenaers, P., Sinturel, C., Janke, A., Pionteck, P., *et al.*, Dynamics of phase separation in poly(acrylonitrile-butadienestyrene)-modified epoxy/ DDS system: kinetics and viscoelastic effects. *J. Phys. Chem. B*, 114, 13271–81, 2010.
16. Jiang, R., Quirk, R.P., White, J.L., Min, K., Polycarbonate-polystyrene block copolymers and their application as compatibilizing agents in polymer blends. *Polym. Eng. Sci.*, 31, 1545, 1991.

17. Jeong, H.M., Song, J.H., Lee, S.Y., Kim, B.K., Miscibility and shape memory property of poly(vinyl chloride)/thermoplastic polyurethane blends. *J. Mater. Sci.*, 36, 5457–63, 2001.
18. Zhang, W., Chen, L., Zhang, Y., Surprising shape-memory effect of polylactide resulted from toughening by polyamide elastomer. *Polymer*, 50, 1311–5, 2009.
19. Mather, P.T., Liu, C., Campo, C.J., *Blends of amorphous and semicrystalline polymers having shape memory properties*, 2007, US Patent and Trademark Office database. 7371799.
20. Liu, C. and Mather, P.T., Thermomechanical characterization of blends of poly (vinyl acetate) with semicrystalline polymers for shape memory applications, in: *Conference Proceedings*, vol. 2, pp. 1962–6, 2003.
21. Arnoult, O.B. and Mather, P.T., Crystallization of shape memory polymer blends, in: *Society of Plastics Engineers Annual Technical Conference, conference proceedings*, Cincinnati, OH, United States, vol. 2, pp. 839–44, 2007.
22. Behl, M., Ridder, U., Feng, Y., Kelch, S., Lendlein, A., Shape-memory capability of binary multiblock copolymer blends with hard and switching domains provided by different components. *Soft Matter*, 5, 676–84, 2009.
23. Li, S., Lu, L., Zeng, W., Thermostimulative shape-memory effect of reactive compatibilized high-density polyethylene/poly(ethylene terephthalate) blends by an ethylene-butyl acrylate-glycidyl methacrylate terpolymer. *J. Appl. Polym. Sci.*, 112, 3341–6, 2009.
24. Weiss, R.A., Izzo, E., Mandelbaum, S., New design of shape memory polymers: mixtures of an elastomeric ionomer and low molar mass fatty acids and their salts. *Macromolecules*, 41, 2978–80, 2008.
25. Zhang, H., Wang, H., Zhong, W., Du, Q., A novel type of shape memory polymer blend and the shape memory mechanism. *Polymer*, 50, 1596–601, 2009.
26. Zhang, S., Feng, Y., Zhang, L., Sun, J., Xu, X., Xu, Y., Novel interpenetrating networks with shape-memory properties. *J. Polym. Sci., Part A: Polym. Chem.*, 45, 768–75, 2007.
27. Zhu, G., Xu, S., Wang, J., Zhang, L., Shape memory behaviour of radiation crosslinked PCL/PMVS blends. *Radiat. Phys. Chem.*, 75, 443–8, 2006.
28. Nagata, M. and Mamoto, Y., Synthesis and characterization of photo crosslinked poly(e-caprolactone) s showing shape-memory properties. *J. Polym. Sci., Part A: Polym. Chem.*, 47, 2422–33, 2009.
29. Zhang, C.S. and Nib, Q.Q., Bending behavior of shape memory polymer based laminates. *Compos. Struct.*, 78, 153–61, 2007.
30. Schultz, M.R., Francis, W.H., Campbell, D., Lake, M.S., Analysis techniques for shape memory composite structures, in: *AIAA/ASME/ASCE/AHS/ASC structures, structural dynamics, and materials conference*, collection of technical papers – 48th Waikiki, HI, United States, 2007, vol. 8, pp. 8393–403, 2007.

31. Gall, K., Tupper, M.L., Munshi, N.A., Mikulas, M., Micro-mechanisms of deformation in fiber reinforced polymer matrix elastic memory composites, in: *AIAA/ASME/ASCE/AHS/ASC structures, structural dynamics and materials conference*, Seattle, WA, United State, 2001, vol. 3, pp. 1773–80, 2001.
32. Keller, P.N., Lake, M.S., Codell, D., Barrett, R., Taylor, R., Schultz, M.R., Development of elastic memory composite stiffeners for a flexible precision reflector, in: *47th AIAA/ASME/ASCE/AHS/ASC structures, structural dynamics, and materials conference, AIAA 2006-2179*, Newport, Rhode Island, 1–4 May 2006.
33. Campbell, D., Lake, M.S., Scherbarth, M.R., Nelson, E., Six, R.W., Elastic memory composite material: an enabling technology for future furable space structures, article no. AIAA 2005-2362, in: *AIAA/ ASME/ASCE/AHS/ASC structures, structural dynamics and materials conference, AIAA*, Reston, VA, 2005.
34. Ohki, T., Ni, Q.Q., Ohsako, N., Iwamoto, M., Mechanical and shape memory behavior of composites with shape memory polymer. *Compos. Part A*, 35, 1065–73, 2004.
35. Liang, C., Rogers, C., Malafeew, E., Investigation of shape memory polymers and their hybrid composites. *J. Intell. Mater. Syst. Struct.*, 8, 380–6, 1997.
36. Ni, Q.Q., Ohsako, N., Ohki, T., Sakaguchi, M., Wang, W., Iwamoto, M., Development of smart composites based on shape memory polymer, in: *The international symposium on smart structures and microsystems*, Hong Kong, 2000.
37. Wei, Z.G. and Sanstrom, R., Shape memory materials and hybrid composites for smart systems. Part II. Shape-memory hybrid composites. *J. Mater. Sci.*, 33, 3763–83, 1998.
38. Gall, K., Mikulas, M., Munshi, N.A., Beavers, F., Tupper, M., Carbon fiber reinforced shape memory polymer composites. *J. Intell. Mater. Syst. Struct.*, 11, 877–86, 2000.
39. Yang, J.H., Chun, B.C., Chung, Y.C., Cho, J.W., Cho, B.G., Vibration control ability of multilayered composite material made of epoxy beam and polyurethane copolymer with shape memory effect. *J. Appl. Polym. Sci.*, 94, 302–7, 2004.
40. Yang, C., Wang, B., Li, D., Tian, X., Modelling and characterisation for the responsive performance of CF/PLA and CF/PEEK smart materials fabricated by 4D printing. *Virtual Phys. Prototyping*, 12, 69–76, 2017.
41. Rodriguez, J.N., Zhu, C., Duoss, E.B., Wilson, T.S., Spadaccini, C.M., Lewicki, J.P., Shape-morphing composites with designed micro-architectures. *Sci. Rep.*, 6, 27933, 2016.
42. Le Duigou, A., Chabaud, G., Scarpa, F., Castro, M., Bioinspired Electro-Thermo-Hygro Reversible Shape-Changing Materials by 4D Printing. *Adv. Funct. Mater.*, 29, 1903280, 2019.

43. Demczyk, B.G., Wang, Y.M., Cumings, J., Hetman, M., Han, W., Zettl, A., *et al.*, Direct mechanical measurement of the tensile strength and elastic modulus of multiwalled carbon nanotubes. *Mater. Sci. Eng., A*, 334, 173–8, 2002.
44. Huang, H.M., Liu, I.C., Chang, Y.U.C., Tsai, H.C., Preparing a polystyrene functionalized multiple-walled carbon nanotubes via covalently linking acyl chloride functionalities with living polystyryllithium. *J. Polym. Sci., Part A: Polym. Chem.*, 42, 5802–10, 2004.
45. Zhang, W.D., Shen, L., Phang, I.Y., Liu, T., Carbon nanotubes reinforced nylon-6 composite prepared by simple melt-compounding. *Macromolecules*, 37, 256–9, 2004.
46. Jung, Y.C., Sahoo, N.G., Cho, J.W., Polymeric nanocomposites of polyurethane block copolymers and functionalized multi-walled carbon nanotubes as crosslinkers. *Macromol. Rapid Commun.*, 27, 126–31, 2006.
47. Cadek, M., Coleman, J., Ryan, K., Nicolosi, V., Bister, G., Fonseca, A., Reinforcement of polymers with carbon nanotubes: the role of nanotube surface area. *Nano Lett.*, 4, 353–6, 2004.
48. Haggenmueller, R., Gommans, HHAGR, Fischer, J.E., Winey, K.I., Aligned singlewall carbon nanotubes composites by melt processing methods. *Chem. Phys. Lett.*, 330, 219–25, 2000.
49. Andrews, R. and Weisenberger, M.C., Carbon nanotube polymer composites. *Curr. Opin. Solid State Mater. Sci.*, 8, 31–7, 2004.
50. Biercuk, M.J., Llaguno, M.C., Radosavljevic, M., Hyun, J.K., Johnson, A.T., Fischer, J.E., Carbon nanotube composites for thermal management. *Appl. Phys. Lett.*, 80, 2767–9, 2002.
51. Rahul, S., Bin, Z., Daniel, P., Mikhail, I., Hui, H., James, L., Preparation of single-walled carbon nanotube reinforced polystyrene and polyurethane nanofibers and membranes by electrospinning. *Nano Lett.*, 4, 459–64, 2004.
52. Gommans, H.H., Alldredge, J.W., Tashiro, H., Park, J., Magnuson, J., Rinzler, A.G., Fibers of aligned single-walled carbon nanotubes: polarized Raman spectroscopy. *J. Appl. Phys.*, 88, 2509–14, 2000.
53. Jimenez, G.A. and Jana, S.C., Polyurethane-carbon nanofiber composites for shape memory effects, in: *Society of Plastics Engineers Annual Technical Conference: plastics encounter at ANTEC 2007, conference proceedings*, Paper presented at the annual technical conference – ANTEC, conference proceedings Cincinnati, OH, United States, vol. 1, pp. 18–22, 2007.
54. Thostenson, E.T. and Chou, T.W., Aligned multi-walled carbon nanotube-reinforced composites: processing and mechanical characterization. *J. Phys. D Appl. Phys.*, 35, L77–80, 2002.
55. Hou, H.Q., Ge, J.J., Zeng, J., Li, Q., Reneker, D.H., Greiner, A., *et al.*, Electrospun polyacrylonitrile nanofibers containing a high concentration of well-aligned multiwall carbon nanotubes. *Chem. Mater.*, 17, 967–73, 2005.
56. Kwon, J. and Kim, H., Comparison of the properties of waterborne polyurethane/multiwalled carbon nanotube and acid-treated multiwalled

carbon nanotube composites prepared by *in situ* polymerization. *J. Polym. Sci., Part A: Polym. Chem.*, 43, 973–3985, 2005.
57. Chen, W. and Tao, X.M., Self-organizing alignment of carbon nanotubes in thermoplastic polyurethane. *Macromol. Rapid Commun.*, 26, 1763–7, 2005.
58. Mondal, S. and Hu, J.L., Shape memory studies of functionalized MWNT-reinforced polyurethane copolymers. *Iran. Polym. J.*, 15, 135–42, 2006.
59. Wan, X., Zhang, F., Liu, Y., Leng, J., CNT-based electro-responsive shape memory functionalized 3D printed nanocomposites for liquid sensors. *Carbon*, 155, 77–87, 2019.
60. Cortés, A., Cosola, A., Sangermano, M., Campo, M., González Prolongo, S., Pirri, C.F., Jiménez-Suárez, A., Chiappone, A., DLP 4D-Printing of Remotely, Modularly, and Selectively Controllable Shape Memory Polymer Nanocomposites Embedding Carbon Nanotubes. *Adv. Funct. Mater.*, 31, 2106774, 2021.
61. Hua, D., Zhang, X., Ji, Z., Yan, C., Yu, B., Li, Y., Wang, X., Zhou, F., 3D printing of shape-changing composites for constructing flexible paper-based photothermal bilayer actuators. *J. Mater. Chem. C*, 6, 2123–2131, 2018.
62. Gunes, I.S., Cao, F., Jana, S.C., Evaluation of nanoparticulate fillers for development of shape memory polyurethane nanocomposites. *Polymer*, 49, 2223–34, 2008.
63. Meng, Q.H., Hu, J.L., Mondal, S., Thermal sensitive shape recovery and mass transfer properties of polyurethane/modified MWNT composite membranes synthesized via *in situ* solution pre-polymerization. *J. Membr. Sci.*, 319, 102–10, 2008.
64. Namathoti, S. and Vakkalagadda, M.R.K., Development of Multiwalled Carbon Nanotubes/Halloysite Nanotubes Reinforced Thermal Responsive Shape Memory Polymer Nanocomposites for Enhanced Mechanical and Shape Recovery Characteristics in 4D Printing Applications. *Polymers*, 15, 1371, 2023.
65. Sahoo, N.G., Jung, Y.C., Yoo, H.J., Cho, J.W., Influence of carbon nanotubes and polypyrrole on the thermal, mechanical and electroactive shape-memory properties of polyurethane nanocomposites. *Compos. Sci. Technol.*, 67, 1920–9, 2007.
66. Kausar, A., Ahmad, I., Zhao, T., Aldaghri, O., Eisa, M.H., Polymer/Graphene Nanocomposites via 3D and 4D Printing—Design and Technical Potential. *Processes*, 11, 868, 2023.
67. Wang, G., Cheng, T., Do, Y., Yang, H., Tao, Y., Gu, J., An, B., Yao, L., Printed paper actuator: A low-cost reversible actuation and sensing method for shape-changing interfaces, in: *Proceedings of the 2018 CHI Conference on Human Factors in Computing Systems*, pp. 1–12, 2018.
68. Li, F., Qi, L., Yang, J., Xu, M., Luo, X., Ma, D., Polyurethane/conducting carbon black composites: structure, electric conductivity, strain recovery behavior, and their relationships. *J. Appl. Polym. Sci.*, 75, 68–77, 2000.

69. Yang, H., Leow, W.R., Wang, T., Wang, J., Yu, J., He, K., Qi, D., Wan, C., Chen, X., 3D Printed Photo responsive Devices Based on Shape Memory Composites. *Adv. Mater.*, 29, 1701627, 2017.
70. Cao, F. and Jana, S.C., Nanoclay-tethered shape memory polyurethane nanocomposites. *Polymer*, 48, 3790–800, 2007.
71. Cao, F. and Jana, S.C., Nano-clay tethered shape memory polyurethane nanocomposites. Society of Plastics Engineers Annual Technical Conference 2006, in: *ANTEC 2006 – conference proceedings*, vol. 2, pp. 646–9, 2006.
72. Bulut, M., Alsaadi, M., Erkliğ, A., The effects of nanosilica and nanoclay particles inclusions on mode II delamination, thermal and water absorption of intraply woven carbon/aramid hybrid composites. *Int. Polym. Proc.*, 35, 367–375, 2020.
73. Eng, H., Maleksaeedi, S., Yu, S., Choong, Y., Wiria, F., Tan, C., Su, P., Wei, J., 3D stereolithography of polymer composites reinforced with orientated nanoclay. *Procedia Eng.*, 216, 17, 2017.
74. Hmeidat, N.S., Kemp, J.W., Compton, B.G., High-strength epoxy nanocomposites for 3D printing. *Compos. Sci. Technol.*, 160, 9–20, 2018.
75. Alsaadi, M., Erkliğ, A., Abbas, M., Effect of clay nanoparticles on the mechanical and vibration characteristics of intraply aramid/carbon fiber reinforced epoxy composite. *Polym. Compos.*, 41, 2704–2712, 2020.
76. Li, Y., Zheng, W., Li, B., Dong, J., Gao, G., Jiang, Z., Double-layer temperature-sensitive hydrogel fabricated by 4D printing with fast shape deformation. *Colloids Surf., A*, 648, 129307, 2022.
77. Mao, Y., Ding, Z., Yuan, C., Ai, S., Isakov, M., Wu, J., Wang, T., Dunn, M.L., Qi, H.J., 3D printed reversible shape changing components with stimuli responsive materials. *Sci. Rep.*, 6, 24761, 2016.
78. Li, X., Yang, Y., Zhang, Y., Wang, T., Yang, Z., Wang, Q., Zhang, X., Dual-method molding of 4D shape memory polyimide ink. *Mater. Des.*, 191, 108606, 2020.
79. Kumar, S., Hofmann, M., Steinmann, B., Foster, E.J., Weder, C., Reinforcement of stereolithographic resins for rapid prototyping with cellulose nanocrystals. *ACS Appl. Mater. Interfaces*, 4, 5399–5407, 2012.
80. Bodkhe, S. and Ermanni, P., 3D printing of multifunctional materials for sensing and actuation: Merging piezoelectricity with shape memory. *Eur. Polym. J.*, 132, 109738, 2020.
81. Choong, Y.Y.C., Maleksaeedi, S., Eng, H., Yu, S., Wei, J., Su, P.-C., High speed 4D printing of shape memory polymers with nanosilica. *Appl. Mater. Today*, 18, 100515, 2020.
82. Ng, C.S., Subramanian, A.S., Su, P.-C., Zinc oxide nanoparticles as additives for improved dimensional accuracy in vat photopolymerization. *Addit. Manuf.*, 59, 103118, 2022.
83. Krivec, M., Roshanghias, A., Abram, A., Binder, A., Exploiting the combination of 3D polymer printing and inkjet Ag-nanoparticle printing for advanced packaging. *Microelectron. Eng.*, 176, 1–5, 2017.

84. Ikram, H., Al Rashid, A., Koç, M., Synthesis, Characterization, and 3D Printing of Silver Nanoparticles/Photopolymer Resin Composites, in: *IOP Conference Series: Materials Science and Engineering*, IOP Publishing, p. 012003, 2022.
85. Chen, K., Kuang, X., Li, V., Kang, G., Qi, H.J., Fabrication of tough epoxy with shape memory effects by UV-assisted direct-ink write printing. *Soft Matter*, 14, 1879–1886, 2018.
86. Wu, A.S., Small IV, W., Bryson, T.M., Cheng, E., Metz, T.R., Schulze, S.E., Duoss, E.B., Wilson, T.S., 3D printed silicones with shape memory. *Sci. Rep.*, 7, 4664, 2017.
87. Zhang, B., Zhang, W., Zhang, Z., Zhang, Y.-F., Hingorani, H., Liu, Z., Liu, J., Ge, Q., Self-Healing Four-Dimensional Printing with an Ultraviolet Curable Double-Network Shape Memory Polymer System. *ACS Appl. Mater. Interfaces*, 11, 10328–10336, 2019.
88. Lin, C., Lv, J., Li, Y., Zhang, F., Li, J., Liu, Y., Liu, L., Leng, J., 4D-Printed Biodegradable and Remotely Controllable Shape Memory Occlusion Devices. *Adv. Funct. Mater.*, 29, 1906569, 2019.
89. Song, M., Liu, X., Yue, H., Li, S., Guo, J., 4D printing of PLA/PCL-based bio-polyurethane via moderate cross-linking to adjust the microphase separation. *Polymer, 256*, 125190, 2022.
90. Lin, C., Liu, L., Liu, Y., Leng, J., 4D printing of shape memory polybutylene succinate/polylactic acid (PBS/PLA) and its potential applications. *Compos. Struct.*, 279, 114729, 2022.
91. Peng, B., Yang, Y., Ju, T., Cavicchi, K., Fused Filament Fabrication 4D Printing of a Highly Extensible, Self-Healing, Shape Memory Elastomer Based on Thermoplastic Polymer Blends. *ACS Appl. Mater. Interfaces*, 13, 1, 12777–12788, 2020.
92. Li, A., Chen, X.-G., Zhang, L.-Y., Zhang, Y.-F., Temperature and Infill Density Effects on Thermal, Mechanical and Shape Memory Properties of Polylactic Acid/Poly(ε-caprolactone) Blends for 4D printing. *Materials*, 15, 8838, 2022.
93. Tekay, E., Şen, S., Korkmaz, M., Nugay, N., Preparation and characterization of thermo-responsive shape memory ester-based polymer blends. *J. Mater. Sci.*, 58, 1–20, 2023. 10.1007/s10853-023-08549-6.
94. Rahmatabadi, D., Ghasemi, I., Baniassadi, M., Abrinia, K., Baghani, M., 4D printing of PLA-TPU blends: effect of PLA concentration, loading mode, and programming temperature on the shape memory effect. *J. Mater. Sci.*, 58, 16, 7227–7243, 2023
95. Ma, S., Jiang, Z., Wang, M., Zhang, L., Liang, Y., Zhang, Z., Ren, L., Ren, L., 4D printing of PLA/PCL shape memory composites with controllable sequential deformation. *Bio-Des. Manuf.*, 4, 1–12, 2021.
96. Shan, W., Chen, Y., Hu, M., Qin, S., Peng, L., 4D printing of shape memory polymer via liquid crystal display (LCD) stereolithographic 3D printing. *Mater. Res. Express*, 7, 10, 105305, 2020.

97. Dong, X., Zhang, F., Wang, L., Liu, Y., Leng, J., 4D printing of electroactive shape-changing composite structures and their programmable behaviors. *Composites, Part A*, 157, 106925, 2022.
98. Liu, H., Wang, F., Wu, W., Dong, X., Sang, L., 4D printing of mechanically robust PLA/TPU/Fe3O4 magneto-responsive shape memory polymers for smart structures. *Composites, Part B*, 248, 110382, 2023.
99. Wan, X., He, Y., Liu, Y., Leng, J., 4D Printing of Multiple Shape Memory Polymer and Nanocomposites with Biocompatible, Programmable and Selectively Actuated Properties. *Addit. Manuf.*, 53, 53, 102689, 2022.
100. Lai, J., Ye, X., Liu, J., Wang, C., Li, J., Wang, X., Ma, M., Wang, M., 4D Printing of Highly Printable and Shape Morphing Hydrogels Consisted of Alginate and Methylcellulose. *Mater. Des., 205*, 109699, 2021.
101. Su, J.-W., Gao, W., Trinh, K., Kenderes, S., Pulatsu, E., Zhang, C., Whittington, A., Lin, M., Lin, J., 4D printing of polyurethane paint-based composites. *Int. J. Smart Nano Mater.*, 10, 1–12, 2019.
102. Liu, Y., Zhang, F., Jinsong, L., Fu, K., Lu, L., Wang, L., Cotton, C., Sun, B., Gu, B., Chou, T.-W., Remotely and Sequentially Controlled Actuation of Electroactivated Carbon Nanotube/Shape Memory Polymer Composites. *Adv. Mater. Technol.*, 4, 2019.
103. Liu, Y., Zhang, W., Zhang, F., Jinsong, L., Pei, S., Wang, L., Jia, X., Cotton, C., Sun, B., Chou, T.-W., Microstructural design for enhanced shape memory behavior of 4D printed composites based on carbon nanotube/polylactic acid filament. *Compos. Sci. Technol.*, 181, 107692, 2019.
104. Herath, M., Islam, M.M., Epaarachchi, J., Zhang, F., Jinsong, L., *4D Printed Shape Memory Polymer Composite Structures for Deployable Small Spacecrafts*, 2019.
105. Chowdhury, J., Anirudh, P.V., Karunakaran, C., Rajmohan, V., Mathew, A.T., Koziol, K., Alsanie, W.F., Kannan, C., Balan, A.S.S., Thakur, V.K., 4D Printing of Smart Polymer Nanocomposites: Integrating Graphene and Acrylate Based Shape Memory Polymers. *Polymers*, 13, 3660, 2021.
106. Wei, H., Zhang, Q., Yao, Y., Liu, L., Liu, Y., Jinsong, L., Direct-Write Fabrication of 4D Active Shape-Changing Structures Based on a Shape Memory Polymer and Its Nanocomposite. *Appl. Mater. Interfaces*, 2016.

7

4D Printing in Micro/Nano Scale: Technologies, Challenges, and Applications

Kaustav Moni Bora[1]*, Anita Jena[2], Ujjal Dey[3], Shailendra Kumar Varshney[4] and Cheruvu SivaKumar[3]

[1]School of Nano Science and Technology, IIT Kharagpur, Kharagpur, India
[2]Advanced Technology Development Centre, IIT Kharagpur, Kharagpur, India
[3]Department of Mechanical Engineering, IIT Kharagpur, Kharagpur, India
[4]Department of Electronics and Electrical Communication Engineering, IIT Kharagpur, Kharagpur, India

Abstract

3D printing of flexible and active shapes, commonly known as 4D printing, has revolutionized the world of additive manufacturing by introducing the concept of self-transforming structures. The added dimension in 4D printing signifies the ability of additively manufactured objects to change their characteristics, such as shape, over time in response to external stimuli. Although significant advancements have been achieved in the creation and production of printing materials on a large scale, the field of 4D printing at the micro-/nano- scale is still in its early stages due to several difficulties in constructing structures with micro/nano-sized features.

This book chapter delves into the emerging field of micro/nano scale 4D printing, highlighting the cutting-edge technologies being developed and exploring their potential applications. The chapter begins by introducing the concept of 4D printing and its significance in advanced manufacturing. It provides an overview of the transition from 3D printing to 4D printing and the potential impact of this evolution. The main focus of the chapter is on presenting various micro/nano scale 4D printing technologies. These techniques include liquid crystal elastomer- based miniaturization printing, two-photon polymerization, micro stereolithography, laser transfer micro additive manufacturing, ink-based micro additive

**Corresponding author*: kstvbora@kgpian.iitkgp.ac.in

Bijaya Bikram Samal, Cheruvu Siva Kumar and Shailendra Kumar Varshney (eds.) 4D Printing Technology: Principles, Materials and Applications, (205–244) © 2025 Scrivener Publishing LLC

manufacturing, hydrogel-based miniaturization printing, micro laser sintering, and beam deposition. Each of these technologies is discussed in detail, highlighting their unique capabilities and limitations.

The chapter presents several compelling case studies that showcase the practical applications of micro/nano scale 4D printing. These examples illustrate how these innovative technologies can be leveraged in diverse fields such as micro-robotics, optics, and biomedicine. The utilization of 4D microprinting has the capacity to demonstrate exceptional shape memory characteristics and activation at minimal stimulation levels; hence, creating new opportunities for the advancement of practical and adaptable micro/nano devices.

Keywords: 4D printing, shape memory alloys, beam deposition, two-photon lithography

Abbreviations

Abbreviation	**Full Name**
AM	additive manufacturing
3DP	3D printing
4DP	4D printing
IR	infrared radiation
CAD	computer-aided design
LCEs	liquid crystal elastomers
SMHs	shape memory hydrogels
SMPs	shape memory polymers
BD	beam deposition
LIFT	laser induced forward transfer
SMAs	shape memory alloys
ESMs	electro-stimuli materials
DEAs	dielectric elastomers actuators
SMPCs	shape memory polymer composites
FDM	fused deposition modeling
UV	ultraviolet
TPL	two-photon lithography
SLA	stereolithography
DIW	direct ink writing
DLW	direct laser writing
DOD	drop-on-demand

MEMS	microelectromechanical systems
SLS	selective laser sintering
DLP	digital light processing
μLS	micro-laser sintering

7.1 Introduction

In response to the imperative for environmental sustainability and society's increasing energy demands, researchers have actively pursued alternative energy sources and technologies capable of harnessing various energy forms, including atomic, magnetic, hydrogen, and piezoelectric energies. The overarching goal is to reduce dependence on fossil fuels and mitigate associated environmental impacts. Addressing a critical bottleneck, the application of 3D printing (3DP) emerges as a promising platform, offering myriad opportunities for designing and producing advanced engineering structures. Such structures exhibit advantageous characteristics, such as lightweight construction, high porosity, intricate detailing, minimal material waste, cost- effectiveness, and the ability to bypass intermediate processing steps by enabling single-step fabrication.

Moreover, these structures demonstrate promising mechanical properties. The implementation of 3D printing extends to various soft materials, encompassing polymers, hydrogels, and elastomers. However, the fabrication of crystalline polymers without resorting to polymers introduces challenges associated with processing complications, including dimensional imprecision, significant shear deformation, potential component damage, warping, and diminished mechanical properties. The additive manufacturing technique for crystalline polymers encounters constraints and limitations such as unresponsive materials, extended processing times, improper adhesion, fluctuating shrinkage, and void formation.

To overcome these constraints, it is necessary to integrate an advanced technology called 4D printing (4DP). This innovative technology allows items to modify their form, structure, dimensions, hue, and composition by harnessing renewable energy sources. Skylar Tibbits introduced the notion of 4D printing during his TED conference presentation in 2013, which led to a surge of study and publications on the subject in the following decade. 4D printing utilizes several construction techniques that involve the layering of materials and programming. It also incorporates intelligent materials that may change shape in reaction to external stimuli.

The technology utilizes shape memory programming, modeling, simulations, and implementation.

The utilization of intelligent materials offers a diverse array of functionalities, including as self-assembly, self-healing, self-adaptability, and self-sensing; hence, significantly broadening the scope of research opportunities across various applications. Figure 7.1 illustrates the upward trajectory of publications in the domain of 4D printing starting from 2013, based on data obtained from Web of Science. 4D printed products have shape-changing abilities in response to various stimuli, akin to the way plants, organs, and animals adapt to changes in their natural environment. Extensive research has examined the diverse shape-modifying capabilities of 4D printed structures, encompassing unidirectional, bidirectional, or multidirectional transformations.

The practical consequences involve the capacity to convert both printed and intermediate structures in a two-way manner, without the need for reprogramming or mechanical intervention [2]. The morphology of intelligent materials is determined by the proportions of the soft segment, chain mobility, and conductive additives. These factors, in turn, affect the features of 4D printing behavior [3]. The two main modeling methodologies for 4DP buildings are categorized as direct heating and indirect heating. Direct heating refers to the process of applying heat directly to the surface of an object in a high-temperature environment, such as immersing the object in boiling water to raise its temperature. This leads to an increase in the transition temperature (known as Tg for polymers) of the respective material or structures [1, 4]. Direct heating refers to the direct transfer of external heat or thermal energy to a structure, such as immersing an item in warm water or placing it in a high-temperature enclosed space. This approach bypasses the requirement for additives or specialized machinery, rendering it very uncomplicated. Nevertheless, difficulties occur in providing significant amounts of heat across a wide expanse and for a prolonged period of time [5]. Notwithstanding these difficulties, electrical current phenomena such as ohmic heating, Joule heating [6–8], and magnetism [9–14] continue to be relevant and important in academic discussions. Scientific studies have found that infrared (IR) radiation between 24-26 units, water, pH levels as specified by references 27 and 28, and numerous biological stimuli cited in 29-32, can significantly impact a wide variety of measurements. These stimulation mediums utilize sustainable energy sources, such as solar and wind power, to cause visible changes in shape. Internal heating methods

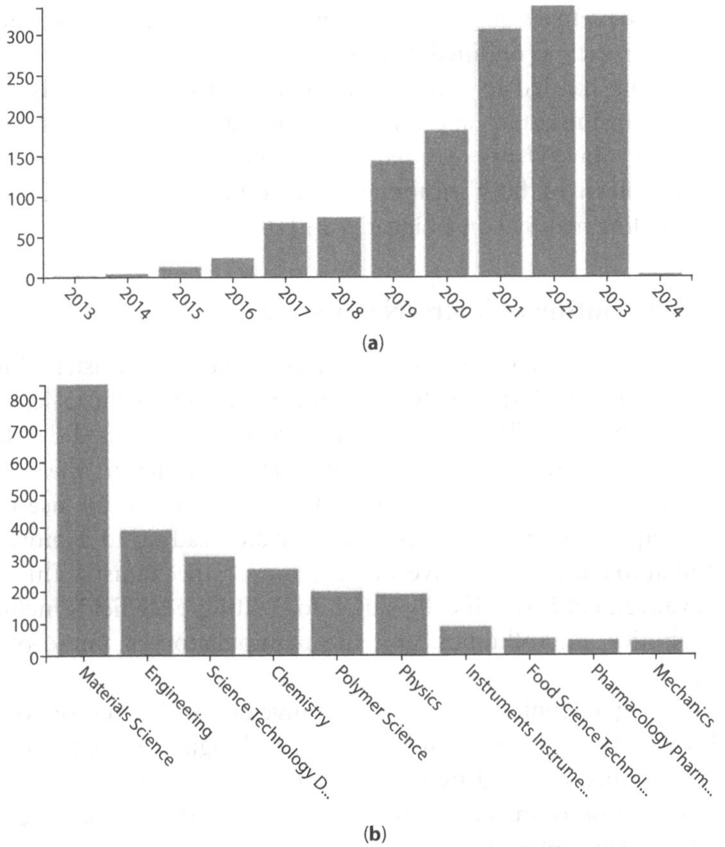

Figure 7.1 (a) 4D printing publications from 2013 to 2024 based on web of science data (collected on 9 November 2023) (b) 4D printing publications in different domains.

include dipole rearrangement, current conduction, photochemical heating, and the absorption of molecules like water and acid-based chemicals [2, 15].

Diverse modern materials utilize a wide spectrum of stimuli, including electrical, magnetic, and light-based stimuli. The materials consist of active electro systems, namely electrically stimulated liquid crystal elastomers (LCEs), shape memory hydrogels (SMHs), and shape memory polymers (SMPs). Furthermore, there have been developments in the creation of non-electroactive systems that include fillers such as carbon-based materials, metals, and metal oxides. When employed in various printing techniques, these materials exhibit prompt, two-way, dynamic, and adaptable characteristics, as outlined in Table 7.1.

3D Printing has led to substantial progress in different fields and is widely used in important applications, acting as a strong basis for the 4DP process. The 3D printing technique, sometimes referred to as fast prototyping, involves the fabrication of three-dimensional things by incrementally adding material on a bigger scale [16, 17].

7.1.1 4D Printing in Micro/Nano Scale

Academic research has extensively investigated the construction of three-dimensional (3D) structures at the micro and nanoscale, using 3D printing methods. The significant progress has stimulated the development of 4D printing, leading to new opportunities in the field of micro and nanostructures [17–19]. At these small sizes, the use of 4D Printing requires very little energy to activate, leading to a more efficient and automatically adaptive shape-changing mechanism. This technical advancement is specifically suited for cutting-edge fields including micro-robotics, microfluidics, and the manipulation of micro-objects [19, 20].

3D printing technology offers a wide range of possibilities for creating complex structures, such as sandwich panels, designs inspired by nature, and architectures inspired by origami. This technology also allows for the incorporation of material functionality [21]. Prior research [22–24] have extensively explored the complex interplay among mechanical and surface properties, functioning behavior, and thermal stability in connection to fabricated parts, design, and geometry. These researches highlight the vulnerability of morphing processes to unforeseen disruptions, nonlinearity, temporal changes, and time delay [25–27]. Shape memory materials primarily observe and analyze changes in form, improving their strength and usefulness in a wide range of applications [28]. The notion of dynamic shift corresponds to the life cycle of intelligent textiles, encompassing the transformations experienced by shape memory polymers (SMPs), shape memory hybrids (SMHs), and liquid crystal elastomers (LCEs) without any noticeable deterioration in their capacity to preserve and recover their initial form.

Nevertheless, a significant obstacle occurs, specifically for 3D printers, when attempting to create features that are smaller than around 50 micrometers. This is mostly due to limitations imposed by the materials used and the technical difficulties involved. To meet the increasing need for making things smaller, especially as modern electronics continue to shrink in size, a new field called microscale 3D printing, or micro-AM, has emerged. This particular aspect of 3D printing focuses on creating small, intricately designed, and accurately engineered parts. To work at microscale dimensions, often within a range of a few micrometers, certain equipment, procedures, and materials are necessary. The utilization of micro-AM encompasses a wide range of areas, such as the advancement of microelectromechanical systems (MEMS) devices, the creation of 3D-printed photonics, the detailed assembly of micro-additively made microfluidics, and the manufacturing of micro stents. The micro-AM domain serves sectors that require the production of small, intricately designed components to facilitate technological advancement.

Since its inception in 2013, a substantial body of literature has explored the vast potential of 4D printing. However, there is a significant lack of comprehensive data that enables the manipulation of materials by indirect heating methods, such as electrical, magnetic, and light- based techniques. This review analyses the prospective trajectory of 4D printing with regards to downsizing the printing process to submicron and nano sizes. Furthermore, it investigates the associated mechanisms implicated in motion, enabling precise adaptability over time, tailored sensory input, and versatility. Subsequently, we delineate the challenges that this technology will encounter in the future, along with its prospects and potential, in tandem with the growing adoption of 4DP in different fields.

7.2 Materials

Electro-stimuli materials (ESMs) have gained significant attention from the scientific community due to their exceptional electromechanical capabilities. They are considered a crucial class of soft multifunctional materials [17]. Electroactive shape memory materials (ESMs) possess a distinct capacity to change their shape when exposed to electrical fields, setting them apart from other soft materials. ESMs exhibit remarkable characteristics such as rapid reaction and spontaneous actuation in biomedical environments, water, or typical air conditions, distinguishing them from thermal or direct heating stimuli [30]. ESMs predominantly employ non-electrosensitive polymers and liquid crystal elastomers (LCEs) as their primary materials. The materials are mixed with other particles, including carbon-based systems, metallic oxides (Fe2O3, Fe3O4, CoFe2O3,

NiFe2O3), graphene oxide, and metal blends [29]. These materials are crucial for actuation since they enable the capacity to alter shape, making them highly ideal for usage in actuators, robotics, sensors, and energy harvesting devices [34]. Electro-responsive particles induce internal resistance and generate warmth by means of electric current, resulting in the production of heat (known as Joule heat) in shape memory polymers (SMPs) [32]. Various forms of 3D computer-aided design (CAD) models and fabrication techniques, including stereolithography (SLS), direct ink writing (DIW), fused deposition modeling (FDM), and drop- on-demand (DOD)/ inkjet printing, can be employed to produce intricate structures using shape memory polymers (SMPs) [36]. Multiple significant studies highlight the potential of ESMs in 4D printing [31, 33]. For example, the combination of carbon fiber-reinforced PLA utilizing FDM results in transient structures that react to changes in temperature [37]. The incorporation of graphene into wiggle structures, together with PLA, demonstrates the ability to sense and control electrical signals in a reversible manner, as reported in reference [35]. Moreover, the utilization of biodegradable carboxymethyl cellulose polymer with graphite particles grafted using DIW emphasizes the significant shear thinning properties necessary for the printing technique [36].The reversible actuators demonstrate remarkable shape-changing abilities triggered by heat and light [38]. An intriguing study involves the integration of an ionic hybrid consisting of elastomer and hydrogels, which is aimed at enhancing the capabilities of soft robotics and bio-inspired machines [39]. Furthermore, 4D printed dielectric elastomers actuators (DEAs) showcase substantial bending in response to electrical fields [40].

The second method involves magnetism, introducing a means of local heating through the influence of a magnetic field [41]. Studies demonstrate the 4D printing of magnetic responsive polymer structures, where materials such as PLA grafted by Fe3O4 composite filament exhibit bone tissue-like shapes responding to a 27.5 kHz magnetic field [42]. Composite systems like PLA/Fe3O4 in DIW printing showcase the recovery of original shapes under a 30 kHz magnetic field [43].

The choice of materials assumes paramount significance, dictating the form, function, and responsiveness of printed structures. This study scrutinizes a diverse array of materials classified into resists and coatings, each contributing unique attributes to the fabricated constructs. The subsequent elucidation provides a comprehensive examination of these materials, encompassing their specific shapes, stimuli responsiveness, dimensions, and applications.

a) **SU-8: A Versatile Resisting Agent:** SU-8, a widely utilized resist material, imparts a helical structure to the printed objects. Leveraging magnetic-driven stimuli, SU-8 structures, ranging from 4 to 65 μm, find application in the realm of transportation, showcasing the material's adaptability to diverse dimensional requirements [1, 2]. Additionally, SU-8, when coated with Ni/Ti, manifests cylindrical structures conducive to magnetic-driven rolling motion, presenting opportunities for innovative locomotion in micro robotic applications [7].

b) **Hydrogel Blend of PEGDA + Irgacure 369: Navigating Medical Microrobots:** The combination of PEGDA and Irgacure 369, constituting a hydrogel resist, synergizes with a Ni+Au coating to yield double-helical structures. Under the influence of magnetic- driven stimuli, these structures, with dimensions of 25 μm, find a niche in medical micro robotics, demonstrating the material's potential for intricate biomedical applications [3].

c) **OrmoComp: Precision in Biocompatible Architectures:** OrmoComp, serving as a resist material with Fe or Fe/Ti coating, imparts a helical structure responsive to magnetic-driven stimuli. With dimensions measuring 15.7 μm, the resulting structures showcase biocompatibility, opening avenues for applications in biomedicine [4].

d) **IP-L 780: Light-Driven Drug Delivery Vessels:** IP-L 780, as a resist, in tandem with a Ti+Au coating, gives rise to vessel structures adorned with spherical handles. Light- driven stimuli guide these structures, sized at 60 μm, towards drug delivery applications, highlighting the material's potential in controlled and targeted therapeutic interventions [5].

e) **IP-S: Magnetic-Driven Scaffold Architectures:** The resist material IP-S, coupled with a Ni/TiO2 coating, engenders helical scaffold structures featuring needles. Responding to magnetic-driven stimuli, these structures, spanning 300 μm, carve a niche in scaffold architecture applications, showcasing the material's suitability for the construction of intricate three-dimensional frameworks [6].

f) **SZ2080: Tailored for Dragging and Tumbling:** SZ2080, a resist material, complemented by MNPs coating, yields concave and tapered structures under magnetic- driven stimuli. Ranging from 35 to 75 μm, these structures find application in dragging and tumbling scenarios, showcasing the material's adaptability for varied micro robotic functionalities [8].

g) **SMP Blends: Thermally-Responsive Color Palette and Shape Morphing:** The class of shape memory polymers (SMPs), encompassing Vero Clear, HPPA, BPA, TPO, and Isobornyl acrylate, PEGDA 575, TcddA, HexA, unveils a spectrum of possibilities. Thermally-driven stimuli bring forth a color palette with dimensions of 20 μm, demonstrating applications in nanophononics [9]. Furthermore, SMPs showcase their versatility in shape morphing, as exemplified by double platform and box-like structures with dimensions of 15 μm, broadening the horizons for adaptive structures [10].

h) **SU-8 50: Precision in Swimming Micro Robotics:** SU-8 50, in conjunction with a Ni+Pt coating, introduces serrated trapezoidal structures under magnetic-driven stimuli. Sized at 65 μm, these structures find application in the domain of swimming microrobots, exemplifying the material's precision in crafting functional and specialized constructs [11].

These materials and some more along with the applications are presented in the Table 7.1.

7.3 Micro-Nano Scale 4D Printing Processes

The escalating demand for precision in various applications has driven the development of additive manufacturing (AM) processes tailored for micro- and nanoscale structures. This niche within AM responds to the imperative need for intricate designs, with features measuring only a few micrometers. The inherent advantages of AM, such as precision and adaptability, make it a compelling choice over traditional fabrication methods. Within this realm, micro-nano AM methods have emerged, presenting a specialized subset of techniques explicitly designed for crafting microscale objects. Notably, this technology holds immense promise across diverse fields, including biomedicine, electronics, and materials science. Researchers are actively engaged in exploring avant-garde materials and fabrication methods to enhance the capabilities of micro-nano AM, making it a focal point for innovative solutions in applications where precision and adaptability in miniaturized structures are paramount. In the subsequent sections, we delve into an in-depth discussion of six major processes, as

Table 7.1 Materials of different structures with its dimensions.

Materials	Shape of the structure	Stimuli	Dimensions (μm)	Application	Ref
Resist:SU-8 Coating: Ni/Ti	helices	magnetic- driven	4-65	transportation	[44, 45]
Resist (Hydrogel): PEGDA + Irgacure 369 Coating: Ni+Au	double-helical	magnetic- driven	25	medical microrobots	[46]
Resist: OrmoComp Coating: Fe or Fe/Ti	helix	magnetic- driven	15.7	biocompatible	[47]
Resist: IP-L 780 Coating: Ti+Au	vessel with spherical handles	light-driven	60	drug delivery	[48]
Resist: IP-S Coating: Ni/TiO2	helical scaffold with needle	magnetic- driven	300	scaffold architecture	[49]
Resist:SU-8 Coating: Ni/Ti	cylindrical	magnetic- driven	144–160	rolling motion	[50]
Resist: SZ2080 Coating: MNPs	concave and tapered	magnetic- driven	35–75	dragging and tumbling	[51]
Resist (SMP): Vero Clear, HPPA, BPA, TPO	color palette	thermo- driven	20	nanophononics	[52]
Resist (SMP): Isobornyl acrylate, PEGDA 575, TcddA, HexA	double platform & box-like structure	thermo- driven	15	shape morphing	[53]

(Continued)

Table 7.1 Materials of different structures with its dimensions. (*Continued*)

Materials	Shape of the structure	Stimuli	Dimensions (μm)	Application	Ref
Resist:SU-8 50 Coating: Ni+Pt	serrated trapezoid	magnetic- driven	65	swimming microrobot	[54]
Resist (Hydrogel): NIPAM, MBA, LAP, SWNTs, EG, TEOA	hollow buckyball, micropillar, micro clamp, artificial aortic valve	light-driven	30	micromachines	[55]
Resist (LCE): ST3866, ST3021, Irgacure 369, DR1A	voxels, line, rectangular frame	thermo- driven	200	optics, and programmable origami	[56]
Resist:SU-8 Coating: Au	triple-barreled tube	photothermal-driven	45	microrobotic applications in biomedical	[57]
Resist (LCE): C6BP, RM257, Irgacure 369	woodpile, hexagonal crystal, Frame structures, microclamp	photothermal-driven	140	microactuators	[58]
Resist: SZ2080, IESL- FORTH Coating: Ni/Ti	hollow conical helices	magnetic- driven	45–75	hybrid drug delivery	[59]
Resist (LCE): PETA, PEGda, DMAEMA, HMPP, Gycerol, TPO	flytrap micro actuator	pH-driven	40	micro actuators	[60]

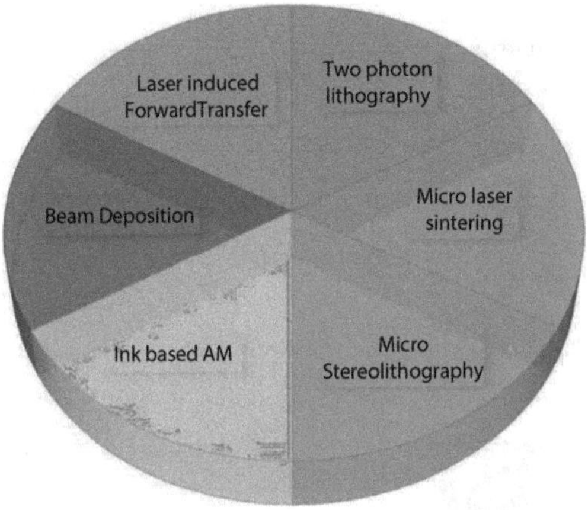

Figure 7.2 Different types of micro/nano scale 4D printing processes.

illustrated in the Figure 7.2, shedding light on their distinctive characteristics and applications.

7.3.1 Two-Photon Lithography/3D Printing by Direct Laser Writing

Two- photon lithography (TPL) stands at the forefront of advanced microscale fabrication, showcasing its versatility in crafting tailored polymeric structures. This technique harnesses the power of two-photon absorption, a nonlinear optical process, to selectively solidify a photosensitive resin, thereby manifesting intricate 3D patterns with unparalleled precision. The appeal of TPL lies in its capacity to generate complex microstructures, boasting high resolution and precision, which has found application across diverse domains, including optics, microfluidics, drug delivery, micro-robotics, and tissue engineering [61–66]. As the technology matures, its potential applications across various industries are expected to burgeon, contributing to the development of highly specialized and bespoke microdevices and surfaces. A pivotal frontier in this domain involves harnessing TPL to print 3D micro-nano structures responsive to external stimuli such as heat, moisture, light, or magnetic fields, thereby paving the way for innovative applications [67].

The operational principle of TPL involves the utilization of a laser source to focus high- intensity femtosecond laser pulses into a photosensitive resin. Through the simultaneous absorption of two lower-energy photons, a photochemical reaction is triggered within the resin. The laser

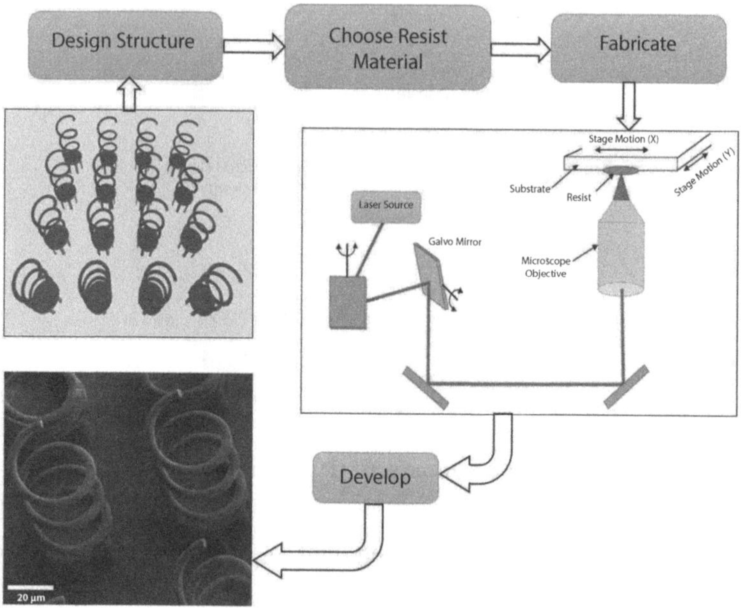

Figure 7.3 Schematic overview of the two-photon polymerization process for fabrication 3D nanostructures.

focus, akin to a precise "pen," navigates in three dimensions, facilitating the meticulous polymerization of the resin point by point. Remarkably, this process achieves feature sizes on the order of nanometers, rendering it suitable for the creation of intricate and small-scale structures. Integration of 4D printing with TPL further elevates the resolution of responsive structures to the micro- nanoscale, unlocking the potential to craft deformable or multifunctional micro-nano structures. Illustrated in Figure 7.3 is the fabrication process of a 3D helical nanostructure using TPL, underscoring its capabilities and adaptability for 4D applications. This synergy between TPL and 4D printing heralds a new era in micro-nano fabrication, offering unprecedented precision and responsiveness for a myriad of cutting-edge applications.

7.3.2 Micro Laser Sintering

Micro Laser Sintering (μLS), as shown in the Figure 7.4 emerges as an advanced additive manufacturing technology, aligning itself within the realm of selective laser sintering (SLS). Its primary domain lies in the high-precision, layer-by-layer fabrication of three-dimensional objects,

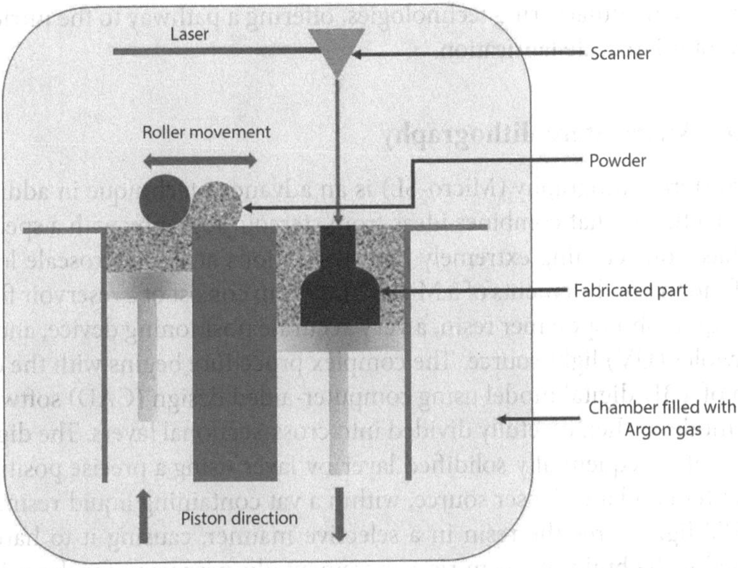

Figure 7.4 Laser sinering.

with a particular emphasis on the microscale. µLS is adept at producing small, intricate, and highly detailed parts and structures, rendering it indispensable for applications demanding meticulous precision.

In the µLS process, a fine powder material, commonly polymers or metals, is delicately spread as a thin layer over a build platform. The key player, a high-powered laser, is meticulously controlled to selectively sinter (heat and fuse) the powdered material in accordance with the 3D model's design in a cross-sectional layer. Upon completing a layer, the build platform descends, and a fresh layer of powder is applied. This layer- by-layer approach persists until the entire 3D object materializes. Remarkably versatile, µLS accommodates various materials, including polymers, metals, and ceramics. Material selection hinges on the specific application and the desired properties of the final part. µLS distinguishes itself with exceptional precision, capable of producing parts with fine details and features, often operating at the sub-micron scale. This technology's precision makes it a potent tool for applications necessitating intricate microscale structures. Although the build volume in µLS tends to be modest, its small- scale capabilities render it ideal for crafting microscale components, prototypes, and intricate structures. The adaptability to a spectrum of materials, coupled with unparalleled precision, positions µLS as a stalwart in the arsenal of additive manufacturing technologies, offering a pathway to the intricate world of microscale fabrication.

7.3.3 Micro Stereolithography

Micro stereolithography (Micro-SL) is an advanced technique in additive manufacturing that combines ideas from stereolithography with a specific emphasis on reaching extremely high resolutions at the microscale level. The fundamental elements of a Micro-SL system consist of a reservoir filled with liquid photopolymer resin, a very accurate positioning device, and an ultraviolet (UV) light source. The complex procedure begins with the creation of a 3D digital model using computer-aided design (CAD) software. This model is then carefully divided into cross-sectional layers. The digital blueprint is sequentially solidified layer by layer using a precise positioning system and a UV laser source, within a vat containing liquid resin. As the UV light cures the resin in a selective manner, causing it to harden gradually, the build platform rises incrementally, preparing for the subsequent layer. This repeating procedure persists until the complete microscale item materializes. The last phase of the process is removing any remaining uncured resin and then subjecting the material to a post-curing procedure to enhance its qualities. Micro-SL is well recognized for its ability to create highly detailed microscale structures with outstanding surface smoothness. It is used in several fields, including microelectronics, microfluidics, and biomedical engineering, where accuracy and complexity at the microscale are crucial. This process is demonstrated in Figure 7.5.

7.3.4 Ink Based AM

Within the field of advanced additive manufacturing, Ink-based micro additive manufacturing (Ink Micro-AM, Figure 7.6) emerges as a state-of-the-art method, offering the potential for precise creation at a microscale level. The fundamental elements of this technology consist of a specific ink, a meticulously regulated dispensing mechanism, and a substrate. The ink, which frequently consists of functional elements like polymers, nanoparticles, or biological components, is carefully released in small amounts onto a substrate. The process progresses sequentially, with each layer being formed under the guidance of a digitally created plan. The regulated dispensing mechanism guarantees accurate positioning of the ink, enabling the creation of elaborate patterns and structures. After being deposited, the ink proceeds through a process of curing or solidification, which is determined by the material qualities. This process leads to the formation of

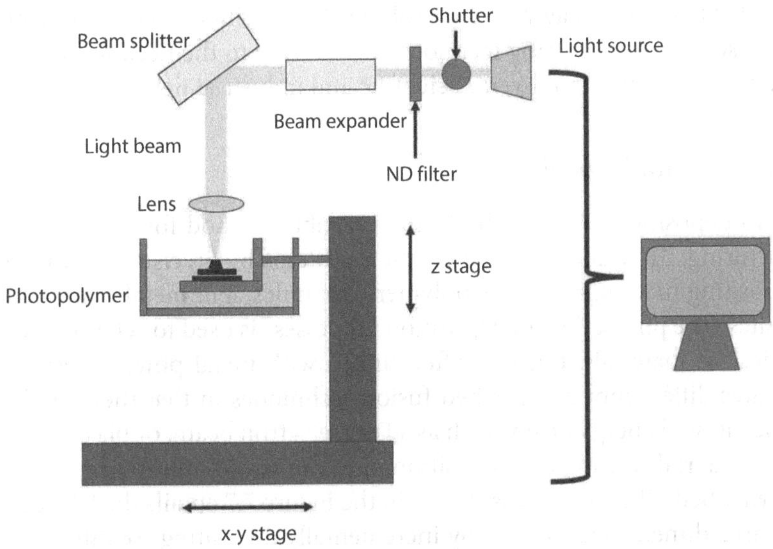

Figure 7.5 Micro stereolithography.

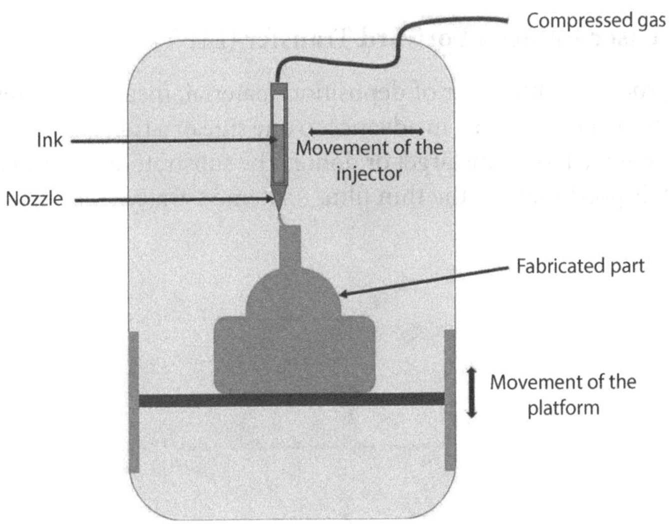

Figure 7.6 Ink based additive manufacturing.

a microscale object. This approach, which is both accurate and adaptable, may be used in several sectors like electronics, photonics, and biology. It provides a potential way to create small-scale objects with specific material compositions and functional qualities. The investigation and improvement of Ink Micro-AM play a crucial role in the continuous advancement of microscale manufacturing technologies, catering to the need for accuracy and personalization in diverse scientific and industrial fields.

7.3.5 Beam Deposition

Beam deposition (BD) methods are a flexible method for additive manufacturing. They allow for the creation of items by precisely melting and depositing materials, such as polymers, ceramics, and metal matrix composites. The phrase "beam deposition processes" is used to refer to a wider variety of materials, but it is often linked with metal powders. BD procedures differ from powder bed fusion techniques in that they employ a concentrated energy beam, such as a laser, electron beam, or plasma arc, to melt materials during the deposition process instead of using pre-existing powder beds. The process as shown in the Figure 7.7 entails the fabrication of three-dimensional objects by incrementally depositing solidified material using a controlled motion of a BD head. To create complex shapes, it is possible to use either support material or a multi axis deposition head.

7.3.6 Laser Induced Forward Transfer (LIFT)

In this process, a thin layer of deposition material, measuring a few hundred nanometers, is placed in advance to one side of a transparent support, which is referred to as the target or donor. The substrate, also known as the receptor, is positioned as the thin film. As demonstrated in the Figure 7.8,

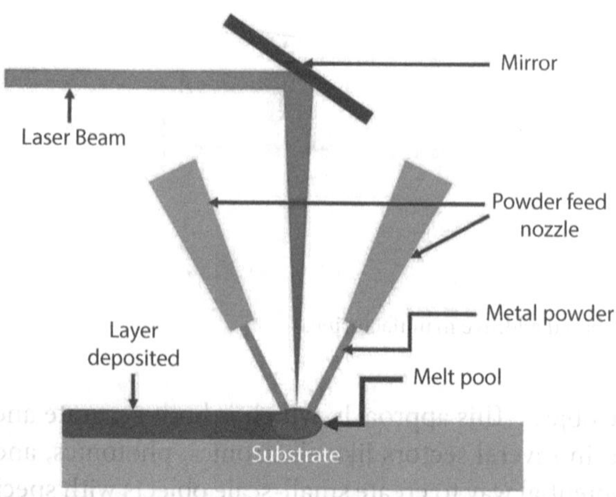

Figure 7.7 Beam deposition.

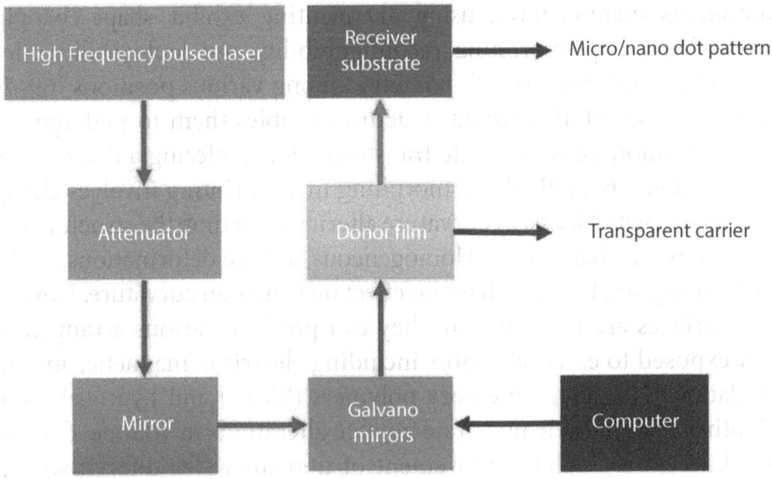

Figure 7.8 Schematic of LIDT.

a laser emitting periodic pulses is directed towards the transparent support from the other side. The pre-coated material absorbs the laser energy, resulting in the generation of heat and the removal of a thin layer of film from the support. Deposition takes place when the removable film is transferred to the substrate. The duration of laser pulses used for ablation and transfer typically ranges from nanoseconds to femtoseconds. The laser beam undergoes relative motion with respect to the substrate to create a pattern. The laser radiation technology employs an iterative process to precisely regulate the thickness of the pattern. It is a technique for the purpose of additive micropatterning of functional particles and films. The usefulness of this process approach is being explored for its ability to provide precise micropatterning, which could potentially revolutionize the manufacture of functional materials. This process is also about how LIFT can be used to regulate and modify micro structures, which will help additive manufacturing technologies get better.

7.4 Applications of 4D Printing

7.4.1 Structural Components

4D printing goes beyond the production of complex mechanical components and explores the development of smart materials that incorporate integrated elements and possess distinctive capabilities [68]. Structural components manufactured using 4D printing exhibit shape-changing, shapeshifting, shape morphing, reconfigurability, and deployability, allowing for the transformation of geometry among various positions [69, 70]. The adaptability of 4D printed structures enables them to undergo non-affine and inhomogeneous state transformations, offering a diverse range of applications [69, 70]. Shape morphing in 4D printing involves changes in curvature, with Gaussian curvature altering each time the structure transitions between shapes [70]. Homogeneous surface deformations, such as pure bending and twisting, have no effect on Gaussian curvature. However, when surfaces are not uniform, they can produce various arrangements when exposed to external factors including electrical, magnetic, and light stimulation [70]. Shape memory polymers (SMPs) and hydrogels, along with other materials, demonstrate notable alterations in volume, deformation, differential swelling, movement of domain walls, uneven stiffness, and remaining stress [71, 72].

Within the field of structural applications, 4D printing presents two widely used approaches for achieving magnetic responsive structural deformation. One approach is to create localized heating using magnetic fields that may be adjusted, resulting in forms that change over time [75–78]. The second approach utilizes components that consist of many layers and materials, where each layer undergoes swelling or morphing at distinct time intervals [75–78]. Studies show that in multi-material composites, the spatial distribution of gradient and concentration is deliberately regulated, leading to the alignment of active magneto particles during the structural 3D printing process [79]. Multi-stimuli ciliary systems offer a promising opportunity for achieving structural morphing via 4D printing. These techniques facilitate the development of cilia that can respond to magnetic fields and light in a remotely-controlled manner, concurrently reacting to several stimuli [74]. In addition, structural components that possess self-healing capabilities are utilized in the fields of tissue engineering and drug delivery systems. A specific instance of a material with the ability to repair itself is a gel that is produced using three-dimensional technology and is responsive to magnetic fields. This gel is composed of oxidized hyaluronate, glycol chitosan, and superparamagnetic iron oxide nanoparticles. This gel has the ability to undergo self-repair when subjected to a reversed magnetic field [80].

While magnetic field activation has been extensively studied for structural components, ohmic energy presents another avenue for reconfiguring shapes with positive and negative Gaussian curvature [81]. Electrically responsive constructions utilize complex methods, such as including many layers of electrodes with different arrangements and concentric rings on polymer surfaces, or controlling dielectric liquid in liquid crystal elastomers by creating patterns [81–83]. Moon Young Choi *et al.* have developed a technique to convert hydrogel copolymer sheets from a two-dimensional (2D) shape into complex and predictable three-dimensional forms by employing electrically responsive hydrogel architectures [73]. These applications demonstrate the versatility and potential of 4D printing in creating intelligent and adaptive structural components for various industries and fields.

7.4.2 Robotics

The field of soft robotics and actuators has witnessed significant advancements through the rapid proliferation of 4D printing technologies. The advancements are evident in the improved actuation stress, reliability, work capacity, and time response [84, 85]. Soft polymers and elastomers, such as silicone and polydimethylsiloxane, have garnered significant attention in scientific research due to their facile processing using additive manufacturing techniques [85].

Magnetic-responsive soft robots, which are a type of 4D printing applications, demonstrate transformation abilities that rely on the features of the materials used, the properties of the base matrix, and the magnetic properties of additive particles such as Fe, GO, Fe3O4, Co, NbFeB, CNT, and CNF [86–88]. A remarkable achievement in this domain is the investigation carried out by Xu Wang and collaborators, whereby they devised simulation- guided, long-lasting, compact, autonomous, pliable robots capable of adapting to diverse circumstances. These robots are capable of performing tasks such as walking, carrying cargo, floating, compressing, and swimming [89]. These ultrafast robots exhibit nonlinear crosslinking behavior, responding to a low magnetic field and achieving specific density at a remarkable rate [89].

Elastomers, which are very promising materials for soft robotics, utilize their exceptional ability to deform, possess high energy density, and elongate by up to 200% to generate Maxwell electro-static pressure and motion within an electrical field [90–93]. A recent advancement involves the implementation of liquid dielectrics in hydraulically amplified electrically sensitive actuators, which possess the ability to self-heal. This innovation was documented in reference [94]. These electro-static soft actuators utilize the electro-static Maxwell stress to enhance their reaction to applied voltage between conductors, allowing them to self-heal [95]. Despite challenges in mechanical strength, these actuators exhibit high life cycles and responsiveness to both magnetic and electric actuation [96, 97].

The research explores the integration of electromagnetic particles, including Ni, Ag, Co, and carbon-based materials, into soft robotics materials at submicron and nanoscale dimensions. This integration aims to improve the mechanical properties of the materials, allowing them to be activated by electrical and magnetic fields [98]. Grippers designed using shape-morphing hydrogels, magnetic field-driven intravascular grippers, and sodium polyacrylate-based gel grippers exemplify the versatility of 4D printing in creating functional soft robotic components [99, 100].

In addition, the utilization of 4D printing in liquid crystal elastomers (LCEs) has been employed to create composite structures for soft robotics and actuators, demonstrating increased work capacity and stress during nematic-isotropic transformation [17]. Layered manufactured liquid elastomers exhibit quick snap-through deformation and repetition, achieving substantial actuation even when lifting five times the weight of the actuator [101]. Researchers, such as Kotikian *et al.*, have successfully 4D printed LCE actuators with two-way and extensive transformation capabilities, lifting loads 1000 times heavier than themselves [102].

7.4.3 Biomedical and Tissue Engineering

In the realm of biomedical and tissue engineering, the application of 4D printing has garnered attention for its potential to create structures with low actuation temperature, biocompatibility, local heating, and tunable shapes. This holds promise for more effective tissue and organ regeneration, expanding replacement or repair procedures for damaged tissues [103]. Conventional techniques for repairing organs have several drawbacks, including limited movement, extended healing periods, inflammation, and the risk of rejection [104]. The emergence of magneto-/electro-responsive polymeric and hydrogel materials provides opportunities for creating novel biomaterials with distinct structural characteristics and inherent signals for altering cell mechanobiology [105]. Electric, magnetic, and light stimuli have the ability to induce cell differentiation, activate mechanotransduction signaling pathways, and impact immunomodulation and proliferation [106]. The electrical conductivity of materials is crucial in the choice of substrates for electro-responsive tissues such as brain and heart tissue. Attempts have been made to adjust the electrical properties of biodegradable materials by including nanoparticles like graphene, GO, CNTs, and CNFs into matrices, resulting in the development of multifunctional biomaterials with electrical responsiveness [107, 108].

Stimuli-responsive hydrogels have been utilized in diverse tissue engineering applications, catering to distinct biological tissues. Xia and colleagues achieved successful development of a hydrogel composite by incorporating single-walled carbon nanotubes (SWCNTs) into a poly(N-isopropylacrylamide) (PNIPAAm) hydrogel matrix, resulting in improved compatibility with living cells [109]. Xu C. and colleagues fabricated electro-responsive hydrogels with notable electromechanical characteristics, demonstrating potential for applications in tissue engineering and biomedical fields, namely in neuro-regeneration [88]. Filippi and their colleagues proposed a new magnetized hydrogel nanocomposite for tissue engineering by incorporating magneto-active components into the underlying hydrogel materials [110]. The scaffold, when subjected to a stationary magnetic field, demonstrated enhanced osteoblastic/vasculogenic characteristics, mineralization, and upregulated expression of osteogenesis-related genes. A study was undertaken by Akbari and the team on the 4D bioprinting of pH-responsive porous sensor composites. The researchers employed alginate fibers and pH-responsive polymers to fabricate these composites, which can be progressively triggered in drug delivery systems. Alginate fibers were utilized in a controlled experiment to create drug-eluting stents. These fibers demonstrated the ability to respond to changes in pH and release drugs at specific locations, so aiding in the elimination of bacteria [111].

Although the utilization of 4D printing in tissue engineering is now limited to research, study, and prototype development, it possesses considerable potential for future therapeutic applications. Advancements in intelligent materials that can detect natural signals and pathological anomalies in the body have facilitated the investigation of 4D bioprinting for the purpose of drug delivery. Researchers are prioritizing a comprehensive comprehension of medicinal chemistry, pathophysiology, and bioengineering technology to create efficient and secure drug delivery systems for diverse applications in the biomedical area.

7.4.4 Electronic Devices and Absorbers

The emerging field of 4D printing is expanding its scope to include the fabrication of sophisticated meta-structures that have promising applications in radar and electronic devices, namely in the area of electromagnetic absorption. Scientists are utilizing 4D printing to create radar absorber materials that fulfil the requirements for structures that are lightweight, thin, all-encompassing, and durable [112]. The need for electromagnetic metamaterials (EMMs) arises from the growing data storage and demand caused by the emergence of 5G and 6G wireless communication, particularly in the frequency range of 100 GHz to 1 THz. These EMMs are essential for the efficient implementation of radar and electronic applications [113, 114]. Scientists have investigated shape memory polymers (SMPs) because of their ability to change shape and recover their original form. The aim is to address the drawbacks of shape memory alloys (SMAs), including difficulties in recovering the desired shape, complicated manufacturing processes, toxicity concerns, and high expenses. Scientists have developed active smart metamaterials (MMs) to address the constraint of the thickness-to-bandwidth ratio, also known as Rozanov's effect. This breakthrough expands the spectrum of possible applications for MM absorbers. This is accomplished by utilizing techniques such as manipulating the material properties to modify electromagnetic characteristics, including varactor diodes, integrating PIN diodes, and employing mechanical adjustments [115, 116].

An experiment of significance entailed the utilization of stereolithography (SLA) to 4D print a shape memory polymer resin, resulting in the creation of a flexible electronic gadget. The polycaprolactone (PCL) resin, containing different levels of methacrylate, was utilized for printing purposes. To improve thermal conductivity, a carbon nanotube (CNT) coating was applied to the resin. The resulting structure exhibited an open-circuit configuration in its initial printed shape, which then transitioned into a closed temporary circuit when a voltage of 50 V was applied. The sandwich structure, employing Kapton tape, demonstrated a substantial reversible metamorphosis with a deflection of 180 degrees [17]. A recent study showcased the microwave shielding capabilities of a flexible shape memory material called thermoplastic polyurethane, which was enhanced by the addition of microwave shielding additives. This material has the potential to be used in various applications, particularly in the development of self-expanding structural designs. The composite, consisting of reduced graphene oxide (rGO), $Fe2O3$, and multi-walled carbon nanotubes (MWCNT), demonstrated an electromagnetic shielding effectiveness of -32 dB. Nevertheless, the shape-changing potential was limited due to the high aspect ratio of the filler and matrix [117].

Furthermore, 4D printing has been employed to create ultrathin metamaterials for terahertz absorption, demonstrating high absorptivity of electromagnetic waves at 5.91 THz. The honeycomb structure of the active metamaterials, which were printed with resistor patches and evaluated within the frequency range of 3.53-24.00 GHz, showed a low level of reflection. The printed active metamaterial portrayed outstanding mechanical properties, including a compressive strength of 10.7 MPa and a low density of 0.294 g/cm3, coupled with an energy absorption per volume of 4.37 MJ/m3 [17].

Within the absorbers category, an alternative method is employing polymeric substrates filled with liquid absorbers. An example of this is a PDMS object with three layers, which was printed and filled with EGaIn liquid metal. This object demonstrated the ability to alter shape in response to variations in absorber frequency. These examples highlight the versatility and potential of 4D printing in the creation of sophisticated structures for electronic devices and absorbers, with applications ranging from radar technology to advanced metamaterials.

7.5 Challenges in 4D Printing Technology

The promising realm of 4D printing, with its potential to eliminate numerous industrial barriers and inspire innovations, is nonetheless faced with significant challenges. Although the technique has numerous advantages, it is now in the proof-of-concept phase and requires addressing various challenges related to input materials, stimulus methods, design/simulations, and printing technology [118]. The challenges outlined below underscore the need for extensive research to overcome obstacles and propel 4D printing into broader practical applications.

1. **Intelligent Material Formulation:** The formulation of intelligent materials is a central challenge in 4D printing. 4D-printed components generally comprise both rigid and flexible sections that possess different capabilities for movement. The rigid segments prioritize superior mechanical qualities, such as rigidity, but frequently lack compatibility and biodegradability. Soft materials offer compatibility suitable for 4D printing but pose challenges in determining their mechanical characteristics. Achieving a balance and enhancing the availability, feasibility, and combined properties of 4D materials are critical challenges, especially considering the diverse requirements of different printing techniques [118, 119].
2. **Material Functionality and Categorization:** The challenge lies in making materials more functional and categorized, tailored to various 4D printing techniques. For instance, materials used in direct ink writing (DIW) must be shape-shifting and capable of reversible transformations. Shape memory polymers (SMPs) and shape memory polymer composites (SMPCs) require programming for each actuation cycle,

limiting their application in soft robotics and mechanical programming. The development of new SMPCs or hybrid SMPs for two-way triggered or interchangeable transformations is a future avenue for exploration [118, 119].

3. **Temperature Challenges in Liquid Crystal Elastomers (LCEs):** In the case of LCEs, challenges arise from shifting temperature (TNI) and temperature distribution discrepancies. The nematic phase, preferred for mesogen alignment in printing, contradicts the higher printing temperature required for a smooth process. Additionally, pure LCEs exhibit higher work capacity and blocking stress compared to the printed part, posing challenges for achieving consistent and desired properties in printed LCE components [120, 121].
4. **Hydrogel Printing Challenges:** Hydrogel 4D printing is crucial in tissue engineering due to its reversible shifting ability in response to water. However, challenges arise in achieving effective actuation without the necessity for a liquid medium. Material selection becomes critical to influence the activation, shape alteration, and mechanical properties of the end part. Materials must also exhibit compatibility with the environment, ensuring resistance to factors such as corrosion, chemicals, humidity, and wear, especially in medical applications [122].
5. **Limited Stimulus Responsiveness:** Many 4D printing materials are responsive to a single stimulus, limiting their versatility. Developing materials that react to various stimuli is complex, and mechanisms for real-time applications, such as controlled heating or constant electricity/magnetic fields over extended life cycles, are yet to be fully realized. Applications involving living cells, where extreme pH values and high temperatures should be avoided, add an additional layer of complexity.
6. **Lack of Reliable Computational Models:** There is a lack of reliable computational models that can accurately predict the improvements in 4D printing. The current design process relies heavily on experimental trials and empirical data, with insufficient control over varying phases of deformation. Efforts are being made to develop finite element simulations and machine learning approaches to track and forecast shape transformations more efficiently.

7. **Printed Structure Quality and Post-Processing Challenges:** Achieving desirable structural, mechanical, and shape memory behavior against stimulation depends on suitable 3D printing processes and materials. However, no standardized parameters exist for material and process combinations to obtain defect-free end objects. While layer-by-layer printing offers an even surface finish, intricate geometries suffer from significant surface roughness, requiring post-processing like machining. Balancing accuracy and surface quality throughout the process is essential for optimal shape-morphing performance. Machine capabilities, build chamber temperature, part design, and printing time are interconnected factors influencing dimensional stability and must be optimized for low- tolerance fabrication [123].
8. **Post-Processing and Industrial Limitations:** Post-processing, which involves the removal of support structures, machining, and coating, is frequently necessary because of the limits of design and 3D printing machines. This leads to higher prices and decreased efficiency. In addition, 3D printing is limited by factors such as the size of the object that can be printed, the time it takes to print, the specific design specifications, and the overall cost for large-scale manufacture. Although micro-/nano printing shows potential in tackling these difficulties, enhancements are necessary to guarantee precise and meticulous printing for seamless and efficient stimulation in advanced applications. Overcoming these challenges is crucial for the continued development and widespread adoption of 4D printing technologies across diverse industries. Researchers and practitioners must collaboratively address these issues to unlock the full potential of this transformative manufacturing approach.

7.6 Future Scope of 4D Printing

The emergence of 4D printing has sparked significant enthusiasm within the research community, opening up possibilities for the creation of dynamic and highly complex structures that can undergo intricate transformations. 4D printing incorporates the use of 3D printing and intelligent materials to bring a new level of responsiveness to external influences. This allows for alterations in geometry, dimension, characteristics, and functioning.

Notwithstanding the advancements achieved, certain obstacles and opportunities for enhancement influence the future potential of 4D printing.

1. **Progress in Printing Technology:** The advancement of appropriate and enduring 3D printing techniques remains essential for the progress of 4D printing. To overcome the restrictions of existing techniques, such as the limited level of detail achievable with fused deposition modeling (FDM) and direct ink writing (DIW), it is necessary to explore new possibilities for manufacturing that offer faster production, superior surface quality, and the ability to work with many materials. Possible future advancements could entail integrating diverse 3D printing methodologies into a unified additive manufacturing (AM) configuration to accomplish multifaceted creation.
2. **Stimuli-Responsive Materials Development:** The development of stimuli-responsive materials is essential for the advancement of 4D printing. Continued research should prioritize the advancement of indirect stimulation polymers, hydrogels, elastomers, and composites specifically designed for 4D printing applications. These materials should have a capacity to react to magnetic, electrical, and light stimuli, allowing for swift implementation in practical situations. Comprehending polymer chemistry, materials science, and 3D printing techniques is essential for maximizing the advancement of these adaptable materials.
3. **Functional Material Characteristics:** Future research should focus on investigating the functional properties of materials used in 4D printing. Factors to be taken into account encompass the compatibility with living organisms, the ability to break down naturally, the ability to be reprogrammed, the efficiency in energy consumption, the quick and reversible transition between different states, and the responsiveness to magnetic, electric, and light stimuli in a directional manner. Attaining a one-step procedure for conceptual design is crucial, necessitating a more profound comprehension of material properties and performance criteria.
4. **Standardization of Testing and Descriptions:** The current state lacks standardized protocols for assessing, specifying, and applying dimensions for 4D printing structures. It is

crucial to establish standardized methods or processes for the surveillance and adjustment of 4D-printed components. The software developed by MIT assembly lab to detect and analyze morphological changes in dynamic 4D printed components shows promise, highlighting the necessity for further development of characterization and analytic tools.

5. **Integration of 5D and 6D Printing Concepts:** The incorporation of 5D and 6D printing concepts is now being explored, which involves the utilization of sophisticated printing techniques. 6D printing is a new concept that combines the concepts of 5D printing, which involves the movement of the printing head and rotation of the print bed at specific angles, with the principles of 4D printing, which involves the use of time-modified smart materials (SMs). This innovative approach has the potential to improve the control and manipulation of printed objects in various dimensions.

The advancement of 4D printing relies heavily on interdisciplinary research in materials science, polymer chemistry, and additive manufacturing methods. The collective endeavors spanning these areas will stimulate creativity, broadening the range of uses for 4D printing in various sectors, such as healthcare, production, and technology. 4D printing's dynamic characteristics offer the potential for a future in which responsive and adaptable structures are seamlessly integrated into our technological environment.

7.7 Conclusion

The exploration of micro/nano scale 4D printing in this chapter unravels the transformative potential of additive manufacturing at the minutest scales. While the macro-scale advancements in 4D printing have garnered significant attention, the micro/nano scale is a burgeoning frontier, laden with challenges and opportunities. The conceptual leap from 3D printing to 4D printing has redefined the landscape of advanced manufacturing, introducing the dynamic dimension of self- transformation in printed structures. The shift toward micro/nano scale 4D printing signifies a nascent yet promising domain that holds the key to unlocking unprecedented possibilities in diverse applications. Our comprehensive examination of various micro/nano scale 4D printing technologies has shed light on cutting-edge methodologies, from hydrogel-based miniaturization printing to laser transfer micro additive manufacturing. Each technique brings unique capabilities and limitations to the fore, providing a nuanced understanding of their potential contributions to the micro/nano fabrication realm. As micro/nano scale 4D printing continues to evolve, addressing challenges in fabricating structures with features at the micro/nano scale, it holds the promise of revolutionizing industries where precision, adaptability, and miniaturization are paramount. The journey from concept to application in this nascent field is marked by rapid advancements, laying the groundwork for a future where micro/nano scale 4D printing becomes an integral facet of advanced manufacturing.

References

1. Leist, S.K. and Zhou, J., Current status of 4D printing technology and the potential of light-reactive smart materials as 4D printable materials. *Virtual Phys. Prototyp.*, 11, 249–262, 2016.
2. Rastogi, P. and Kandasubramanian, B., Breakthrough in the printing tactics for stimuli-responsive materials: 4D printing. *Chem. Eng. J.*, 366, 264–304, 2019.
3. Oladapo, B.I., Oshin, E.A., Olawumi, A.M., Nanostructural computation of 4D printing carboxymethylcellulose (CMC) composite. *Nano-Struct. Nano-Objects*, 21, 100423, 2020.
4. Liu, T., Zhou, T., Yao, Y., Zhang, F., Liu, L., Liu, Y., Leng, J., Stimulus methods of multi-functional shape memory polymer nanocomposites: A review. *Compos. Part A Appl. Sci. Manuf.*, 100, 20–30, 2017.
5. Yang, C., Boorugu, M., Dopp, A., Ren, J., Martin, R., Han, D., Choi, W., Lee, H., 4D printing reconfigurable, deployable and mechanically tunable metamaterials. *Mater. Horiz.*, 6, 1244–1250, 2019.
6. Romasanta, L.J., Lopez-Manchado, M.A., Verdejo, R., Increasing the performance of dielectric elastomer actuators: A review from the materials perspective. *Prog. Polym. Sci.*, 51, 188–211, 2015.
7. Leng, J.S., Huang, W.M., Lan, X., Liu, Y.J., Du, S.Y., Significantly reducing electrical resistivity by forming conductive Ni chains in a polyurethane shape-memory polymer/carbon-black composite. *Appl. Phys. Lett.*, 92, 20, 204101, 2008.
8. Brochu, P. and Pei, Q., Dielectric elastomers for actuators and artificial muscles. *Electroact. Polym. Mater.*, 9781461408, 1–56, 2012.
9. Wu, S., Hu, W., Ze, Q., Sitti, M., Zhao, R., Multifunctional magnetic soft composites: A review. *Multifunct. Mater.*, 3, 4, 042003, 2020.

10. Ze, Q., Kuang, X., Wu, S., Wong, J., Montgomery, S.M., Zhang, R., Kovitz, J.M., Yang, F., Qi, H.J., Zhao, R., Magnetic Shape Memory Polymers with Integrated Multifunctional Shape Manipulation. *Adv. Mater.*, 32, 1–8, 2020.
11. Testa, P., Style, R.W., Cui, J., Donnelly, C., Borisova, E., Derlet, P.M., Dufresne, E.R., Heyderman, L.J., Magnetically Addressable Shape-Memory and Stiffening in a Composite Elastomer. *Adv. Mater.*, 31, 29, 1900561, 2019.
12. Lee, Y.W., Ceylan, H., Yasa, I.C., Kilic, U., Sitti, M., 3D-Printed Multi-Stimuli- Responsive Mobile Micromachines. *ACS Appl. Mater. Interfaces*, 13, 11, 12759–12766, 2020.
13. Yarali, E., Baniasadi, M., Bodaghi, M., Baghani, M., 3D constitutive modeling of electro-magneto-visco- hyperelastic elastomers: A semi-analytical solution for cylinders under large torsion-extension deformation. *Smart Mater. Struct.*, 29, 8, 085031, 2020.
14. Baniasadi, M., Yarali, E., Bodaghi, M., Zolfagharian, A., Baghani, M., Constitutive modeling of multi- stimuli-responsive shape memory polymers with multi-functional capabilities. *Int. J. Mech. Sci.*, 192, 106082, 2021.
15. Subash, A. and Kandasubramanian, B., 4D printing of shape memory polymers. *Eur. Polym. J.*, 134, 109771, 2020.
16. Jin, B., Song, H., Jiang, R., Song, J., Zhao, Q. and Xie, T., Programming a crystalline shape memory polymer network with thermo-and photo-reversible bonds toward a single-component soft robot. *Sci. Adv.*, 4, 1, eaao3865, 2018.
17. Zarek, M., Layani, M., Cooperstein, I., Sachyani, E., Cohn, D., Magdassi, S., 3D Printing of Shape Memory Polymers for Flexible Electronic Devices. *Adv. Mater.*, 28, 4449–4454, 2016.
18. del Barrio, J. and S´anchez-Somolinos, C., Light to Shape the Future: From Photolithography to 4D Printing. *Adv. Opt. Mater.*, 7, 1900598, 2019.
19. Spiegel, C.A., Hippler, M., Münchinger, A., Bastmeyer, M., Barner-Kowollik, C., Wegener, M., Blasco, E., 4D Printing at the Microscale. *Adv. Funct. Mater.*, 30, 1907615, 2020.
20. Wan, X., Luo, L., Liu, Y., Leng, J., Direct Ink Writing Based 4D Printing of Materials and Their Applications. *Adv. Sci.*, 7, 2001000, 2020.
21. Rastogi, P., Gharde, S., Kandasubramanian, B., *Thermal Effects in 3D Printed Parts*, pp. 43– 68, Springer, 2020.
22. Wu, W., Hu, W., Qian, G., Liao, H., Xu, X., Berto, F., Mechanical design and multifunctional applications of chiral mechanical metamaterials: A review. *Mater. Des.*, 180, 107950, 2019.
23. Safarabadi, M., Haghighi-Yazdi, M., Sorkhi, M.A., Yousefi, A., Experimental and numerical study of buckling behavior of foam-filled honeycomb core sandwich panels considering viscoelastic effects. *J. Sandw. Struct. Mater.*, 23, 3985–4015, 2021.
24. Ge, L., Zheng, H., Li, H., Liu, B., Su, H., Fang, D., Compression behavior of a novel sandwich structure with bi-directional corrugated core. *Thin Walled Struct.*, 161, 107413, 2021.

25. Serjouei, A., Yousefi, A., Jenaki, A., Bodaghi, M., Mehrpouya, M., 4D printed shape memory sandwich structures: experimental analysis and numerical modeling. *Smart Mater. Struct.*, 31, 5, 055014, 2022.
26. Ji, Q., Chen, M., Wang, X.V., Wang, L., Feng, L., Optimal shape morphing control of 4D printed shape memory polymer based on reinforcement learning. *Robot. Comput. Integr. Manuf.*, 73, 102209, 2022.
27. Zolfagharian, A., Durran, L., Gharaie, S., Rolfe, B., Kaynak, A., Bodaghi, M., 4D printing soft robots guided by machine learning and finite element models. *Sens. Actuators A Phys.*, 328, 112774, 2021.
28. Samal, S., Kosjakova, O., Vokoun, D., Stachiv, I., Shape Memory Behaviour of PMMA-Coated NiTi Alloy under Thermal Cycle. *Polym. (Basel)*, 14, 2932, 2022.
29. Nguyen, T.T. and Kim, J., 4D-Printing — Fused Deposition Modeling Printing and PolyJet Printing with Shape Memory Polymers Composite. *Fibers Polym.*, 21, 2364–2372, 2020.
30. Shao, L.H., Zhao, B., Zhang, Q., Xing, Y., Zhang, K., 4D printing composite with electrically controlled local deformation. *Extrem. Mech. Lett.*, 39, 100793, 2020.
31. Wang, Y. and Li, X., 4D printing reversible actuator with strain self-sensing function via structural design. *Compos. Part B Eng.*, 211, 108644, 2021.
32. Liu, Y., Zhang, F., Leng, J., Fu, K., Lu, X.L., Wang, L., Cotton, C., Sun, B., Gu, B., Chou, T., Remotely and Sequentially Controlled Actuation of Electroactivated Carbon Nanotube/Shape Memory Polymer Composites. *Adv. Mater. Technol.*, 4, 1900600, 2019.
33. Cortés, A., Cosola, A., Sangermano, M., Campo, M., González Prolongo, S., Pirri, C.F., Jiménez-Suárez, A., Chiappone, A., Dlp, 4D-Printing of Remotely, Modularly, and Selectively Controllable Shape Memory Polymer Nanocomposites Embedding Carbon Nanotubes. *Adv. Funct. Mater.*, 31, 2106774, 2021.
34. Wei, H., Cauchy, X., Navas, I.O., Abderrafai, Y., Chizari, K., Sundararaj, U., Liu, Y., Leng, J., Therriault, D., Direct 3D Printing of Hybrid Nanofiber-Based Nanocomposites for Highly Conductive and Shape Memory Applications. *ACS Appl. Mater. Interfaces*, 11, 24523–24532, 2019.
35. Khan, S.A. and Lazoglu, I., Development of additively manufacturable and electrically conductive graphite–polymer composites. *Prog. Addit. Manuf.*, 5, 153–162, 2020.
36. Dong, K., Panahi-Sarmad, M., Cui, Z., Huang, X., Xiao, X., Electro-induced shape memory effect of 4D printed auxetic composite using PLA/TPU/CNT filament embedded synergistically with continuous carbon fiber: A theoretical & experimental analysis. *Compos. Part B Eng.*, 220, 108994, 2021.
37. Kang, M., Pyo, Y., Jang, J.Y., Park, Y., Son, Y.-H., MyungChan Choi, J.W., W. Chang, Y.-, Lee, C.S., Design of a shape memory composite(SMC) using 4D printing technology. *Sens. Actuators A Phys.*, 283, 187–195, 2018.

38. Ula, S.W., Traugutt, N.A., Volpe, R.H., Patel, R.R., Yu, K., Yakacki, C.M., Liquid crystal elastomers: an introduction and review of emerging technologies. *Liq. Cryst. Rev.*, 6, 78–107, 2018.
39. Haghiashtiani, G., Habtour, E., Park, S.-H., Gardea, F., McAlpine, M.C., 3D printed electrically-driven soft actuators. *Extrem. Mech. Lett.*, 21, 1–8, 2018.
40. Ambulo, C.P., Ford, M.J., Searles, K., Majidi, C., Ware, T.H., 4D-Printable Liquid Metal-Liquid Crystal Elastomer Composites. *ACS Appl. Mater. Interfaces*, 13, 12805–12813, 2021.
41. Le Duigou, A., Castro, M., Bevan, R., Martin, N., 3D printing of wood fibre biocomposites: From mechanical to actuation functionality. *Mater. Des.*, 96, 106–114, 2016.
42. Mohr, R., Kratz, K., Weigel, T., Lucka-Gabor, M., Moneke, M., Lendlein, A., Initiation of shape-memory effect by inductive heating of magnetic nanoparticles in thermoplastic polymers. *Proc. Natl. Acad. Sci.*, 103, 3540–3545, 2006.
43. Wan, X., He, Y., Liu, Y., Leng, J., 4D printing of multiple shape memory polymer and nanocomposites with biocompatible, programmable and selectively actuated properties. *Addit. Manuf.*, 53, 102689, 2022.
44. Tottori, S., Zhang, L., Qiu, F., Krawczyk, K.K., Franco-Obregón, A., Nelson, B.J., Magnetic helical micromachines: Fabrication, controlled swimming, and cargo transport. *Adv. Mater.*, 24, 6, 811–816, 2012. doi: https://doi.org/10.1002/adma.201103818.
45. Huang, T.Y., Sakar, M.S., Mao, A., Petruska, A.J., Qiu, F., Chen, X.B., Nelson, B.J., 3D printed microtransporters: Compound micromachines for spatiotemporally controlled delivery of therapeutic agents. *Adv. Mater.*, 27, 42, 6644–6650, 2015.
46. Yasa, I.C., Ceylan, H., Bozuyuk, U., Wild, A.M., Sitti, M., Elucidating the interaction dynamics between microswimmer body and immune system for medical microrobots. *Sci. Rob.*, 5, 43, eaaz3867, 2020.
47. Qiu, F., Zhang, L., Peyer, K.E., Casarosa, M., Franco-Obregón, A., Choi, H., Nelson, B.J., Noncytotoxic artificial bacterial flagella fabricated from biocompatible ORMOCOMP and iron coating. *J. Mater. Chem. B*, 2, 4, 357–362, 2014.
48. Villangca, M.J., Palima, D., Banas, A.R., Glückstad, J., Light-driven microtool equipped with a syringe function. *Light Sci. Appl.*, 5, 9, e16148–e16148, 2016.
49. Lee, S., *et al.*, A needle-type microrobot for targeted drug delivery by affixing to a microtissue. *Adv. Healthc. Mater.*, 9, 7, 1901697, 2020.
50. Kim, S., *et al.*, Fabrication and characterization of magnetic microrobots for three-dimensional cell culture and targeted transportation. *Adv. Mater.*, 25, 41, 5863–5868, 2013.
51. Yang, L., *et al.*, Targeted single-cell therapeutics with magnetic tubular micromotor by one-step exposure of structured femtosecond optical vortices. *Adv. Funct. Mater.*, 29, 45, 1905745, 2019.

52. Zhang, W., Wang, H., Wang, H., Chan, J.Y.E., Liu, H., Zhang, B., Yang, J.K., Structural multi-colour invisible inks with submicron 4D Printing of shape memory polymers. *Nat. Commun., 12*, 1, 112, 2021.
53. Spiegel, C.A., Hackner, M., Bothe, V.P., Spatz, J.P., Blasco, E., 4D Printing of shape memory polymers: from macro to micro. *Adv. Funct. Mater., 32*, 51, 2110580, 2022.
54. Li, D., Liu, Y., Yang, Y., Shen, Y., A fast and powerful swimming microrobot with a serrated tail enhanced propulsion interface. *Nanoscale, 10*, 42, 19673–19677, 2018.
55. Deng, C., Liu, Y., Fan, X., Jiao, B., Zhang, Z., Zhang, M., Xiong, W., Femtosecond Laser 4D Printing of Light-Driven Intelligent Micromachines. *Adv. Funct. Mater., 33*, 11, 2211473, 2023.
56. Guo, Y., Zhang, J., Hu, W., Khan, M.T.A., Sitti, M., Shape-programmable liquid crystal elastomer structures with arbitrary three-dimensional director fields and geometries. *Nat. Commun., 12*, 1, 5936, 2021.
57. Li, D., Liu, C., Yang, Y., Wang, L., Shen, Y., Micro-rocket robot with all-optic actuating and tracking in blood. *Light Sci. Appl., 9*, 1, 84, 2020.
58. Chen, L., Dong, Y., Tang, C.Y., Zhong, L., Law, W.C., Tsui, G.C., Xie, X., Development of direct-laser-printable light-powered nanocomposites. *ACS Appl. Mater. Interfaces, 11*, 21, 19541–19553, 2019.
59. Xin, C., Yang, L., Li, J., Hu, Y., Qian, D., Fan, S., Chu, J., Conical hollow microhelices with superior swimming capabilities for targeted cargo delivery. *Adv. Mater., 31*, 25, 1808226, 2019.
60. Wang, J.Y., Jin, F., Dong, X.Z., Liu, J., Zheng, M.L., Flytrap inspired pH-driven 3D hydrogel actuator by femtosecond laser microfabrication. *Adv. Mater. Technol., 7*, 8, 2200276, 2022.
61. Wang, X., Qin, X.-H., Hu, C., Terzopoulou, A., Chen, X.-Z., Huang, T.-Y., Maniura-Weber, K., Pané, S., Nelson, B.J., 3D Printed Enzymatically Biodegradable Soft Helical Microswimmers. *Adv. Funct. Mater.*, 28, 1804107, 2018.
62. Lin, Y. and Xu, J., Microstructures fabricated by two-photon polymerization and their remote manipulation techniques: toward 3D printing of micromachines. *Adv. Opt. Mater.*, 6, 8, 1701359, 2018.
63. Xing, J.-F., Zheng, M.-L., Duan, X.-M., Two-photon polymerization microfabrication of hydrogels: an advanced 3D printing technology for tissue engineering and drug delivery. *Chem. Soc. Rev.*, 44, 5031, 2015.
64. Rad, Z.F., Prewett, P.D., Davies, G.J., High-resolution two-photon polymerization: the most versatile technique for the fabrication of microneedle arrays. *Microsyst. Nanoeng.*, 7, 1, 71, 2021.
65. Li, J.H. and Pumera, M., 3D printing of functional microrobots. *Chem. Soc. Rev.*, 50, 2794–838, 2021.

66. Weiß, T., Hildebrand, G., Schade, R., Liefeith, K., Two-photon polymerization for microfabrication of three-dimensional scaffolds for tissue engineering application. *Eng. Life Sci.*, 9, 384–90, 2009.
67. Jian, B., Li, H., He, X., Wang, R., Yang, H.Y., Ge, Q., Two-Photon Polymerization-Based 4D Printing and Its Applications. *Int. J. Extreme Manuf.*, 6, 1, 012001, 2024.
68. Zolfagharian, A., Denk, M., Bodaghi, M., Kouzani, A.Z., Kaynak, A., Topology- Optimized 4D Printing of a Soft Actuator. *Acta Mech. Solida Sin.*, 33, 418–430, 2020.
69. Nath, U., Crawford, B.C.W., Carpenter, R., Coen, E., Genetic Control of Surface Curvature. *Science*, 299, 5611, 1404–1407, 2003.
70. Hajiesmaili, E. and Clarke, D.R., Reconfigurable shape-morphing dielectric elastomers using spatially varying electric fields. *Nat. Commun.*, 10, 1, 183, 2019.
71. Guo, H., Cheng, J., Yang, K., Demella, K., Li, T., Raghavan, S.R., Nie, Z., Programming the Shape Transformation of a Composite Hydrogel Sheet via Erasable and Rewritable Nanoparticle Patterns. *ACS Appl. Mater. Interfaces*, 11, 45, 42654–42660, 2019.
72. Zhang, Y.S. and Khademhosseini, A., Advances in engineering hydrogels. *Science*, 356, 6337, eaaf3627, 2017.
73. Choi, M.-Y., Shin, Y., Lee, H.S., Kim, S.Y., Na, J.-H., Multipolar spatial electric field modulation for freeform electroactive hydrogel actuation. *Sci. Rep.*, 10, 2482, 2020.
74. Liu, J.-C., Evans, B.A., Tracy, J.B., Photothermally Reconfigurable Shape Memory Magnetic Cilia. *Adv. Mater. Technol.*, 5, 7, 2000147, 2020.
75. Lum, G.Z., Ye, Z., Dong, X., Marvi, H., Erin, O., Hu, W., Sitti, M., Shape-programmable magnetic soft matter. *Proc. Natl. Acad. Sci. U.S.A.*, 113, E6007–E6015, 2016.
76. Ren, Z., Hu, W., Dong, X., Sitti, M., Multi-functional soft-bodied jellyfish-like swimming. *Nat. Commun.*, 10, 1, 2703, 2019.
77. Yu, X., Xie, Z., Yu, Y., Lee, J., Vazquez-Guardado, A., Luan, H., Ruban, J., Ning, X., Akhtar, A., Li, D., Ji, B., Liu, Y., Sun, R., Cao, J., Huo, Q., Zhong, Y., Lee, C.M., Kim, S.Y., Gutruf, P., Zhang, C., Xue, Y., Guo, Q., Chempakasseril, A., Tian, P., Lu, W., Jeong, J.Y., Yu, Y.J., Cornman, J., Tan, C.S., Kim, B.H., Lee, K.H., Feng, X., Huang, Y., Rogers, J.A., Skin-integrated wireless haptic interfaces for virtual and augmented reality. *Nature*, 575, 473–479, 2019.
78. Tang, J., Yin, Q., Qiao, Y., Wang, T., Shape Morphing of Hydrogels in Alternating Magnetic Field. *ACS Appl. Mater. Interfaces*, 11, 21194–21200, 2019.
79. Kokkinis, D., Schaffner, M., Studart, A.R., Multimaterial magnetically assisted 3D printing of composite materials. *Nat. Commun.*, 6, 8643, 2015.
80. Zolfagharian, A., Kaynak, A., Bodaghi, M., Kouzani, A.Z., Gharaie, S., Nahavandi, S., Control-based 4D printing: Adaptive 4D-printed systems. *Appl. Sci.*, 10, 9, 3020, 2020.

81. Chen, F., Liu, K., Zhu, X., Buckling-induced Shape Morphing using Dielectric Elastomer Actuators Patterned with Spatially-varying Electrodes, in: *IEEE Int. Conf. Intell. Robot. Syst*, pp. 8306–8311, 2019.
82. Davidson, Z.S., Shahsavan, H., Aghakhani, A., Guo, Y., Hines, L., Xia, Y., Yang, S., Sitti, M., Monolithic shape-programmable dielectric liquid crystal elastomer actuators. *Sci. Adv.*, 5, 11, eaay0855, 2019.
83. Hajiesmaili, E., Khare, E., Chortos, A., Lewis, J., Clarke, D.R., Voltage-controlled morphing of dielectric elastomer circular sheets into conical surfaces. *Extrem. Mech. Lett.*, 30, 100504, 2019.
84. Zolfagharian, A., Kouzani, A.Z., Khoo, S.Y., Moghadam, A.A.A., Gibson, I., Kaynak, A., Evolution of 3D printed soft actuators. *Sens. Actuators A Phys.*, 250, 258–272, 2016.
85. Yee Ling, Y., Swee Leong, S., Wai Yee, Y., A review of 3D printing processes and materials for soft robotics. *Rapid Prototyp. J.*, 26, 8, 1345–1361, 2020.
86. Xu, T., Zhang, J., Salehizadeh, M., Onaizah, O., Diller, E., Millimeter-scale flexible robots with programmable three-dimensional magnetization and motions. *Sci. Robot.*, 4, 2019.
87. Li, X., Zhang, Z., Sun, M., Wu, H., Zhou, Y., Wu, H., Jiang, S., A magneto-active soft gripper with adaptive and controllable motion. *Smart Mater. Struct.*, 30, 1, 015024, 2021.
88. Cao, X., Xuan, S., Sun, S., Xu, Z., Li, J., Gong, X., 3D Printing Magnetic Actuators for Biomimetic Applications. *ACS Appl. Mater. Interfaces*, 13, 30127–30136, 2021.
89. Wang, X., Mao, G., Ge, J., Drack, M., Cañón Bermúdez, G.S., Wirthl, D., Illing, R., Kosub, T., Bischoff, L., Wang, C., Fassbender, J., Kaltenbrunner, M., Makarov, D., Untethered and ultrafast soft-bodied robots. *Commun. Mater.*, 1, 67, 2020.
90. Gao, X., Cao, C., Guo, J., Conn, A., Elastic Electroadhesion with Rapid Release by Integrated Resonant Vibration. *Adv. Mater. Technol.*, 4, 1, 1800378, 2019.
91. Cacucciolo, V., Shintake, J., Kuwajima, Y., Maeda, S., Floreano, D., Shea, H., Stretchable pumps for soft machines. *Nature*, 572, 516–519, 2019.
92. Shintake, J., Rosset, S., Schubert, B., Floreano, D., Shea, H., Versatile Soft Grippers with Intrinsic Electroadhesion Based on Multifunctional Polymer Actuators. *Adv. Mater.*, 28, 231–238, 2016.
93. Yarali, E., Noroozi, R., Yousefi, A., Bodaghi, M., Baghani, M., Multi-trigger thermo-electro-mechanical soft actuators under large deformations. *Polym. (Basel)*, 12, 2, 489, 2020.
94. Rothemund, P., Kellaris, N., Mitchell, S.K., Acome, E., Keplinger, C., HASEL Artificial Muscles for a New Generation of Lifelike Robots—Recent Progress and Future Opportunities. *Adv. Mater.*, 33, 19, 2003375, 2021.
95. Wang, X., Mitchell, S.K., Rumley, E.H., Rothemund, P., Keplinger, C., High-Strain Peano-HASEL Actuators. *Adv. Funct. Mater.*, 30, 1908821, 2020.

96. Bodaghi, M., Damanpack, A.R., Liao, W.H., Self-expanding/shrinking structures by 4D printing. *Smart Mater. Struct.*, 25, 10, 105034, 2016.
97. Zolfagharian, A., Kouzani, A.Z., Khoo, S.Y., Noshadi, A., Kaynak, A., 3D printed soft parallel actuator. *Smart Mater. Struct.*, 27, 4, 045019, 2018.
98. Park, J.M., Kim, S.J., Jang, J.H., Wang, Z., Kim, P.G., Yoon, D.J., Kim, J., Hansen, G., DeVries, K.L., Actuation of electrochemical, electro-magnetic, and electro-active actuators for carbon nanofiber and Ni nanowire reinforced polymer composites. *Compos. Part B Eng.*, 39, 1161–1169, 2008.
99. Yuk, H., Lin, S., Ma, C., Takaffoli, M., Fang, N.X., Zhao, X., Hydraulic hydrogel actuators and robots optically and sonically camouflaged in water. *Nat. Commun.*, 8, 14230, 2017.
100. Palleau, E., Morales, D., Dickey, M.D., Velev, O.D., Reversible patterning and actuation of hydrogels by electrically assisted ionoprinting. *Nat. Commun.*, 4, 2257, 2013.
101. Ambulo, C.P., Burroughs, J.J., Boothby, J.M., Kim, H., Shankar, M.R., Ware, T.H., Four-dimensional Printing of Liquid Crystal Elastomers. *ACS Appl. Mater. Interfaces*, 9, 37332–37339, 2017.
102. Kotikian, A., Truby, R.L., Boley, J.W., White, T.J., Lewis, J.A., 3D Printing of Liquid Crystal Elastomeric Actuators with Spatially Programed Nematic Order. *Adv. Mater.*, 30, 1706164, 2018.
103. Sultana, N., *Biodegradable Polymer-Based Scaffolds for Bone Tissue Engineering*, Springer, Berlin Heidelberg, Berlin, Heidelberg, 2013.
104. Roberts, T.T. and Rosenbaum, A.J., Bone grafts, bone substitutes and orthobiologics the bridge between basic science and clinical advancements in fracture healing. *Organogenesis*, 8, 114–124, 2012.
105. Ahadian, S., Rahal, R., Ramón-Azcón, J., Obregón, R., Hasan, A., Biomaterials in Tissue Engineering. *Tissue Eng. Artif. Organs Regen. Med. Smart Diagn. Pers. Med.*, 1–2, 35–83, 2016.
106. Ribeiro, S., Ribeiro, C., Carvalho, E.O., Tubio, C.R., Castro, N., Pereira, N., Correia, V., Gomes, A.C., Lanceros-M´endez, S., Magnetically Activated Electroactive Microenvironments for Skeletal Muscle Tissue Regeneration. *ACS Appl. Bio Mater.*, 3, 4239–4252, 2020.
107. Ahadian, S., Obregón, R., Ramón-Azcón, J., Salazar, G., Shiku, H., Ramalingam, M., Matsue, T., Carbon nanotubes and graphene-based nanomaterials for stem cell differentiation and tissue regeneration. *J. Nanosci. Nanotechnol.*, 16, 8862–8880, 2016.
108. Ashtari, K., Nazari, H., Ko, H., Tebon, P., Akhshik, M., Akbari, M., Alhosseini, S.N., Mozafari, M., Mehravi, B., Soleimani, M., Ardehali, R., Ebrahimi Warkiani, M., Ahadian, S., Khademhosseini, A., Electrically conductive nanomaterials for cardiac tissue engineering. *Adv. Drug Deliv. Rev.*, 144, 162–179, 2019.

109. Wang, H., Totaro, M., Blandin, A.A., Beccai, L., A wireless inductive sensing technology for soft pneumatic actuators using magnetorheological elastomers, in: *RoboSoft 2019 - 2019 IEEE Int. Conf. Soft Robot*, pp. 242–248, 2019, https://doi.org/10.1109/ROBOSOFT.2019.8722800.
110. Filippi, M., Dasen, B., Guerrero, J., Garello, F., Isu, G., Born, G., Ehrbar, M., Martin, I., Scherberich, A., Magnetic nanocomposite hydrogels and static magnetic field stimulate the osteoblastic and vasculogenic profile of adipose-derived cells. *Biomaterials*, 223, 119468, 2019.
111. Mirani, B., Pagan, E., Currie, B., Siddiqui, M.A., Hosseinzadeh, R., Mostafalu, P., Zhang, Y.S., Ghahary, A., Akbari, M., An Advanced Multifunctional Hydrogel-Based Dressing for Wound Monitoring and Drug Delivery. *Adv. Healthc. Mater.*, 6, 1700718, 2017.
112. Xie, Z., Du, Q., Wu, Y., Hao, X., Liu, C., Full-band UV shielding and highly daylight luminescent silane- functionalized graphene quantum dot nanofluids and their arbitrary polymerized hybrid gel glasses. *J. Mater. Chem. C*, 4, 9879–9886, 2016.
113. Castro-Camus, E., Koch, M., Hernandez-Serrano, A.I., Additive manufacture of photonic components for the terahertz band. *J. Appl. Phys.*, 127, 21, 210901, 2020.
114. Landy, N. and Smith, D.R., A full-parameter unidirectional metamaterial cloak for microwaves. *Nat. Mater.*, 12, 25–28, 2013.
115. Qi, K., Li, L., Su, J., Liu, Y., Chen, J., Design of a Tunable Absorber Based on Active Frequency- Selective Surface for UHF Applications. *Mater. (Basel)*, 12, 3989, 2019.
116. Yang, D., Mei, H., Yao, L., Yang, W., Yao, Y., Cheng, L., Zhang, L., Dassios, K.G., 3D/4D printed tunable electrical metamaterials with more sophisticated structures. *J. Mater. Chem. C*, 9, 12010–12036, 2021.
117. Menon, A.V., Madras, G., Bose, S., Shape memory polyurethane nanocomposites with porous architectures for enhanced microwave shielding. *Chem. Eng. J.*, 352, 590–600, 2018.
118. Zhang, Z., Demir, K.G., Gu, G.X., Developments in 4D-printing: a review on current smart materials, technologies, and applications. *Int. J. Smart Nano Mater.*, 10, 205–224, 2019.
119. Tumbleston, J.R., Shirvanyants, D., Ermoshkin, N., Janusziewicz, R., Johnson, A.R., Kelly, D., Chen, K., Pinschmidt, R., Rolland, J.P., Ermoshkin, A., Samulski, E.T., DeSimone, J.M., Continuous liquid interface production of 3D objects. *Science*, 347, 1349–1352, 2015. (80- .).
120. Kim, H., Boothby, J.M., Ramachandran, S., Lee, C.D., Ware, T.H., Tough, Shape- Changing Materials: Crystallized Liquid Crystal Elastomers. *Macromolecules*, 50, 4267–4275, 2017.
121. Ge, F. and Zhao, Y., Microstructured Actuation of Liquid Crystal Polymer Networks. *Adv. Funct. Mater.*, 30, 2, 1901890, 2020.

122. Pei, E., Loh, G.H., Harrison, D., Almeida, HA., Monzón Verona, M.D., Paz, R., A study of 4D printing and functionally graded additive manufacturing. *Assem. Autom.*, 37, 2, 147–153, 2017.
123. Thompson, M.K., Moroni, G., Vaneker, T., Fadel, G., Campbell, R.I., Gibson, I., Bernard, A., Schulz, J., Graf, P., Ahuja, B., Martina, F., Design for Additive Manufacturing: Trends, opportunities, considerations, and constraints. *CIRP Ann.*, 65, 737–760, 2016.

8

Characterization Techniques for Four Dimensional (4D) Printed Parts

Bijaya Bikram Samal[1]*, Bharat Charan Goud Marupalli[2], Pranabjyoti Talukdar[3], Anita Jena[1], Roja Rani Korrayi[4], Tapasendra Adhikary[2], Shailendra Kumar Varshney[3] and Cheruvu Siva Kumar[5]

[1]*Advanced Technology Development Centre, Indian Institute of Technology, Kharagpur, West Bengal, India*
[2]*Metallurgical and Materials Engineering, Indian Institute of Technology, Kharagpur, West Bengal, India*
[3]*Department of Electronics and Electrical Communication, Indian Institute of Technology, Kharagpur, West Bengal, India*
[4]*Institute of Material Research, Helmholtz-Zentrum Hereon, Geesthacht, Germany*
[5]*Department of Mechanical Engineering, Indian Institute of Technology, Kharagpur, West Bengal, India*

Abstract
Characterization techniques are essential for evaluating the properties and behavior of smart materials involved in four-dimensional (4D) printing. Smart materials possess unique properties that respond to external stimuli such as temperature, pressure, electric fields, or magnetic fields. Characterization helps in understanding these properties thoroughly, including how they change under different conditions. Thermal characterization provides insights into thermal properties, transition temperatures, heat resistance, and thermal conductivity, influencing the performance of 4D printed components. Mechanical characterization techniques, including tensile, compressive, and shear testing, assess the mechanical behavior and recovery of shape memory materials. Surface roughness analysis evaluates the quality and aesthetics of 4D printed parts. Microstructure analysis techniques analyze the internal structures, defects, and surface roughness of printed components. Dimensional accuracy and precision play a crucial role in guaranteeing the fidelity of printed parts to the intended design. Non-destructive testing methods allow for

**Corresponding author*: bijaybikram3991@gmail.com

Bijaya Bikram Samal, Cheruvu Siva Kumar and Shailendra Kumar Varshney (eds.) 4D Printing Technology: Principles, Materials and Applications, (245–280) © 2025 Scrivener Publishing LLC

quality inspection without damaging the parts. Shape memory testing techniques evaluate the response of materials to stimuli. These characterization techniques enable optimization of design parameters, part quality, and performance, driving advancements in four dimesnional additive manufacturing.

Keywords: Characterization techniques, four-dimensional printing, 4D printed parts, additive manufacturing, shape memory materials, part quality, shape memory effect

8.1 Introduction

Among various modern manufacturing processes, additive manufacturing (AM), has transformed the landscape of fabrication with its ability to create very complicated structures in less time without material waste [1]. Among the various AM technologies, four-dimensional (4D) printing, introduced in 2013, has a unique feature of changing the property of the printed structures according to the choice of the user [2]. Due to this dynamic nature of 4D printing, the applications and use cases are very interesting, starting from shape responsive biomedical scaffold to deployable structures in space. These breakthroughs have unlocked a world of possibilities, allowing the manufacturing of structures that not only imitate the intricacy and functioning seen in natural systems [3], but also have the potential to adapt and respond to external fillip over time [4]. The predetermined changes in the shape, structure, or functionality are possible via temperature variations, humidity levels, light exposure, or mechanical stress. The innovative capabilities of 4D printing extend beyond mere creation of intricate geometries with tailored functionalities [5]. This cutting-edge technology transcends traditional manufacturing by imbuing fabricated components with adaptive and autonomous properties. This shift from static and rigid structures to dynamic and responsive materials heralds a new era in design, production, and utilization across diverse sectors such as aerospace, biomedical engineering, robotics, architecture, and consumer products [6].

4D printed structures with stimuli induced variable properties over time, complex material design, and variable mechanism of actuation requires detailed material characterization techniques utilizing various methods and instruments [7]. Comprehending 4D printed components requires analyzing properties such as material characteristics, stress response, heat resistance, and aesthetic qualities. Each of these factors is essential in determining the functionality and reliability of the printed parts. Studying how these components respond to stimulus allows us to forecast their performance [8]. There is also a need to examine surface roughness and visually scrutinize the printed parts to ensure they adhere to required quality requirements. Various post processing techniques can be utilized further to improve the surface finish [9]. For determining the effect of various parameters during printing and postprocess, macroscopic characterization and high-resolution imaging are required. The dimensional stability and accuracy are highly needed; therefore, it needs to be evaluated and also optimized for ensuring consistency and reliability in the layered 3D printed part [10].

Non-destructive testing (NDT) methods is an important method of characterization which can evaluate the integrity and quality of the 4D materials and components without causing any physical damage during the process. These techniques utilize various types of methodologies, including ultrasonic testing, radiography, magnetic particle inspection, and eddy current testing, among others. NDTs help in thorough evaluation of the material properties and can also be used to induce various stimuli, thereby enabling utilization of some other characterization along with it [11]. On the contrary, mechanical testing methods including tensile (pulling force), compression (opposite of tensile), flexural (bending), impact (notched), etc. damages the 4D printed part to study the fracture mechanics of the damage sample through utilization of digital image correlation (DIC) or optical microscopy to analyze the crack generation and its propagation [12]. Thermal analysis techniques like differential scanning calorimetry (DSC), thermogravimetric analysis (TGA), and thermal mechanical analysis (TMA), helps in providing the details about the thermal properties like transition temperature (T_g), melting temperature (T_m), degradation temperature (T_d), coefficient of thermal expansion etc. [13]. The combination of mechanical and thermal properties can be studied in detail utilizing dynamic mechanical analysis (DMA), providing the important information about the creep, storage modulus and loss modulus of a material as a function of temperature, time, or applied force. DMA is particularly useful in characterizing the viscoelastic properties of the smart materials or its composites [14]. Spectroscopy characterization techniques like optical microscopy, including visible, ultraviolet or near infrared helps in study of absorption, scattering or emission of light though the material reveling its chemical composition, molecular structure, and physical properties. Infrared spectroscopy includes fourier transform infrared spectroscopy (FTIR). It works in the infrared area of the electromagnetic spectrum and is used to determine the vibrational modes of molecules in a sample, measuring the energy absorption or emission as required [15]. This absorption pattern reveals the chemical composition, molecular structure, and functional groups present in the material. Other important spectroscopic techniques involve nuclear magnetic resonance (NMR), utilizing magnetic properties of the atomic nuclei to find out the chemical and molecular structure of the material. Another spectroscopic method to find out the molecular structure of the material is mass spectrometry. It is a technique used to analyze the mass-to-charge ratio of ions generated from a sample [16]. X-ray spectroscopy involves the study of the interaction between X-rays and matter to determine the elemental composition and crystal structure of materials. Techniques in this category include X-ray photoelectron spectroscopy (XPS) and X-ray diffraction (XRD). Electron spectroscopy encompasses techniques that use the interaction of electrons with matter to study the electronic structure and properties of materials. Examples include Auger electron spectroscopy (AES) and X-ray photoelectron spectroscopy (XPS) [17]. Raman spectroscopy is a powerful analytical technique used to study the vibrational, rotational, and other low-frequency modes of molecules. It is based on the Raman scattering phenomenon, which occurs when light interacts with a sample and undergoes a change in energy due to molecular vibrations [18]. These various characterization techniques along with its applicability to study the properties of smart materials and 4D printed parts have been described in detail in this chapter.

8.2 Characterization Techniques Overview

8.2.1 Introduction to Characterization and its Importance for 4D Printing

Characterization is a process of analyzing the properties of a material to determine its detailed properties, chemical structures, and behavior at various scales. The methods involved in the characterization provides important information for material selection, process optimization, and quality. These techniques of characterization have widespread applications in various industries like aerospace, automotive, biomedical, electronics, etc. Characterization processes include a broad variety of techniques starting from simple inspection by visualization to complex analytical method like spectroscopy, microscopy and other detailed characterization techniques requiring sophisticated machineries and instruments [19]. Characterization methods have been quite useful for the engineers and researchers for determining the material properties and various attributes and take informed decisions regarding design, fabrication, effect of process parameters and its various applications. Characterization is also very crucial and propels the new material research thorough out the world [20]. Characterization techniques help in understanding the properties of the newly developed or synthesized material or composites providing more information to help discover better materials and create robust composites which will be application centric. The performance evaluation of various parts and materials also helps in re-engineering to create better products to be used widely [21]. In this era of more customization and fast product life cycle, additive manufacturing has played a widespread role for various applications. Therefore, characterization of the AMed parts and newly developed materials is crucial for creating better parts that will help the AM fabricated parts to perform at par with the conventionally built parts but with much more functionality and complexity which was not possible before [22]. Understanding the material properties that affect its behavior in various environments or stimulus in case of 4D printed part is an important aspect of characterization. Material qualities include a number of physical and chemical characteristics that is often dependent on the environmental conditions. Mechanical qualities such as strength, stiffness, toughness, and ductility and its recovery during the 4D motions determine how a material responds to applied forces and loads combined with the external fillip [23]. Thermal qualities such as conductivity, expansion coefficient, and specific heat capacity influence how a thermo- responsive material reacts to temperature fluctuations [24]. Electrical qualities such as conductivity, resistivity, and dielectric constant also affect a material's ability to conduct or resist the passage of electricity. Chemical qualities including reactivity, corrosion resistance, and chemical stability determine how a smart material reacts to different chemical environments. Understanding these properties is crucial for choosing appropriate materials for specific applications of 4D printing [25], designing components with desired performance characteristics, and anticipating material behavior in use. For example, in the aerospace industry, where lightweight materials with high strength-to-weight ratios along with an additional feature of morphing air wings are critical for fuel economy and performance which were not possible in conventional manufactured components, selecting suitable smart materials for aircraft components requires a thorough understanding of material properties under various conditions [26]. This knowledge is especially important in four-dimensional additive manufacturing, where material characteristics may vary greatly based on printing conditions, material composition, and post-processing treatments, and external stimulus.

Therefore, 4D-printed objects are intended to demonstrate dynamic behavior and property change, it is critical to have a complete knowledge of the underlying material properties that regulate their reactions to stimulus [27]. This encompasses not only the fundamental material's mechanical, thermal, electrical, and chemical characteristics, but also how these qualities interact and develop in response to environmental changes.

Moreover, 4D-printed structures designed for self-assembly, shape-changing mechanisms, or adaptive structures, the material must have certain qualities that allow for controlled deformations or transformations over time [28]. Understanding the material's reaction to various stimuli is critical for building and optimizing 4D-printed items that have the appropriate functionality and performance. Furthermore, since 4D printing is still a new and emerging technology, a thorough knowledge of material properties is critical for expanding the capabilities and possible applications of 4D-printed things [29]. Shape memory polymers show distinctive properties of shape memory when they undergo thermo-mechanical programming and recovery cycles as shown in Figure 8.1, which can also be used by various instruments like DMTA or rheometer. Researchers and engineers may open up new avenues for innovation in sectors such as health, robotics, architecture, and consumer items by better understanding how shape memory qualities impact the dynamic behavior of 4D-printed objects. As a result, in the field of 4D printing, knowing material properties

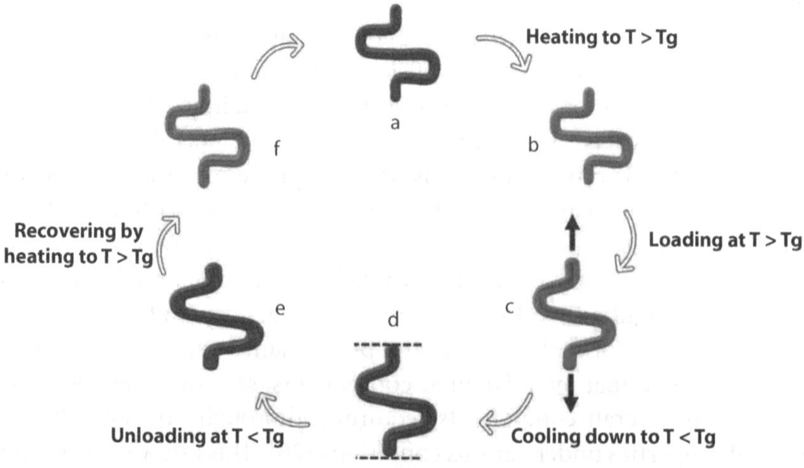

Figure 8.1 Shape memory behavior of a shape memory polymer: (a) original state; (b) heated to temperature above transition; (c) loaded; (d) quenched/cooled while keeping the strain constant; (e) unloaded; (f) shape recovery after being heated [30].

is critical to ensuring the effective creation and deployment of dynamic and responsive printed products.

8.2.2 Characterization Strategies for Ensuring Quality in Additive Manufacturing and 4D Printing

AM with its layer-by-layer building methodology, has broken down the barriers of conventional subtractive technologies. It has transformed manufacturing by providing unprecedented design flexibility, enabling the fabrication of previously unthinkable complicated geometries and sophisticated internal systems. The layer-by-layer structure of additive manufacturing methods effects material uniformity and integrity, offering unique issues that must be carefully considered and mitigated [31]. Material homogeneity refers to the uniform distribution of material qualities inside a printed component, whereas material integrity relates to the item's structural soundness and reliability. The impact of additive manufacturing on material uniformity and integrity may be investigated from a variety of angles:

- **Microstructural Variances:** Because of the layer-by-layer deposition of material, additive manufacturing technologies frequently produce microstructural variations in printed items. These fluctuations can be caused by local temperature gradients, fast solidification rates, or layer-to-layer bonding processes. As a result, printed objects may have diverse microstructures with various grain morphologies, orientations, and flaws like porosity and interlayer interfaces [32].
- **Anisotropic Characteristics:** The layer-by-layer deposition process introduces anisotropic qualities into printed objects, which means that material properties change with direction. This anisotropy results from changes in grain structure, orientation, and residual stresses across neighboring layers. As a result, printed components may exhibit direction-dependent mechanical behavior, including differences in strength, stiffness, and ductility along different axes [33].
- **Porosity and Flaws:** Additive manufacturing technologies can produce porosity and other defects, compromising the material integrity and mechanical qualities of printed products. Porosity is caused by trapped gas bubbles, imperfect fusion of deposited layers, or insufficient energy input

during the printing process. These flaws can weaken the material and act as possible locations for fracture initiation and propagation, resulting in decreased part performance and dependability [34].

- **Residual Strains:** The layer-wise deposition and quick cooling rates used in additive manufacturing procedures can cause residual stresses in printed items [33]. These residual strains are caused by temperature gradients, differential shrinkage, and non-uniform cooling rates between neighboring layers. High residual stresses can harm material integrity by causing distortion, warping, and cracking during the printing process or after post-processing treatments [34].
- **Material Homogeneity:** Material homogeneity in additive manufacturing is a substantial difficulty due to the intricate interaction of processing settings, material characteristics, and component shape. In addition to microstructural differences and anisotropic features, powder quality, powder bed parameters, and process control are important elements in determining material homogeneity [35]. Deviations from the specified material composition, density, and mechanical qualities can occur if material homogeneity is not properly managed, affecting part performance and dependability [36].

Characterization procedures are critical in evaluating and optimizing material characteristics, microstructural features, and defect generation to assure the quality and dependability of printed components in a variety of applications [37]. Addressing the issues of material uniformity and integrity is critical for expanding additive manufacturing capabilities and promoting wider usage in a variety of sectors. However, in the dynamic world of 4D printing, objects go beyond their static state, displaying time-dependent form changes and functions that react to environmental inputs [38]. Characterization is fundamental foundation for ensuring the quality, functionality, and transformative potential of 4D printed parts across a wide range of industries, from shape morphing aerospace to life-saving applications in biomedical engineering, from intricate robotic movements to architectural structures, and even the consumer goods we interact with every day [39]. Characterization in 4D printing is not a single approach; rather, it is a combination of several analytical procedures, each of which plays an important part in understanding the materials' unique features. Spectroscopic methods like infrared (FTIR) and Raman spectroscopy serve as chemical detectives. They investigate the material's composition, showing the complex interaction of atoms and molecules that comprise its structure. Understanding how these structures are structured may reveal how the material reacts to external stimuli such as temperature, light, and pressure. This information is useful in altering the material's shape-shifting properties [40]. For example, consider a 4D printed medicine delivery capsule that dissolves at a precise pH level inside the body [41]. Spectroscopic characterization would allow us to better understand the molecular interactions between the capsule material and the surrounding environment, ensuring that it dissolves precisely enough to deliver the drug successfully. Microscopic methods like as scanning electron microscopy (SEM) and optical microscopy serve as high-resolution characterization tools. They enable us to focus on the precise microstructural characteristics and faults present in the 4D printed object [42]. Understanding these characteristics is critical to comprehending the mechanics behind the material's dynamic behavior. For example, a 4D printed self-healing material may include tiny passageways incorporated in its structure [43]. Using SEM, we can see the size, form, and distribution of these channels, which are critical in enabling the material to heal by carrying healing chemicals throughout its structure. Mechanical testing procedures such as dynamic mechanical analysis (DMA) and shape memory testing are important for evaluating its dynamic mechanical characteristics. These tests give critical information about the material's appropriateness for functional applications and capacity to respond to changing environments. DMA can disclose how a material's stiffness and elasticity vary with temperature, which is useful information for applications where heat causes shape alteration [44]. Shape memory testing, using DMA, focuses on a material's capacity to remember a temporary form and return to it in response to a particular stimulus. It can also be repeated to check the functional fatigue behavior over number of shape recovery cycles [45]. For example, consider a 4D printed, deployable part of a satellite, where DMA testing would guarantee that the part can work up to a number of cycles, thereby improving the reliability of the developed 4D product. Characterization of the 4D printed parts goes beyond merely material analysis. It takes a more comprehensive approach, assessing the following:

- **Dynamic Response:** How does a 4D printed item respond to external stimuli? Does it change properties as quickly and precisely as expected? For example, consider a 4D printed microscale gripper that can hold items of the scale of micro and nano for various applications [46]. Characterization would determine the gripper's response to an electrical signal, guaranteeing that it opens and shuts with the accuracy and speed required to handle the micro/nano object without damage [47].

- **Shape-Changing Capabilities:** Can the component morph into various forms consistently and repeatedly? How well does it retain its programmed shape upon transformation? A 4D printed shape memory polymer actuator may need characterization to guarantee that it consistently unfolds into its operational shape when exposed to a certain environmental state (like temperature) and retains its shape memory over time and over various cycles [48].
- **Performance Validation:** How does the 4D printed item function in different operating conditions? Does it fulfill the functional criteria for the planned application? Characterization is critical in determining the overall performance of a 4D printed object. For example, a 4D printed biomimetic heart valve meant to emulate the natural movement of a human heart valve may be characterized to determine its capacity to endure the pressures and flow dynamics of the circulatory system [49].

8.3 Mechanical Characterization

Mechanical characterization is indeed critical in AM as it provides invaluable insights into the structural integrity and performance of printed objects [50]. In this comprehensive review, we will explore numerous mechanical characterization techniques employed to examine mechanical characteristics, including tensile, compressive, shear, fatigue, and impact testing. These techniques offer a detailed understanding of the mechanical behavior of AM materials and structures, allowing engineers and researchers to optimize design parameters, enhance part performance, and ensure the reliability of printed components [51].

In AM, mechanical characterization plays a vital role in assessing the structural integrity, strength, stiffness, and durability of printed items. However, it is equally important to consider how these techniques apply to the emerging field of 4D printing. Unlike traditional 3D printing, 4D printing introduces an additional dimension of time-dependent shape changes and functionalities in response to external stimuli. Therefore, mechanical characterization techniques tailored to the unique properties of 4D printed materials are essential for gaining deeper insights into their mechanical behavior and performance under various conditions [52]. Table 8.1 provides the standards available for mechanical characterization of additive manufactured parts.

While conventional mechanical characterization techniques remain relevant in the context of 4D printing, such as tensile and compressive testing to evaluate material strength and stiffness, additional considerations are necessary to assess the dynamic nature of 4D printed objects. Techniques such as custom-made tensile testing with hot water testing or heated chamber [30] as shown in Figure 8.2, DMTA [53], and other methods of shape memory testing become crucial in understanding how 4D printed materials respond to stimuli and how they maintain their programmed shape over time.

8.3.1 Tensile, Compressive, and Shear Testing

Tensile, compressive, and shear testing techniques are not only crucial for evaluating the mechanical properties of materials in traditional manufacturing but also hold significant importance in the realm of 4D printing [55]. In the context of 4D printed parts, these mechanical tests serve several important purposes. Firstly, tensile testing can be utilized to assess the dynamic mechanical behavior of 4D printed parts as they undergo shape transformation. By subjecting the printed parts to tensile forces during the testing process, researchers can observe how the material responds to stretching and deformation, providing insights into its tensile strength and elongation properties in different configurations [56]. Similarly, compressive testing is valuable for evaluating the structural integrity and stability of 4D printed parts when subjected to compressive forces. This is particularly relevant for applications where the printed parts need to support or withstand external loads, such as in architectural structures or mechanical components. Compressive testing helps ensure that 4D printed parts maintain their intended form and structural integrity under compression [57]. Furthermore, shear testing can be employed to assess the shear strength and deformation behavior of 4D printed parts, especially in scenarios where these parts experience shear forces during their functional operation [58]. Understanding how 4D printed materials respond to shear forces is essential for applications such as deployable structures, robotic mechanisms, or morphing components where shear stresses play a significant role in their performance. Various mechanical testing methods of additively manufactured parts are shown in Figure 8.3.

Tensile testing in AM is often carried out with standardized test specimens, such as ASTM D638 for polymer materials [60], and ASTM E8 (sometimes miniature test specimens) for metallic materials [61]. The test

Table 8.1 Available standards for mechanical characterization of AM parts.

Mechanical characterization methodologies	Relevant standards for additive manufactured parts
Tensile Strength Assessment	• ASTM F2924: Additive Manufacturing Titanium-6-Aluminum-4-Vanadium with Powder Bed Fusion • ASTM F2971: Outlines specific requirements for reporting various types of data associated with AM test specimens, including dimensional measurements, material properties, test conditions, and manufacturing parameters. • ASTM E8/E8M: Standard Test Methods for Tension Testing of Metallic Materials • ASTM D638: Standard Test Method for Tensile Properties of Plastics • ISO 527-1: Plastics, determination of tensile properties • ISO 6892-1: Metallic materials, Tensile testing, Part 1: Method of test at room temperature
Compressive Strength Evaluation	• ASTM F2924: Additive Manufacturing Titanium-6 Aluminum-4 Vanadium with Powder Bed Fusion • ASTM F2971: Outlines specific requirements for reporting various types of data associated with AM test specimens • ASTM D695: Standard Test Method for Compressive Properties of Rigid Plastics • ISO 604: Plastics: Determination of compressive properties
Shear Testing Techniques	• ASTM F2924: Standard Specification for Additive Manufacturing Titanium-6 Aluminum-4 Vanadium with Powder Bed Fusion • ASTM F2971: Standard Practice for Reporting Data for Test Specimens Prepared by Additive Manufacturing • ASTM D732: Standard Test Method for Shear Strength of Plastics by Punch Tool • ISO 1827: Plastics: Determination of shear strength

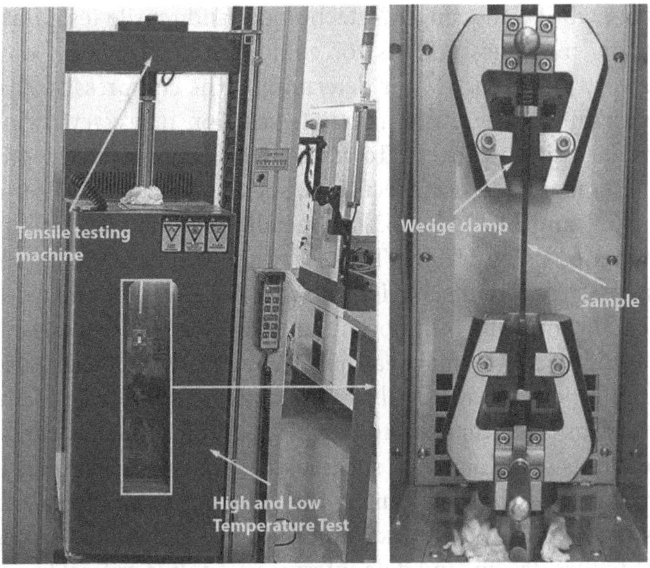

Figure 8.2 Tensile testing apparatus with temperature controller [54].

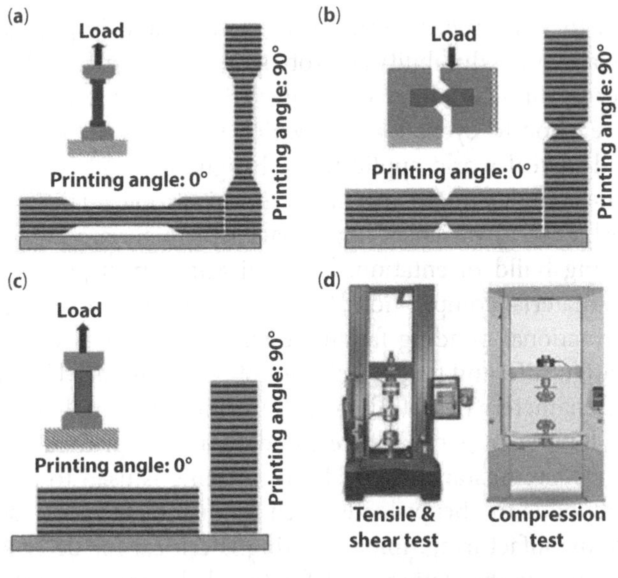

Figure 8.3 Various mechanical testing of AMed parts (a) tensile, (b) shear, (c) compression, (d) machine for tensile and shear tests (D638), and compression test (D695) [59].

coupons are fabricated using AM techniques and tensile tested using universal testing machines (UTMs) or equivalent equipment. Compressive testing is very useful in AM for determining the compressive strength of layered structures, where the material's reaction may vary owing to the layer-wise deposition process. To assess the compressive strength of AM materials, standardized test specimens, such as ASTM D695 for polymer materials [62], and ASTM E9 [63] for metallic materials, are manufactured and compressed using UTMs. The test results are used to determine the compressive stress-strain behavior, compressive strength, and deformation properties. Shear testing is important in AM because it evaluates the interlayer bonding strength and shear qualities of printed objects. Depending on the application and material system, many shear testing methods can be used, such as lap shear tests, torsion tests, and Iosipescu shear tests [58], [64]. These tests entail applying shear loads to the test specimen's interface or shear plane and then measuring the resulting shear stress and strain. Shear testing in AM takes into account specimen geometry, loading conditions, test fixture design, and data interpretation. It is critical to verify that the shear force is delivered uniformly and that the test equipment appropriately simulates the expected application circumstances.

8.3.2 Fatigue and Impact Testing

In addition to static mechanical testing, fatigue and impact testing are required to assess the durability and robustness of AM materials and structures under dynamic loading conditions [65]. Fatigue testing evaluates a material's reaction to cyclic loading, which replicates the recurrent pressures found in real-world applications. Fatigue testing is critical in AM owing to the complex microstructures and anisotropic qualities of printed objects [66]. Several variables can impact AM material's fatigue properties, including build orientation, layer thickness, post-processing treatments, and material composition [67]. Fatigue testing procedures, such as axial and rotational bending fatigue tests, are used to assess the fatigue strength, fatigue life, and fatigue behavior of AM materials. Understanding the fatigue characteristics of 3D and 4D printed materials is critical for forecasting long-term performance and durability of printed components under service circumstances [68]. Impact testing is used to assess how a material responds to abrupt loads, such as an impact or collision event. In additive manufacturing, impact testing is critical for determining the toughness, energy absorption, and fracture behavior of printed objects under impact loading [69]. Charpy and Izod impact tests use a pendulum or hammer to strike a notched or unnotched test specimen and measure the energy absorbed during fracture. The impact energy, absorbed energy, and fracture properties are utilized to determine the material's impact resistance and fracture toughness [70]. Impact testing in additive manufacturing takes into account specimen shape, notch arrangement, loading velocity, and test temperature [71]. The impact loading conditions must be carefully selected to effectively replicate the desired application situations, such as impact occurrences in aircraft, automotive, or consumer electronics applications.

8.4 Thermal Characterization

In applications where thermal properties have an impact on performance, thermal characterization is essential for comprehending the behavior of materials during additive manufacturing (AM) processes [72]. It is imperative to comprehend the manner in which materials react to thermal stimuli in both 3D and 4D printing. Thermal properties are a factor in the selection of materials, design, and ultimate product application in AM, whereas thermal stimuli are frequently employed in 4D printing to induce shape transformations or other dynamic behaviors [73]. The thermal properties of printed materials differ considerably from those of bulk materials due to the layer-by-layer characteristics of AM. The complexity of time-dependent material responses to these properties is further enhanced by 4D printing [74].

8.4.1 Evaluation of Thermal Properties

It is essential to assess the critical thermal properties of materials in both 3D and 4D printing. The following properties are included:

- Thermal conductivity is a critical factor in applications that necessitate heat transfer or insulation. The timing and precision of shape transformations induced by heat can be influenced by thermal conductivity in 4D printing, which impacts the way heat propagates through the material [75].
- Specific heat capacity indicates the quantity of heat that a substance can absorb or release prior to experiencing a temperature change. This influences the energy consumption during the printing process in 3D printing, and in 4D printing, it determines the amount of heat required to trigger a

shape change in stimuli-responsive materials when printing [76].

- Thermal expansion coefficient is a critical property for precision applications, as it indicates the extent to which a material expands or contracts in response to temperature fluctuations. This is especially crucial in 4D printing, as materials are frequently programmed to undergo shape changes when heated. Therefore, a precise comprehension of thermal expansion is required to regulate these shape changes [77].
- Thermal diffusivity is a measure of the rate at which heat is transferred through a material, and it is calculated by combining thermal conductivity, specific heat capacity, and density. The timing of shape changes can be influenced by the rate of heat diffusion in 4D printing, which can affect the speed and uniformity with which a material responds to thermal stimuli [78].

8.4.2 Measurement of Thermal Conductivity in Printed Parts

The anisotropic structure of printed objects makes measuring the heat conductivity of materials used in AM difficult [79]. Commonly utilized techniques include the guarded hot plate method and laser flash analysis. These approaches must be carefully tweaked or chosen to account for the directionality provided by layer-by-layer manufacturing, which can result in large differences in heat conductivity along a printed part's axis. The standards for analysis of thermal properties in additively manufactured parts are shown in Table 8.2.

8.4.3 Heat Resistance Testing for Various Printing Materials

Heat resistance is crucial for AM parts, especially for high-temperature applications, and even more so in 4D printing where materials must not only withstand heat but also retain their functional properties to respond to thermal stimuli.

- Differential Scanning Calorimetry (DSC): This method analyzes how materials behave under thermal transitions, such as melting or crystallization. For 4D printing, DSC helps in determining the precise temperature range at which materials will change shape [80].

Table 8.2 ASTM and ISO standards for thermal analysis in different materials.

Material type	ASTM standards	ISO standards
Polymers	• ASTM E1269: Specific Heat Capacity by Differential Scanning Calorimetry. • ASTM E289: Linear Thermal Expansion of Rigid Solids • ASTM E473: Thermal Conductivity of Powders. • ASTM E1461: Thermal Diffusivity by the Flash Method. • ASTM E1131: Compositional Analysis by Thermogravimetry.	• ISO 11357: Differential Scanning Calorimetry (DSC). • ISO 22007-4: Thermal Conductivity and Diffusivity - Dynamic Laser Method. • ISO 22007-3: Thermal Conductivity and Diffusivity - Temperature Wave Analysis. • ISO 22007-2: Thermal Conductivity and Diffusivity - Transient Plane Heat Source (Hot Disc) Method. • ISO 11357-6: Determination of Oxidative Induction Time (Isothermal OIT) and Oxidation Induction Temperature (Dynamic OIT).
Metals/ Ceramics	• ASTM E1461: Thermal Diffusivity by the Flash Method. • ASTM E1463: Thermal Diffusivity of Advanced Ceramics by Laser Flash Method. • ASTM E1468: Index of Refraction of Transparent Organic Plastics.	• ISO 22007-1: Thermal Conductivity and Diffusivity - Temperature Wave Analysis Method.

- Thermogravimetric Analysis (TGA): It is used to analyze material stability at elevated temperatures. In 4D printing, this is essential to ensure that materials can repeatedly undergo thermal cycles without degrading or losing their ability to transform [81].

8.4.4 Impact of Layering on Thermal Performance

The layering process in both 3D and 4D printing significantly affects thermal performance. Layer interfaces, material density variations, and bonding quality all contribute to how heat moves through a part [82]. In 4D printing, these factors are even more critical because thermal stimuli must effectively penetrate the entire structure to induce shape changes.

- Interlayer Bonding: Weak interlayer bonds can disrupt heat flow, especially in 4D printing, where uniform heat distribution is essential for activating shape transformations across the entire structure.
- Layer Orientation: The orientation of layers relative to the heat source plays a major role in thermal conductivity. In 4D printing, this orientation must be optimized to ensure that heat can trigger predictable shape changes throughout the structure.

8.5 Surface Finish and Roughness

The surface quality and roughness of produced components are important factors that influence their performance, usefulness, and attractiveness. The layer-by-layer building process in additive manufacturing (AM) presents unique problems for attaining a high-quality surface finish [83]. This section discusses methods for analyzing surface quality, the significance of post-processing procedures, and their function in improving AM component surface properties.

8.5.1 Assessing Surface Quality

Surface quality of manufactured components is a multidimensional property that includes roughness, waviness, and the presence of flaws, all of which have a significant impact on the parts' mechanical performance, fatigue resistance, and frictional qualities. The industry follows standards like ISO 4287 and ASME B46.1, which provide a framework for surface roughness measurements.

8.5.2 Quantitative Analysis of Surface Finish

Understanding and quantifying roughness metrics such as R_a (Arithmetic Average) and R_z (Average Maximum Height) is critical for quantitative surface finish analysis since they provide insights into the surface's topography. Various measuring techniques, such as contact profilometry for direct surface contact measurements and non-contact optical profilometry for high-resolution surface mapping, are used in this study [84]. The analysis of this data is critical for directing production and post-processing decisions and achieving the required surface quality.

8.5.3 Visual Inspection Techniques

Visual examination is the first step in determining surface quality, using magnification equipment and surface replicas to discover flaws and evaluate overall polish. Despite its use, the subjective nature of visual inspection requires experienced workers to assure consistent and trustworthy assessment, emphasizing the significance of using quantitative approaches for a full review.

8.5.4 Importance of Post-Processing

Post-processing arises as a necessary step in AM, with the goal of not only increasing surface polish but also eliminating support structures and improving material characteristics. The precise needs of the material, desired finish, and intended use determine the technique used, which ranges from mechanical processes such as sanding and polishing to chemical and heat treatments [85].

8.5.5 Role of Post-Processing in Achieving Desired Surface Finishes

The choice of a suitable post-processing procedure is crucial to attaining the required surface finish. This selection procedure is influenced by elements such as the material used, the level of polish required, and the part's functional use. Detailed case studies demonstrate the transforming influence of various post-processing approaches on surface quality, offering significant insights into optimizing surface finishes in AM components [86].

8.6 Microstructure Analysis

Microstructure analysis in AM is essential for understanding the internal characteristics of materials that directly influence mechanical, thermal, and chemical properties of the final product. AM techniques, such as powder bed fusion, directed energy deposition, and fused deposition modeling, involve processes where materials are built layer by layer, leading to unique microstructural formations that differ significantly from conventionally manufactured parts [87].

8.6.1 Microscopic Examination

Microscopic examination plays a fundamental role in understanding the surface and internal microstructures of 3D and 4D printed parts. In 3D printing, defects like voids, incomplete layer bonding, and porosity can severely affect the mechanical properties of the final product [88]. In 4D printing, where time-dependent changes in shape or function are important, it becomes critical to investigate how the microstructure evolves when exposed to external stimuli.

Optical microscopy is commonly used for basic surface inspections, but high-resolution tools like scanning electron microscopy (SEM) and transmission electron microscopy (TEM) offer detailed insights at the nano and microscale [89]. SEM, in particular, allows researchers to observe surface topology, defects, and the interface between printed layers [90], while TEM provides atomic-level views, which are essential when investigating how grain structure and phase compositions change during deformation or stimuli response in 4D printing [42]. By leveraging these microscopic examination techniques, it is possible to detect surface cracks, layer misalignments, and grain boundary defects, all of which influence the final properties of printed parts.

8.6.2 Techniques for Microstructure Analysis

Different types of microscopies are used based on the analysis goal:

- **Optical Microscopy:** Used for lower magnification analysis of surface features and defect detection.
- **SEM (Scanning Electron Microscopy):** Provides detailed images of surface topology and is ideal for analyzing the quality of interlayer bonding and defect distribution.
- **TEM (Transmission Electron Microscopy):** Offers atomic resolution imaging, useful for understanding grain structure, defect density, and phase transformations in smart materials.

- **Confocal Microscopy:** Used to create 3D reconstructions of the microstructure, helping understand volume and internal structure without cutting the sample.

The following techniques are utilized in the 4D printed parts to study and analyse the complicated internal structure at the micro and nano levels. They also enable the detection of defects, layer bonding quality, and compositional uniformity, which are essential for understanding the performance of parts.

- **X-ray Diffraction (XRD):** Used to determine the crystalline structure of the material. It helps identify phase compositions and any structural changes that occur when the material is exposed to different stimuli in 4D printing. This is essential for predicting how a material will perform over time.
- **Energy-Dispersive X-ray Spectroscopy (EDS):** Often coupled with SEM, this technique provides elemental composition analysis. It helps in identifying impurities or variations in material composition that may influence mechanical properties or stimuli responses in smart materials.
- **Focused Ion Beam (FIB) Milling:** This technique is used in combination with SEM or TEM to slice through printed materials, enabling cross-sectional analysis of internal structures. It allows for detailed views of the bonding between layers, microcracks, and interfacial defects.

8.6.3 Printing Parameter Influence

The influence of printing parameters on the microstructure is significant in both 3D and 4D printing. Important variables such as print speed, layer thickness, temperature, and cooling rates directly affect the final properties of the printed part as shown in Figure 8.4 and Figure 8.5.

- **Layer Thickness:** Thicker layers reduce print time but often result in poor surface finish and weaker interlayer adhesion, which can lead to microstructural defects like porosity or layer misalignment. Thinner layers improve surface quality and strength but increase print time.
- **Print Speed:** A higher print speed can lead to poor bonding between layers, incomplete fusion, or defects like voids. However, a very slow print speed can result in thermal

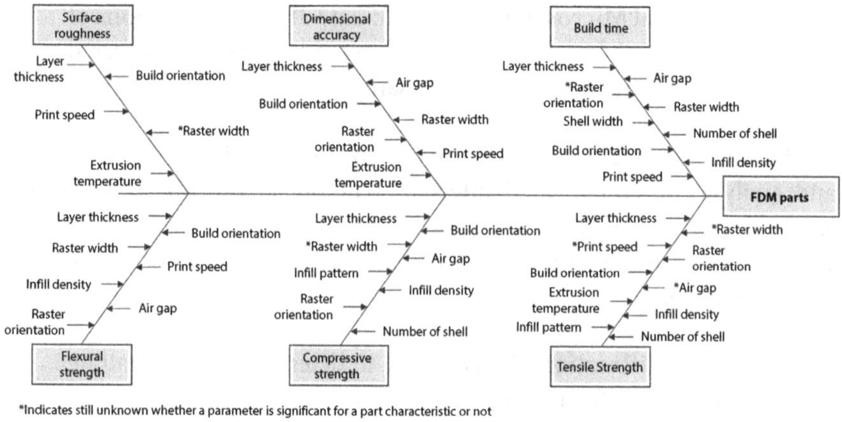

Figure 8.4 Fish bone diagram showing important process parameters affecting the FDM part qualities [93].

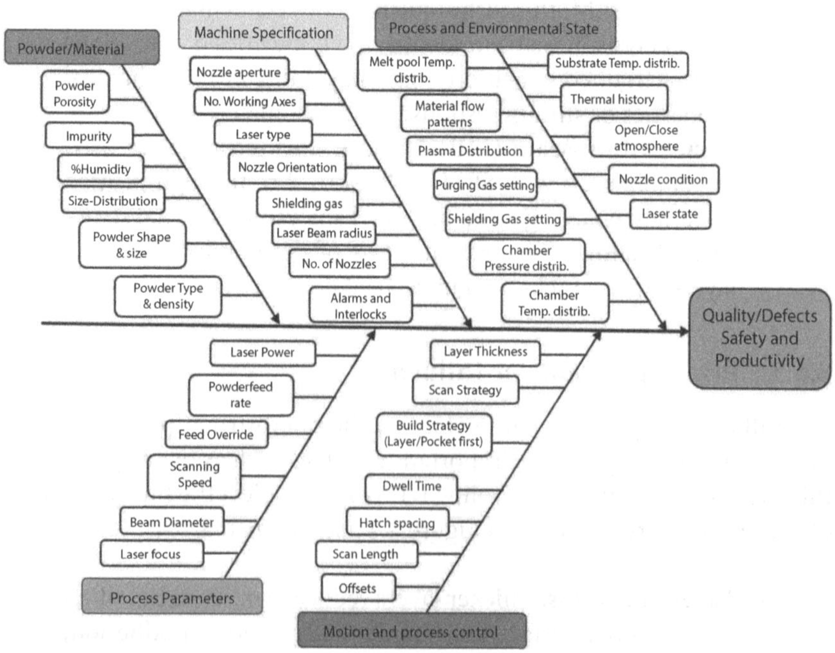

Figure 8.5 Fish bone diagram showing important process parameters affecting the quality/defects safety and productivity of metal AM.

deformation or overheating, affecting the material's internal structure.

- **Nozzle Temperature:** For smart materials in 4D printing, maintaining an optimal nozzle temperature is crucial for ensuring that the material retains its shape-memory properties or its ability to transform under external stimuli [91]. Too high a temperature can cause phase distortions, while a very low a temperature might lead to incomplete fusion between layers.
- **Bed Temperature:** Bed temperature is a critical parameter in both 3D and 4D printing processes, directly influencing the microstructure of the printed part and its subsequent properties. In 3D printing, especially with thermoplastics like polylactide (PLA), bed temperature helps improve layer adhesion and reduce warping, which impacts the mechanical properties of the final component [92]. In 4D printing, bed temperature plays an even more significant role, not only affecting the microstructure but also influencing the shape memory properties that are crucial for the material's ability to respond to external stimuli such as heat or moisture [4].
- **Laser Power:** Influence on Microstructure: Laser power in DED dictates the energy input to the material, which directly affects melting and solidification behavior. Higher laser power can lead to deeper penetration and larger melt pools, resulting in coarse grain structures and increased heat-affected zones. On the other hand, lower laser power can lead to insufficient melting, poor layer bonding, and increased porosity.
- **Powder Feed Rate:** Powder feed rate plays a significant role in determining the deposition rate and the amount of material added per layer. A high powder feed rate can result in excessive material accumulation, leading to inconsistent layers and defects like voids.

8.7 Dimensional Accuracy and Precision

Dimensional accuracy and precision are critical features in the manufacturing industry, defining how closely a produced product follows its original design. This is especially important in additive manufacturing (AM), where the layer-by-layer production approach presents unique problems for retaining certain properties [94]. This study goes into the complexities of analyzing, measuring, and improving dimensional accuracy and precision in AM processes, identifying problems and suggesting ways to solve them.

While AM provides enormous prospects for innovation in a variety of industries, the difficulties of dimensional accuracy and precision must be thoroughly addressed [95]. AM components' dimensional consistency may be considerably improved by combining modern measuring tools, a thorough grasp of layered manufacturing issues, and the adoption of tailored solutions. As the technology matures, continued research and development in this field will improve the dependability and application span of additive printing, pushing the limits of what is possible in precise production.

8.7.1 Assessing Accuracy and Precision

In manufacturing, accuracy refers to how close a measured dimension is, to its true value or the intended design; whereas, precision refers to the repeatability of the manufacturing process, or how closely multiple items produced are to one another, regardless of their proximity to the intended design. In AM, accuracy and precision are critical for the effective application of printed components in sectors ranging from aerospace to medical devices, where even tiny variations can result in failure [96].

8.7.2 Measurement Techniques for Dimensional Accuracy

Dimensional accuracy in AM components is assessed using a variety of measurement techniques, which range from standard calipers and micrometers to modern metrology equipment like as coordinate measuring machines (CMM) and optical scanning technologies. CMMs provide great precision in measuring geometric dimensions and may be used to generate comprehensive maps of a part's surface, revealing deviations from the CAD model [97]. Optical scanning, which includes laser scanning and structured light methods, is a non-contact approach for swiftly capturing the geometry of complicated parts, albeit post-processing may be required to effectively interpret the data.

8.7.3 Precision Considerations in Layered Manufacturing

Layered manufacturing, a feature of AM, necessitates unique accuracy concerns. Layer thickness, orientation, and the direction of material deposition or sintering all have a substantial impact on a part's ultimate dimensions and surface polish. Variations in these parameters can cause stair-stepping effects, warping, and other problems that reduce accuracy [98]. Understanding and regulating process parameters is critical to reducing these impacts and obtaining high accuracy.

8.8 Non-Destructive Testing (NDT)

8.8.1 Introduction to NDT Techniques

Non-Destructive Testing (NDT) refers to a set of testing procedures that enable the examination and analysis of materials, components, and assemblies without affecting or destroying their future usage. These approaches are essential in businesses where structural integrity and dependability are critical. NDT is crucial to maintaining safety and dependability, from the earliest production phases to maintenance and troubleshooting of in-service components [11].

8.8.2 Overview of Non-Destructive Testing Methods

NDT techniques vary greatly, with each having benefits, limits, and uses. Techniques include:

- **Ultrasonic Testing (UT):** is the process of detecting internal faults or characterizing materials using high-frequency sound waves. UT is extremely effective for detecting internal faults in solid materials because to its deep penetration.
- **Radiographic Testing (RT):** uses X-rays or gamma rays to create pictures of a component's interior structure, which is beneficial for detecting density changes caused by fractures, voids, or porosity.
- **Magnetic Particle Testing (MPT):** is the application of magnetic fields to ferromagnetic materials to identify surface and near-surface discontinuities, as demonstrated by the use of small magnetic particles.
- **Eddy Current Testing (ECT):** Uses electromagnetic induction to discover surface and subsurface faults in conductive materials, notably surface fractures.

8.8.3 NDT Suitability for Layered Structures

Layered manufacture, such as 3D printing, presents new issues for NDT because to the anisotropic qualities and internal structures of additive manufacturing [99]. For example, ultrasonic testing must be precisely calibrated to account for the heterogeneous density and complicated geometries of printed components, which might alter sound wave propagation.

8.8.4 Techniques for Identifying Defects in 3D and 4D Prints

With the advancement of 3D and 4D printing, NDT techniques have evolved to meet the particular problems offered by these production technologies. Advanced ultrasonic phased array technologies and computed tomography (CT) scans have become critical in evaluating the intricate interior structures of printed things, allowing for a thorough evaluation of their integrity.

8.9 Conclusions

Characterization techniques are integral to advance the field of additive manufacturing, particularly in the context of four-dimensional (4D) printing technology. These techniques enable the analysis of material properties, mechanical behavior, thermal characteristics, surface morphology, microstructural features, and dimensional accuracy of 4D printed parts is vital for optimizing design parameters, ensuring part quality, and validating performance under various operating conditions. Thermal characterization techniques such as DSC, DMA, and TGA, are essential for evaluating thermal properties, including glass transition, melting, phase transition temperature, and thermal stability. The understanding of these thermal characteristics enables the selection of appropriate materials and design parameters for specific applications. Mechanical characterization techniques, including tensile, compressive, and shear testing, are essential for assessing the mechanical properties, and shape recovery of 4D printed parts. These tests provide critical information about the strength, stiffness, ductility, and fatigue resistance of the printed components. Surface finish and roughness analysis are crucial for optimizing the aesthetics and functionality of 4D printed parts. Quantitative analysis of surface finish helps evaluate the topography and roughness parameters, while visual inspection techniques provide a preliminary assessment of the overall surface quality. Microstructure analysis techniques such as microscopic examination and high-resolution imaging offer insights into the internal structures, defects, and surface roughness of 4D printed parts. These analyses help optimize the printing parameters and improve the microstructural characteristics of the printed components. Dimensional accuracy and precision are fundamental attributes in additive manufacturing, and their evaluation ensures the fidelity of the printed parts to the intended design. Advanced measurement techniques and process control methods help address the challenges associated with maintaining dimensional consistency in layered manufacturing. Non-destructive testing (NDT) methods play a significant role in inspecting and analyzing the structural integrity and reliability of 4D printed parts without damaging them. These techniques, such as ultrasonic testing, radiographic testing, magnetic particle testing, and eddy current testing, help identify defects and ensure quality throughout the manufacturing process. Shape memory testing techniques, such as differential scanning calorimetry and thermomechanical analysis, allow for the evaluation of the shape memory effect and mechanical properties of shape memory materials used in 4D printing. These tests help assess the materials' response to various stimuli and optimize their functional properties. Through comprehensive characterization techniques, researchers and practitioners can optimize design parameters, ensure part quality, and validate performance under various operating conditions, realizing the full potential of 4D printing technologies. These efforts are paramount in advancing the field of additive manufacturing, fostering innovation, and driving transformative advancements in diverse industrial domains.

References

1. Gibson, I., Rosen, D.W., Stucker, B., Springer, U.S., *Additive manufacturing technologies: Rapid prototyping to direct digital manufacturing*, Springer, Springer New York, USA, 2010, doi: 10.1007/978-1-4419-1120-9/COVER.
2. Tibbits, S., *The emergence of '4D printing,'*, TED Conferences, LLC (Technology, Entertainment, Design), New York, USA 2013, Accessed: Jan. 24, 2021. [Online]. Available: https://www.ted.com/talks/skylar_tibbits_the_emergence_of_4d_printing.
3. Samal, B.B., Jena, A., Varshney, S.K., Kumar, C.S., FDM 4D printing: A low-cost approach of shape programming and assessing the shape memory properties using angle measurement methods in hot water actuation testing apparatus. *Mater. Today Proc.*, 101, 7–14, 2023. doi: 10.1016/j.matpr.2023.02.038.
4. Samal, B.B., Varshney, S.K., Kumar, C.S., Four-dimensional (4D) printing through FDM: Effect of infill density and bed temperature on shape memory

properties in different thermo-mechanical programming conditions. *J. Mech. Sci. Technol.*, 38, 8, 4313–4319, 2024. doi: 10.1007/s12206-024-0727-3.
5. Kuang, X., Yue, L., Qi, H.J., Introduction to 4D Printing: Concepts and Material Systems, in: *Additive Manufacturing Technology*, pp. 1–42, John Wiley & Sons, Ltd, Weinheim, Germany, 2023, doi: https://doi.org/10.1002/9783527833931.ch1.
6. Bodaghi, M. and Zolfagharian, A., 1 - 4D printing principles and manufacturing, in: *Smart Materials in Additive Manufacturing*, M. Bodaghi and X.X.X.A. Zolfagharian (Eds.), pp. 1–17, Elsevier, Amsterdam, Netherlands, 2022, doi: https://doi.org/10.1016/B978-0-12-824082-3.00014-3.
7. Rayate, A. and Jain, P.K., A Review on 4D Printing Material Composites and Their Applications. *Mater. Today Proc.*, 5, 9, Part 3, 20474–20484, 2018. doi: https://doi.org/10.1016/j.matpr.2018.06.424.
8. Ahmed, A., Arya, S., Gupta, V., Furukawa, H., Khosla, A., 4D printing: Fundamentals, materials, applications and challenges. *Polymer*, 228, 123926, 2021. doi: https://doi.org/10.1016/j.polymer.2021.123926.
9. Kumar, A., Ramadas, H., Samal, B.B., Kumar, C.S., Nath, A.K., Laser polishing of cobalt chrome alloy fabricated by laser powder bed fusion process: design of experiment-based approach for reducing surface roughness. *Int. J. Adv. Manuf. Technol.*, 1245–1264, 2024. doi: 10.1007/s00170-024-14101-w.
10. Kumar, A., Samal, B.B., Nath, A.K., Kumar, C.S., Laser surface modification of metal additive manufactured parts : A case study of *ex-situ* and *in-situ* methodology, in: *Surface Engineering*, pp. 121–144, CRC Press, Boca Raton, Florida, USA, 2022, doi: 10.1201/9781003319375-5.
11. Segovia Ramírez, I., García Márquez, F.P., Papaelias, M., Review on additive manufacturing and non-destructive testing. *J. Manuf. Syst.*, 66, 260–286, 2023. doi: https://doi.org/10.1016/j.jmsy.2022.12.005.
12. Nezhad, I.S., Golzar, M., Hossein Behravesh, A., Zare, S., Comprehensive study on shape shifting behaviors in FDM-based 4D printing of bilayer structures. *Int. J. Adv. Manuf. Technol.*, 120, 1, 959–974, 2022. doi: 10.1007/s00170-022-08741-z.
13. Ebeid, E.-Z.M. and Zakaria, M.B., Chapter 3 - Thermal analysis in materials science, in: *Thermal Analysis*, E.-Z.M. Ebeid and M.B. Zakaria (Eds.), pp. 129–166, Elsevier, Amsterdam, Netherlands, 2021, doi: https://doi.org/10.1016/B978-0-323-90191-8.00003-8.
14. Azra, C., Plummer, C.J.G., Månson, J.-A.E., Dynamic mechanical analysis for rapid assessment of the time-dependent recovery behavior of shape memory polymers. *Smart Mater. Struct.*, 22, 7, 75037, Jun. 2013. doi: 10.1088/0964-1726/22/7/075037.
15. Bergström, J., 2 - Experimental Characterization Techniques, in: *Mechanics of Solid Polymers*, J. Bergström (Ed.), pp. 19–114, William Andrew Publishing, Amsterdam, Netherlands, 2015, doi: https://doi.org/10.1016/B978-0-323-31150-2.00002-9.

16. Kaliva, M. and Vamvakaki, M., Chapter 17 - Nanomaterials characterization, in: *Polymer Science and Nanotechnology*, R. Narain (Ed.), pp. 401–433, Elsevier, Amsterdam, Netherlands, 2020, doi: https://doi.org/10.1016/B978-0-12-816806-6.00017-0.
17. Skjærvø, S.L., Juelsholt, M., Jensen, K.M.Ø., 10.10 - *In situ* scattering studies of material formation during wet-chemical syntheses, in: *Comprehensive Inorganic Chemistry III (Third Edition)*, Third, J. Reedijk and K.R. Poeppelmeier (Eds.), pp. 248–272, Elsevier, Oxford, 2023, doi: https://doi.org/10.1016/B978-0-12-823144-9.00023-6.
18. Bumbrah, G.S. and Sharma, R.M., Raman spectroscopy – Basic principle, instrumentation and selected applications for the characterization of drugs of abuse. *Egypt. J. Forensic Sci.*, 6, 3, 209–215, 2016. doi: https://doi.org/10.1016/j.ejfs.2015.06.001.
19. Cross, J.O., Opila, R.L., Boyd, I.W., Kaufmann, E.N., Materials characterization and the evolution of materials. *MRS Bull.*, 40, 12, 1019–1034, 2015. doi: 10.1557/mrs.2015.271.
20. Zhang, S., Li, L., Kumar, A., *Materials characterization techniques*, CRC press, Boca Raton, Florida, USA, 2008.
21. Sharma, S.K., Verma, D.S., Khan, L.U., Kumar, S., Khan, S.B., *Handbook of materials characterization*, Springer, Switzerland, 2018.
22. Kern, J., Additive Manufacturing, in: *The Digital Transformation of Logistics*, pp. 41–60, John Wiley & Sons, Ltd, Hoboken, New Jersey, USA, 2021, doi: https://doi.org/10.1002/9781119646495.ch4.
23. Subramaniyan, M., Karuppan, S., Eswaran, P., Anandhamoorthy, A., Review on 4D printing materials, stimuli and its future application. *Int. J. Mech. Prod. Eng. Res. Dev.*, 9, 316–327, 2019. [Online]. Available: https://www.scopus.com/inward/record.uri?eid=2-s2.0-85063477827&partnerID=40&md5=-85fac0842c2e7d3bc832de01af3888ea.
24. Shin, S. and So, H., Effect of 3D printing raster angle on reversible thermo-responsive composites using PLA/paper bilayer. *Smart Mater. Struct.*, 29, 10, 105016, Oct. 2020. doi: 10.1088/1361-665X/aba490.
25. Xin, X., Liu, L., Liu, Y., Leng, J., Mechanical Models, Structures, and Applications of Shape-Memory Polymers and Their Composites. *Acta Mech. Solida Sin.*, 535–565, 2019. doi: 10.1007/s10338-019-00103-9.
26. Barbarino, S., Saavedra Flores, E.L., Ajaj, R.M., Dayyani, I., Friswell, M.I., A review on shape memory alloys with applications to morphing aircraft. *Smart Mater. Struct.*, 23, 6, 063001, Jun. 2014. doi: 10.1088/0964-1726/23/6/063001.
27. Chae, M.P., Hunter-Smith, D.J., De-Silva, I., Tham, S., Spychal, R.T., Rozen, W.M., Four-Dimensional (4D) Printing: A New Evolution in Computed Tomography-Guided Stereolithographic Modeling Principles and Application. *J. Reconstructive Microsurgery*, 31, 6, 458–463, 2015. doi: 10.1055/s-0035-1549006.

28. Mao, Y., Yu, K., Isakov, M.S., Wu, J., Dunn, M.L., Jerry Qi, H., Sequential Self-Folding Structures by 3D Printed Digital Shape Memory Polymers. *Sci. Rep.*, 5, 13616, 2015. doi: 10.1038/srep13616.
29. Jeong, H.Y., An, S.-C., Lim, Y., Jeong, M.J., Kim, N., Jun, Y.C., 3D and 4D Printing of Multistable Structures. *Appl. Sci.*, 10, 20, 7254, 2020. doi: 10.3390/app10207254.
30. Nabavian Kalat, M., *et al.*, Investigating a shape memory epoxy resin and its application to engineering shape-morphing devices empowered through kinematic chains and compliant joints. *Mater. Des.*, 233, 112263, 2023. doi: https://doi.org/10.1016/j.matdes.2023.112263.
31. Ngo, T.D., Kashani, A., Imbalzano, G., Nguyen, K.T.Q., Hui, D., Additive manufacturing (3D printing): A review of materials, methods, applications and challenges. *Composites, Part B*, 143, 172–196, 2018. doi: https://doi.org/10.1016/j.compositesb.2018.02.012.
32. Han, Y., Griffiths, R.J., Yu, H.Z., Zhu, Y., Quantitative microstructure analysis for solid-state metal additive manufacturing via deep learning. *J. Mater. Res.*, 35, 15, 1936–1948, 2020. doi: 10.1557/jmr.2020.120.
33. Somireddy, M. and Czekanski, A., Anisotropic material behavior of 3D printed composite structures – Material extrusion additive manufacturing. *Mater. Des.*, 195, 108953, 2020. doi: https://doi.org/10.1016/j.matdes.2020.108953.
34. Du, C., *et al.*, Pore defects in Laser Powder Bed Fusion: Formation mechanism, control method, and perspectives. *J. Alloys Compd.*, 944, 169215, 2023. doi: https://doi.org/10.1016/j.jallcom.2023.169215.
35. Li, H., Brodie, E.G., Hutchinson, C., Predicting the chemical homogeneity in laser powder bed fusion (LPBF) of mixed powders after remelting. *Addit. Manuf.*, 65, 103447, 2023. doi: https://doi.org/10.1016/j.addma.2023.103447.
36. Li, M., Liu, Z., Ho, J.Y., Wong, T.N., Improving Homogeneity of 3D-Printed Cementitious Material Distribution for Radial Toolpath. *Fluids*, 8, 3, 87, 2023. doi: 10.3390/fluids8030087.
37. Sanaei, N., Fatemi, A., Phan, N., Defect characteristics and analysis of their variability in metal L-PBF additive manufacturing. *Mater. Des.*, 182, 108091, 2019. doi: https://doi.org/10.1016/j.matdes.2019.108091.
38. Wan, X., *et al.*, Recent Advances in 4D Printing of Advanced Materials and Structures for Functional Applications. *Adv. Mater.*, 36, 34, 2312263, 2024. doi: https://doi.org/10.1002/adma.202312263.
39. Yu, Y., *et al.*, Material characterization and precise finite element analysis of fiber reinforced thermoplastic composites for 4D printing. *Comput.-Aided Des.*, 122, 102817, 2020. doi: https://doi.org/10.1016/j.cad.2020.102817.
40. Bril, M., Fredrich, S., Kurniawan, N.A., Stimuli-responsive materials: A smart way to study dynamic cell responses. *Smart Mater. Med.*, 3, 257–273, 2022. doi: https://doi.org/10.1016/j.smaim.2022.01.010.
41. Gazzaniga, A., *et al.*, Towards 4D printing in pharmaceutics. *Int. J. Pharm.: X*, 5, 100171, 2023. doi: https://doi.org/10.1016/j.ijpx.2023.100171.

42. Lefkowitch, J.H., Chapter 17 - Electron Microscopy and Other Techniques, in: *Scheuer's Liver Biopsy Interpretation (Tenth Edition)*, Tenth, J.H. Lefkowitch (Ed.), pp. 410–430, Elsevier, London, 2021, doi: https://doi.org/10.1016/B978-0-7020-7584-1.00017-6.
43. Cañada, J. and Velásquez-García, L.F., Monolithically 3D-Printed, Self-Heating Microfluidics, in: *2023 IEEE 22nd International Conference on Micro and Nanotechnology for Power Generation and Energy Conversion Applications (PowerMEMS)*, pp. 210–213, 2023, doi: 10.1109/PowerMEMS59329.2023.10417177.
44. Li, H., *et al.*, Dual and triple shape memory properties of poly(ε-caprolactone)-based cross-linked polymer elastomers. *Polym. Test.*, 115, 107738, 2022. doi: https://doi.org/10.1016/j.polymertesting.2022.107738.
45. Hornat, C.C., *et al.*, Quantitative predictions of maximum strain storage in shape memory polymers (SMP). *Polymer*, 186, 122006, 2020. doi: https://doi.org/10.1016/j.polymer.2019.122006.
46. Korrayi, R.R., Jena, A., Marupalli, B.C.G., Samal, B.B., Varshney, S.K., Kumar, C.S., pp. 429–457, John Wiley & Sons, Ltd, River Street, USA & Beverly, MA, USA, 2024, doi: https://doi.org/10.1002/9781394198085.ch14.
47. Adam, G., Benouhiba, A., Rabenorosoa, K., Clévy, C., Cappelleri, D.J., 4D Printing: Enabling Technology for Microrobotics Applications. *Adv. Intell. Syst.*, 3, 5, 2000216, 2021. doi: https://doi.org/10.1002/aisy.202000216.
48. Samal, B.B., Jena, A., Varshney, S.K., Kumar, C.S., 4D Printing of Shape Memory Polymers: A Comparative Study of Programming Methodologies on Various Material Properties. *Smart Mater. Struct.*, 32, 7, 074003, May 2023. doi: 10.1088/1361-665X/acda6e.
49. Capulli, A.K., *et al.*, JetValve: Rapid manufacturing of biohybrid scaffolds for biomimetic heart valve replacement. *Biomaterials*, 133, 229–241, 2017. doi: https://doi.org/10.1016/j.biomaterials.2017.04.033.
50. Dizon, J.R.C., Espera, A.H., Chen, Q., Advincula, R.C., Mechanical characterization of 3D-printed polymers. *Addit. Manuf.*, 20, 44–67, 2018. doi: https://doi.org/10.1016/j.addma.2017.12.002.
51. Riva, L., Ginestra, P.S., Ceretti, E., Mechanical characterization and properties of laser-based powder bed–fused lattice structures: a review. *Int. J. Adv. Manuf. Technol.*, 113, 3, 649–671, 2021. doi: 10.1007/s00170-021-06631-4.
52. C., Z., *et al.*, Mechanical properties and shape memory behavior of 4D printed functionally graded cellular structures. *Sci. China Technol. Sci.*, 66, 12, 3522–3533, 2023. doi: 10.1007/s11431-023-2475-3.
53. Xia, Y., He, Y., Zhang, F., Liu, Y., Leng, J., A Review of Shape Memory Polymers and Composites: Mechanisms, Materials, and Applications. *Adv. Mater.*, 33, 6, 2000713, 2021. doi: https://doi.org/10.1002/adma.202000713.
54. Wang, Q., Wang, J., Wang, A., Zhou, C., Hu, J., Pan, F., Effect of Strain Rate and Temperature on the Tensile Properties of Long Glass Fiber-Reinforced Polypropylene Composites. *Polymers*, 15, 15, 3260, 2023. doi: 10.3390/polym15153260.

55. Zeng, C., Liu, L., Lin, C., Xin, X., Liu, Y., Leng, J., 4D printed continuous fiber reinforced shape memory polymer composites with enhanced mechanical properties and shape memory effects. *Composites, Part A*, 180, 108085, 2024. doi: https://doi.org/10.1016/j.compositesa.2024.108085.
56. Safai, L., Cuellar, J.S., Smit, G., Zadpoor, A.A., A review of the fatigue behavior of 3D printed polymers. *Addit. Manuf.*, 28, 87–97, 2019. doi: https://doi.org/10.1016/j.addma.2019.03.023.
57. Xu, P., *et al.*, Compression behavior of 4D printed metamaterials with various Poisson's ratios. *Int. J. Mech. Sci.*, 264, 108819, 2024. doi: https://doi.org/10.1016/j.ijmecsci.2023.108819.
58. Sieberer, S., Savandaiah, C., Leßlhumer, J., Schagerl, M., Shear property measurement of additively manufactured continuous fibre reinforced plastics by in-plane torsion testing. *Addit. Manuf.*, 55, 102805, 2022. doi: https://doi.org/10.1016/j.addma.2022.102805.
59. Lim, T., Cheng, H., Song, W., Lee, J., Kim, S., Jung, W., Simulated and Experimental Investigation of Mechanical Properties for Improving Isotropic Fracture Strength of 3D-Printed Capsules. *Materials*, 14, 16, 4677, 2021. doi: 10.3390/ma14164677.
60. Phillips, C., Kortschot, M., Azhari, F., Towards standardizing the preparation of test specimens made with material extrusion: Review of current techniques for tensile testing. *Addit. Manuf.*, 58, 103050, 2022. doi: https://doi.org/10.1016/j.addma.2022.103050.
61. Wawrzyniak, N., Provencher, P.R., Brochu, M., Brochu, M., Tensile properties of SS316L produced by LPBF: Influence of specimen dimensions and surface condition. *Mater. Charact.*, 203, 113117, 2023. doi: https://doi.org/10.1016/j.matchar.2023.113117.
62. Popescu, D., Zapciu, A., Amza, C., Baciu, F., Marinescu, R., FDM process parameters influence over the mechanical properties of polymer specimens: A review. *Polym. Test.*, 69, 157–166, 2018. doi: https://doi.org/10.1016/j.polymertesting.2018.05.020.
63. Sun, J., Yang, Y., Wang, D., Mechanical properties of a Ti6Al4V porous structure produced by selective laser melting. *Mater. Des.*, 49, 545–552, 2013. doi: https://doi.org/10.1016/j.matdes.2013.01.038.
64. James, S. and Dang, C., Investigation of shear failure load in ultrasonic additive manufacturing of 3D CFRP/Ti structures. *J. Manuf. Processes*, 56, 1317–1321, 2020. doi: https://doi.org/10.1016/j.jmapro.2020.04.026.
65. Sarkar, S., Siva Kumar, C., Kumar Nath, A., Effect of mean stresses on mode of failures and fatigue life of selective laser melted stainless steel. *Mater. Sci. Engineering A*, 700, 92–106, 2017. doi: https://doi.org/10.1016/j.msea.2017.05.118.
66. Mouritz, A.P. (Ed.), 5 - Mechanical and durability testing of aerospace materials, in: *Introduction to Aerospace Materials*, pp. 91–127, Woodhead Publishing, Cambridge, UK, 2012, doi: https://doi.org/10.1533/9780857095152.91.

67. Sarkar, S., Kumar, C.S., Nath, A.K., Effects of heat treatment and build orientations on the fatigue life of selective laser melted 15-5 PH stainless steel. *Mater. Sci. Engineering A*, 755, 235–245, 2019. doi: https://doi.org/10.1016/j.msea.2019.04.003.
68. Speirs, M., Van Hooreweder, B., Van Humbeeck, J., Kruth, J.-P., Fatigue behaviour of NiTi shape memory alloy scaffolds produced by SLM, a unit cell design comparison. *J. Mech. Behav. Biomed. Mater.*, 70, 53–59, 2017. doi: https://doi.org/10.1016/j.jmbbm.2017.01.016.
69. Waqas, A., Xiansheng, Q., Jiangtao, X., Chaoran, Y., Fan, L., Impact toughness of components made by GMAW based additive manufacturing. *Procedia Struct. Integrity*, 13, 2065–2070, 2018. doi: https://doi.org/10.1016/j.prostr.2018.12.207.
70. Popa, C.-F., Mărghitaş, M.-P., Galaţanu, S.-V., Marşavina, L., Influence of thickness on the IZOD impact strength of FDM printed specimens from PLA and PETG. *Procedia Struct. Integrity*, 41, 557–563, 2022. doi: https://doi.org/10.1016/j.prostr.2022.05.064.
71. Ty, A., Mokhtari, M., Balcaen, Y., Votié, A., Cloué, J.-M., Alexis, J., Performance evaluation of Charpy impact tests to investigate and detect process defects of a Ni-based superalloy elaborated by laser powder bed fusion. *J. Mater. Res. Technol.*, 25, 6644–6659, 2023. doi: https://doi.org/10.1016/j.jmrt.2023.07.064.
72. Compton, B.G., Post, B.K., Duty, C.E., Love, L., Kunc, V., Thermal analysis of additive manufacturing of large-scale thermoplastic polymer composites. *Addit. Manuf.*, 17, 77–86, 2017. doi: https://doi.org/10.1016/j.addma.2017.07.006.
73. Shanmugam, V., Babu, K., Kannan, G., Mensah, R.A., Samantaray, S.K., Das, O., The thermal properties of FDM printed polymeric materials: A review. *Polym. Degrad. Stab.*, 228, 110902, 2024. doi: https://doi.org/10.1016/j.polymdegradstab.2024.110902.
74. Goo, B., Hong, C.-H., Park, K., 4D printing using anisotropic thermal deformation of 3D-printed thermoplastic parts. *Mater. Des.*, 188, 108485, 2020. doi: https://doi.org/10.1016/j.matdes.2020.108485.
75. Shemelya, C., *et al.*, Anisotropy of thermal conductivity in 3D printed polymer matrix composites for space based cube satellites. *Addit. Manuf.*, 16, 186–196, 2017. doi: https://doi.org/10.1016/j.addma.2017.05.012.
76. Blanco, I., Cicala, G., Recca, G., Tosto, C., Specific Heat Capacity and Thermal Conductivity Measurements of PLA-Based 3D-Printed Parts with Milled Carbon Fiber Reinforcement. *Entropy*, 24, 5, 654, 2022. doi: 10.3390/e24050654.
77. Rădulescu, B., *et al.*, Thermal Expansion of Plastics Used for 3D Printing. *Polymers*, 14, 15, 3061, 2022. doi: 10.3390/polym14153061.
78. Farzinazar, S., Wang, Y., Abdol-Hamid Owens, C., Yang, C., Lee, H., Lee, J., Thermal transport in 3D printed shape memory polymer metamaterials. *APL Mater.*, 10, 8, 81105, Aug. 2022. doi: 10.1063/5.0094036.

79. Prajapati, H., Ravoori, D., Woods, R.L., Jain, A., Measurement of anisotropic thermal conductivity and inter-layer thermal contact resistance in polymer fused deposition modeling (FDM). *Addit. Manuf.*, 21, 84–90, 2018. doi: https://doi.org/10.1016/j.addma.2018.02.019.
80. Doostmohammadi, H., Kashmarizad, K., Baniassadi, M., Bodaghi, M., Baghani, M., 4D printing and optimization of biocompatible poly lactic acid/ poly methyl methacrylate blends for enhanced shape memory and mechanical properties. *J. Mech. Behav. Biomed. Mater.*, 160, 106719, 2024. doi: https://doi.org/10.1016/j.jmbbm.2024.106719.
81. Ghalayaniesfahani, A., Oostenbrink, B., van Kasteren, H., Gibson, I., Mehrpouya, M., 4D printing of biobased shape memory sandwich structures. *Polymer*, 307, 127252, 2024. doi: https://doi.org/10.1016/j.polymer.2024.127252.
82. Alqahtani, S., Ali, H.M., Farukh, F., Silberschmidt, V.V., Kandan, K., Thermal performance of additively manufactured polymer lattices. *J. Build. Eng.*, 39, 102243, 2021. doi: https://doi.org/10.1016/j.jobe.2021.102243.
83. Markovits, T. and Varga, L.F., Investigating the surface roughness of 3D printed metal parts in case of thin 20 μm build layer thickness. *J. Mater. Res.*, 39, 13, 1841–1851, 2024. doi: 10.1557/s43578-023-01254-9.
84. Jin, Y., Li, H., He, Y., Fu, J., Quantitative analysis of surface profile in fused deposition modelling. *Addit. Manuf.*, 8, 142–148, 2015. doi: https://doi.org/10.1016/j.addma.2015.10.001.
85. Mahmood, M.A., Chioibasu, D., Ur Rehman, A., Mihai, S., Popescu, A.C., Post-Processing Techniques to Enhance the Quality of Metallic Parts Produced by Additive Manufacturing. *Metals*, 12, 1, 77, 2022. doi: 10.3390/met12010077.
86. Boban, J., Ahmed, A., Jithinraj, E.K., Rahman, M.A., Rahman, M., Polishing of additive manufactured metallic components: retrospect on existing methods and future prospects. *Int. J. Adv. Manuf. Technol.*, 121, 1, 83–125, 2022. doi: 10.1007/s00170-022-09382-y.
87. Liu, Z., *et al.*, Additive manufacturing of metals: Microstructure evolution and multistage control. *J. Mater. Sci. Technol.*, 100, 224–236, 2022. doi: https://doi.org/10.1016/j.jmst.2021.06.011.
88. Le Duigou, A., Correa, D., Ueda, M., Matsuzaki, R., Castro, M., A review of 3D and 4D printing of natural fibre biocomposites. *Mater. Des.*, 194, 108911, 2020. doi: https://doi.org/10.1016/j.matdes.2020.108911.
89. Wang, P., *et al.*, Scanning optical microscopy for porosity quantification of additively manufactured components. *Addit. Manuf.*, 21, 350–358, 2018. doi: https://doi.org/10.1016/j.addma.2018.03.019.
90. Zhou, W., Apkarian, R., Wang, Z.L., Joy, D. Fundamentals of Scanning Electron Microscopy (SEM). In: Zhou, W., Wang, Z.L. (eds) *Scanning Microscopy for Nanotechnology*. Springer, New York, NY., 2006, https://doi.org/10.1007/978-0-387-39620-0_1.

91. Samal, B.B., Jena, A., Varshney, S.K., Kumar, C.S., 4D Printing Technology: Experimental investigation of thermoresponsive shape-changing honeycomb and negative honeycomb structures for different proposed programming strategies, in: 대한기계학회 춘추학술대회, p. 10, 2023, [Online]. Available: https://www.dbpia.co.kr/Journal/articleDetail?nodeId=NODE11708468.
92. Aliheidari, N., Tripuraneni, R., Hohimer, C., Christ, J., Ameli, A., Nadimpalli, S., The impact of nozzle and bed temperatures on the fracture resistance of FDM printed materials, in: *Proc. SPIE*, vol. 10165, p. 1016512, Apr. 2017, doi: 10.1117/12.2260105.
93. Dey, A. and Yodo, N., A Systematic Survey of FDM Process Parameter Optimization and Their Influence on Part Characteristics. *J. Manuf. Mater. Process.*, 3, 3, 64, 2019. doi: 10.3390/jmmp3030064.
94. Lee, P.-H., Chung, H., Lee, S.W., Yoo, J., Ko, J., *Review: Dimensional Accuracy in Additive Manufacturing Processes*, International Manufacturing Science and Engineering Conference (MSEC), Jun. 09, 2014, American Society of Mechanical Engineers, Detroit, Michigan, USA. doi: 10.1115/MSEC2014-4037.
95. Al-Ahmari, A., Ashfaq, M., Mian, S.H., Ameen, W., Evaluation of additive manufacturing technologies for dimensional and geometric accuracy. *Int. J. Mater. Prod. Technol.*, 58, 2–3, 129–154, Jan. 2019. doi: 10.1504/IJMPT.2019.097665.
96. Berger, U., Aspects of accuracy and precision in the additive manufacturing of plastic gears. *Virtual Phys. Prototyping*, 10, 2, 49–57, Apr. 2015. doi: 10.1080/17452759.2015.1026127.
97. Mansi, H.K. and Singholi, A.K.S., Additive Manufacturing Metrology, in: *Handbook of Metrology and Applications*, D.K. Aswal, S. Yadav, T. Takatsuji, P. Rachakonda, H. Kumar (Eds.), pp. 1165–1180, Springer Nature Singapore, Singapore, 2023, doi: 10.1007/978-981-99-2074-7_60.
98. Paul, R. and Anand, S., Optimization of layered manufacturing process for reducing form errors with minimal support structures. *J. Manuf. Syst.*, 36, 231–243, 2015. doi: https://doi.org/10.1016/j.jmsy.2014.06.014.
99. Helsel, M.A., Popovics, J.S., Stynoski, P.B., Kreiger, E., Non-destructive testing to characterize interlayer bonds of idealized concrete additive manufacturing products. *NDT & E Int.*, 121, 102443, 2021. doi: https://doi.org/10.1016/j.ndteint.2021.102443.

9

4D Printing Applications in Photonics

Bijaya Bikram Samal[1*†], Shubhanshi Sharma[2†], Monica Pradhan[3†] and Shailendra K. Varshney[2]

[1]*Advanced Technology Development Centre, Indian Institute of Technology, Kharagpur, West Bengal, India*
[2]*Department of Electronics and Electrical Communication, Indian Institute of Technology, Kharagpur, West Bengal, India*
[3]*School of Nanoscience and Technology, Indian Institute of Technology, Kharagpur, India*

Abstract
This chapter reviews the progress and possible applications of 4D printing in the field of photonics, highlighting its ability to produce adaptable components for advanced optical characteristics. The combination of shape memory materials with 3D printing technology allows for the development of tunable photonic devices that respond to changes in the environment, leading to groundbreaking applications. This review presents and discusses various smart mate- rials and their applications in the field of photonics. Different additive manufacturing processes utilized for photonics applications have also been reviewed. The main areas of interest are re- configurable devices with dynamically changing colors, infrared (IR) detectors, soft robotics, biomedical devices, and adaptive optical components. By harnessing the distinctive features of 4D printed materials, it is possible to create advanced photonic devices for dynamic displays, anti-counterfeiting measures, fano resonant structures, smart optical fibers, sensors and other applications. The chapter also emphasizes the need for future research to enhance the performance and properties that can expand the actual applications of these advanced materials and structures.

**Corresponding author*: bijaybikram3991@gmail.com
†These authors contributed equally to this work

Bijaya Bikram Samal, Cheruvu Siva Kumar and Shailendra Kumar Varshney (eds.) 4D Printing Technology: Principles, Materials and Applications, (281–314) © 2025 Scrivener Publishing LLC

Keywords: 4D printing, photonics, optics, adaptable optics, actuators, 4D printed fibers, optical resonators, sensors

9.1 Introduction

Photonics is the key area of science and technology that deals with manipulating, transmitting, and detecting photons. Photons are the basic particles of light. The term 'photonics' was first introduced in 1954, in a letter from John W. Campbell to Gotthard Gunther, which is restated as, "Incidentally, I've decided to invent a new science — photonics. It bears the same relationship to optics that electronics does to electrical engineering. Photonics, like electronics, will deal with the individual units; optics and EE deal with the group phenomena! And note that you can do things with electronics that are impossible in electrical engineering!" Studies in photonics started with the creation of Microwave Amplification by Stimulated Emission of Radiation (MASER) and Light Amplification by Stimulated Emission of Radiation (LASER) between 1958 and 1960 [1]. Some other important developments were the invention of the laser diode in 1970, which revolutionized various industries with its compact size and efficiency, and the introduction of optical fibers for long- distance communication in the late 1970s, by transmitting light signals through thin strands of glass or plastic fibers [2]. These important technological inventions served as the basic foundation for the revolution in communication, and the internet. Though the word photonics was developed earlier, it became widely used in the 1980s when telecommunications network operators used fiber-optical data transmission. At that time, the word was frequently used at Bell Labs. Its usage became more famous when the Institute of Electrical and Electronics Engineers (IEEE) Lasers and Electro- optic Society created an archival Journal called Photonics Technology Letters towards the end of the 1980s [3]. Photonics then has encompassed the study of various physical phenomena related to optical devices like optical fibers, photonic integrated circuits, photodetectors, lasers, etc., used for harnessing the characteristics of light for various applications. The application domain of photonics is quite broad, covering telecommunications, information processing, and imaging sensing to micro- fluidics, biomedical, manufacturing, and environmental monitoring [4]. Introduced in the 1980s, additive manufacturing (AM), also referred to as 3D printing or rapid prototyping, has transformed numerous industries, making possible innovations that were once deemed unattainable [5]. Additive manufacturing has also become a key driver of the fourth industrial revolution (Industry 4.0).

It involves creating a three-dimensional object from a digital model, typically called computer-aided design (CAD), by progressively layering materials. Compared to conventional manufacturing techniques, it provides numerous advantages such as the creation of complex structures, reduced material waste, increased design flexibility, faster prototyping, and the facilitation of mass customization [6]. In the domain of photonics, 3D printing have lately enabled the construction of intricate structures unattainable using traditional techniques [7]. Researchers have utilized 3D printing to create metaphotonic structures. Metaphotonics provides unparalleled control over electromagnetic fields in designed materials and structures, which is not possible with traditional optics. As a result, metaphotonics often demand unconventional, complicated 3D structures requiring AM techniques to fabricate [8]. Moreover, 3D printing can be utilized to create complex structures with very high precision. This has led to the design and fabrication of custom optical elements like lenses, waveguides, diffractive optics elements, etc. Apart from building macrostructures, recent developments in 3D microprinting have further enabled the creation of compact, lightweight, miniaturized structures, reducing costs. The advancements in AM, like new material developments, miniaturization, the ability to create high-resolution parts, etc., have enabled researchers and engineers to explore and develop novel designs and structures [9]. This has catered to solutions in various industries and has propelled the creation of new photonic devices across diverse domains, including telecommunications, biomedicine, sensing, and imaging. 3D printing has become a fundamental fabrication technique for various photonics applications, owing to its countless benefits. However, a revolutionary development in AM emerged in 2013, with the introduction of the concept of four-dimensional (4D) printing [10]. This innovative technological advancement not only inherits the benefits of 3D printing but also introduces the ability to create dynamic structures and components that can alter their properties when exposed to specific stimuli [11]. Therefore, this transformative technological advancement in AM can further revolutionize photonic device design and fabrication [12]. This chapter presents a comprehensive examination of 4D printing in the field of photonics, illustrating how this cutting-edge technology can transform the design and manufacturing of photonic devices. It enables the development of tunable optical devices, reconfigurable photonic circuits, adaptive optical materials, metamaterials, and various other applications. The chapter seeks to provide insights into the potential of 4D printing to alter the landscape of photonics science and technology, discussing the current state of the art, existing challenges, and future prospects.

9.2 Smart Materials in Photonics

Smart materials can react to various environmental stimuli, such as temperature, humidity, and particular triggers like light or magnetic fields. This dynamic responsiveness allows 4D-printed objects to modify their characteristics over time. This degree of control is crucial when producing items that need precise alterations at certain times or under specified situations. In photonics applications, smart materials can be pre-programmed with high accuracy to change at a particular condition for manipulating light in various ways to achieve novel applications. Utilizing smart materials for certain applications like actuation and sensing doesn't require any external energy source. Various types of smart materials are mentioned in Figure 9.1. Recent advancements in novel types of smart materials and their composites have fueled the growth of 4D printing technology. Smart materials, including shape memory polymers, magnetic shape memory alloys, hydrogels, electro-responsive polymers, liquid crystal elastomers, and smart inorganic polymers, have been crucial in advancing 4D printing technology and its diverse applications in photonics.

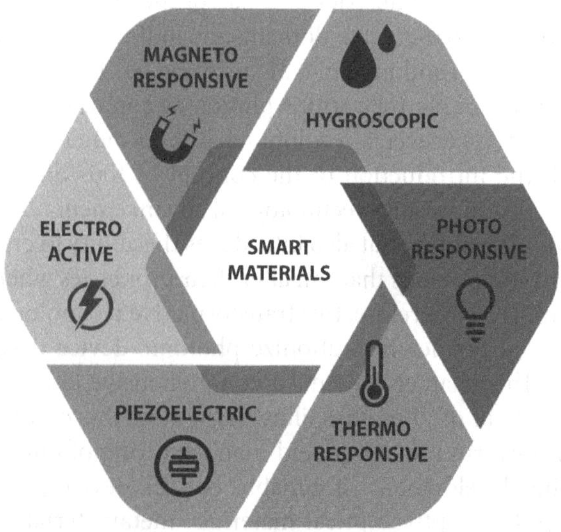

Figure 9.1 Various types of smart materials.

9.2.1 Piezoelectric Materials

Piezoelectric materials have been used extensively in photonics because of their capacity to convert mechanical energy into electrical signals and vice versa. Such materials can be used to precisely manipulate light signals, allowing for functional applications such as optical modulation, beam steering, and frequency conversion. Mechanical stress applied to piezoelectric crystals or films can change their refractive index, controlling light propagation and modifying optical signals. This can be especially useful in telecommunications, laser technology, and optical imaging systems, where precise control and manipulation of light are required to achieve desired functions. Additionally, piezoelectric materials are used in devices such as acousto-optic modulators and piezoelectric scanners, which improve the performance and adaptability of photonics systems in a variety of applications. Studies have shown that AlN has excellent piezoelectric and mechanical properties, which may open up new possibilities for manipulating phonons and photons at the wavelength scale utilizing nanophotonic frameworks [17]. These properties make AlN a better choice for addressing imperfections in traditional silicon and silicon nitride systems. In 2013, researchers developed an AlN photonic crystal nano-beam with contactless electrodes for piezoelectric actuation, achieving a high Qm of around 1.0×105 at 2.5 K [18]. In another work, researchers designed a fully integrated AlN piezo- optomechanical crystal with on-chip electrodes to generate mechanical waves at 4.2 GHz [19]. The piezoelectric manipulation of soliton microcombs is shown by AlN actuators on ultralow-loss Si3N4 waveguides, facilitating voltage-controlled soliton initiation, adjustment, and stabilization, hence enhancing current applications and lidar systems. Moreover, the resonant buildup of bulk acoustic energy allows a 14-fold reduction in the required driving voltage by tuning the micro resonator to the HBAR frequency [15] as shown in Figure 9.2.

9.2.2 Thermo-Responsive Materials

Thermo-responsive materials are a kind of smart material known for their unique ability to change shape or properties subject to changes in surrounding temperature. A distinguishing property of thermo-responsive materials is their capacity to demonstrate a shape memory effect (SME), which allows them to revert to their original configuration after being deformed by a certain temperature. For example, Nitinol, a well-known shape memory alloy, demonstrates this property. At ambient temperature, Nitinol remains in a low-temperature martensitic structure, however, it

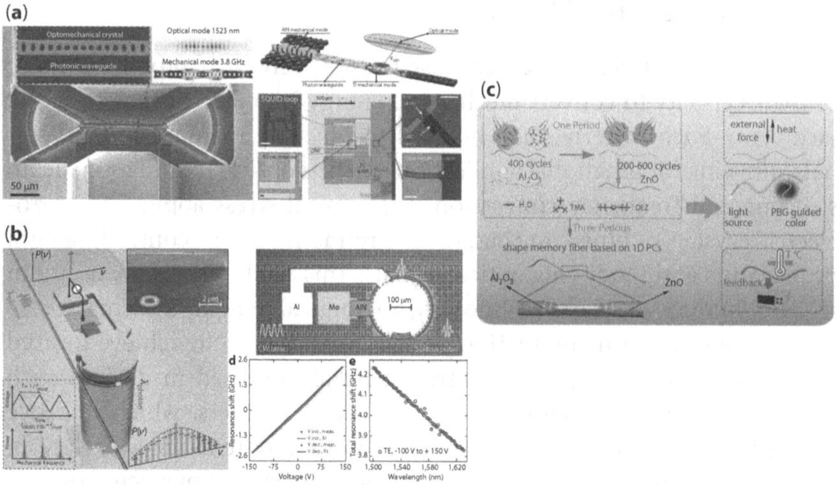

Figure 9.2 (a) Bidirectional microwave–optical transducer comprising a pair of radially symmetric IDTs coupled to an AlN optomechanical crystal (left) [13], and Qubit-to-optical transducer in a hybrid AlN-on-Si platform with the simulated resonant modes and fabricated devices (right) [14]. (b) Schematic illustrating the utilization of AlN to activate the soliton microcomb, using either the HBAR modes or quasi-static stress-optic effect [15] (c) [16].

transforms into a twinned martensite structure upon application of stress. When heated, it reverts to its original austenitic structure, restoring its original form. A novel shape memory photonic crystal fiber (PCF) architecture has been created, including a transparent polyester fiber core and an Al2O3/ZnO photonic shell [16]. The Al2O3/ZnO photonic shell is evenly deposited onto the fiber core surface via an atomic layer deposition technique. The shape memory polymer fibers (SMPFs) exhibit improved shape memory properties, facilitating light transmission and conforming to diverse curved surfaces as shown in Figure 9.2. The pyroelectric action of ZnO further augments the SMPF's temperature-sensing proficiency, enabling the detection of oral respiration, nasal breathing, and infrared radiation even post-deformation.

9.2.3 Magneto-Responsive Materials

Magneto-responsive materials, also known as magnetostrictive or magnetoelastic materials, are a fascinating family of substances known for their ability to change form or mechanical properties when exposed to a magnetic field. This phenomenon, known as magnetostriction, results from the

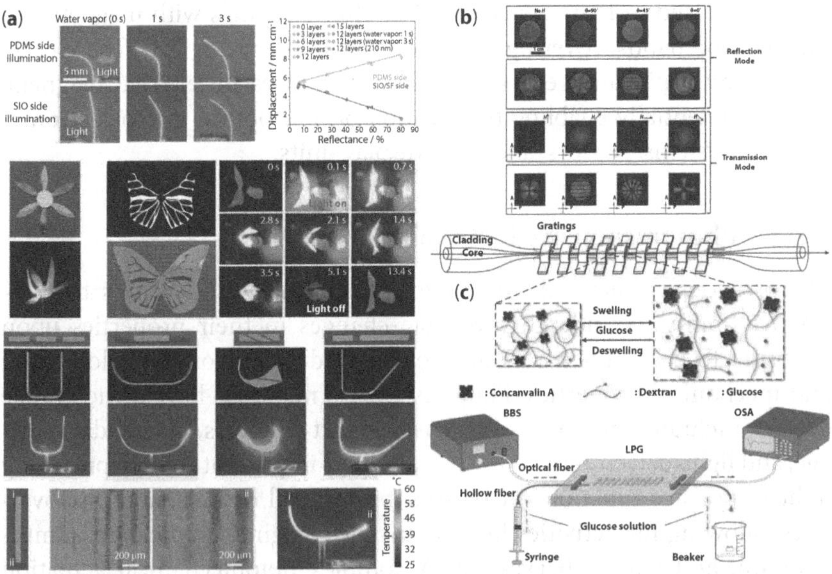

Figure 9.3 (a) Pictures showing how the same photonic actuator was distorted across different time periods both before and after being exposed to water vapor $(I = 35mcm_2)$ [20] (b) Digital photos showing the magneto-optical response of a bifunctional display and anticounterfeiting device based on the nanorod CCA [21] (c) Setup for testing the LPG glucose sensor [22].

reorientation of magnetic domains inside a material in the presence of a magnetic field. As these domains align, the material's form or mechanical characteristics change, which is useful in a variety of applications. Magneto-responsive materials, which are often used in sensors and actuators, provide an adaptable framework for the creation of novel devices and systems. Their capacity to properly transform magnetic energy into mechanical motion highlights their relevance in a variety of technical developments, ranging from precision equipment to sophisticated robots. Magnetostrictive material's potential uses are expanding as a result of continuing research and development, with more advances expected in disciplines ranging from biomedical engineering to industrial automation. In photonics applications like smart windows and magnetic tuning of polarization, researchers have developed a new type of smart optical materials with multiple magneto-optical responses, combining magnetically responsive photonic crystals and magnetically responsive liquid crystals [21] as shown in Figure 9.3. The system allows the modulation of optical properties by a single stimulus, with a switching rate of 50 Hz using a low magnetic field. Superparamagnetic magnetite colloidal nanocrystal clusters (CNCs) have been used to fabricate colloidal photonic crystals with magnetically tunable stop bands, covering the entire visible spectrum [23]. These crystals exhibit rapid and reversible optical responses to external magnetic stimuli, potentially enabling the creation of new optical microelectromechanical systems, sensors, and color display units.

9.2.4 Photoresponsive Materials

Photoresponsive materials, or photochromic materials, possess the distinctive ability to undergo reversible changes in their properties upon exposure to light. These modifications include variations in color, form, and mechanical properties. Photoresponsive materials have a wide range of uses, including photochromic lenses, smart windows, optical data storage, and light-driven actuators. They also play important roles in providing light-responsive functions in sensors, biomedical devices, and photovoltaics, allowing for activities like as sensing, imaging, medication administration, and energy harvesting. A complex optomechanical actuation mode has been developed that enables controlled spectrum distribution at the micro scale by combining silk fibroin with a programmable photonic crystal structure and with an elastomer [20]. This strategy has the potential for micro-robotic technologies and photonic micro actuators with complex deformation for angle-dependent light harvesting. A new strategy uses light to control the optical response of photonic materials (liquid crystal network), using mechanical deformation as an intermediate step. The materials consist of photoresponsive polymers structure by direct laser writing [24]. A 2D grating was prepared, demonstrating a large optical response at sub-millisecond time scales. This technology can be integrated into various systems, including integrated photonic circuits, and engineer photonic structures with optomechanical bi stability or shape memory effects. A unique method of photothermal actuation is achievable by two-photon absorption of concentrated light, resulting in an increase in temperature inside the focal area. Researchers have altered Poly(N-isopropylacrylamide) (pNIPAM) for the creation of functional three-dimensional hetero-microstructures. Experimental findings indicate that actuation response and spatially regulated response are attained using two-photon absorption of focused light [25].

9.2.5 Hygroscopic Materials

Hygroscopic materials possess a unique ability to absorb and retain moisture from their surroundings. Hydrogels, for instance, are a prime example, capable of altering their shape and size by absorbing or desorbing water molecules. These materials play a crucial role in moisture sensors, offering precise detection and regulation of humidity levels in diverse industrial settings and environmental conditions. One way to make a printable hybrid hydrogel is to put PNIPAm microparticles into a silica-alumina gel matrix that is rich in water [26]. This gel can withstand a water content of up to 70% and can be created in layers using a commercial 3D printer. A conductive electrode composed of silver nanoparticles is printed onto a translucent and flexible substrate using the hybrid ink. The patterning of flexible devices, such as smart windows and 3D optical waveguide valves, is made possible by this. An optical shift occurs when water molecules bound to PNIPAm particles are released into the Si/Al sol-gel matrix. Here we show the first example of a 3D printing approach called projection-suspended stereolithography (PSS) for making hydrogel fibers that are uniform and axially continuous [27]. The method optimizes light-guiding efficiency by optimizing fiber diameter, core-cladding thickness, and refractive index distribution. The method offers high-resolution printing, allowing for sub-micrometer-sized to millimeter-sized diameters, flexible core-cladding structures, and modulation of refractive index. In another study, digital light processing (DLP) 3D printed long-period fiber gratings (LPG) devices are fabricated containing a glucose-sensitive hydrogel [22]. It has a high sensitivity of 0.206 nm/mM, a short response time, and a large detection range. The sensor's reversibility and time-response behavior are also studied.

9.2.6 Electroactive Materials

Electroactive materials are a family of compounds that undergo major structural and property changes when exposed to an electric field. Numerous structural modifications are possible in these materials by adjusting parameters such as voltage, current, and pulse length. These dynamic reactions make electroactive materials important in a wide range of applications, from sensors and actuators to transducers, including robots, medical devices, energy harvesting, and photonics. Electrochromic materials may change their optical characteristics, such as color and transparency, in response to an electrical stimulus. To control the color of stretchable photonic crystals, an electrochromic device based on shape memory alloy composite (SMAC) was recently introduced. This device transforms electrical energy into mechanical energy [28]. With a total deformation need of less than 30%, the device is capable of achieving color shifts over the whole visible light spectrum under 1.0V-1.5V voltage.

An effective color-changing area, low driving voltage, excellent compliance, and a wide color-changing range are all features of this device. It also has a rapid reaction time. Additionally, a pattern-specific soft display device is being constructed.

9.3 4D Printing Processes in Photonics

AM techniques have been categorized and standardized into seven groups based on Standard Terminology for Additive Manufacturing Technologies F2792-12, (American Society for Testing and Materials (ASTM)) International Committee F42 on Additive Manufacturing. As illustrated in Figure 9.4, material vat photopolymerization, material jetting, material extrusion, lamination, powder bed fusion, directed energy deposition, and binder jetting are the seven primary types of AM technologies. These methods of AM, enable the creation of very complicated 4D structures in combination with various smart materials.

9.3.1 Liquid-Based

Liquid-based additive manufacturing methods for 4D printing involve the use of materials that can undergo phase transitions in response to external stimuli such as temperature, light, or pH changes. These methods typically utilize liquid resin which may or may not be photo polymerized, shape memory polymers or hydrogel materials that can be printed and then the desired shapes are achieved through controlled exposure to stimuli.

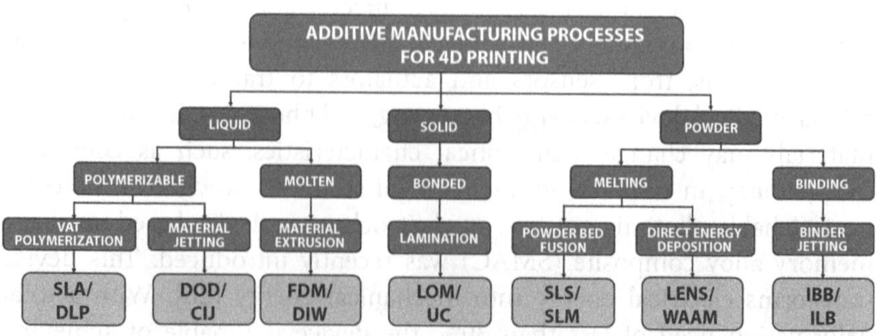

Figure 9.4 Additive manufacturing processes for 4D printing.

9.3.1.1 Molten

The process used in this kind of classification is called extrusion additive manufacturing or fused deposition modeling (FDM) or fused filament fabrication (FFF). It is a method of additive manufacturing wherein a plastic filament is melted within a heated nozzle of a 3D printer. The molten plastic is then extruded onto a platform where it quickly solidifies to form a thin layer. Subsequently, the printer incrementally moves upwards, repeating the process to add successive layers atop one another until the entire object is constructed [11]. Through this layer-by-layer approach, each new layer adheres to the preceding one, gradually building a solid 3D object. This technique enables the fabrication of complex shapes and structures using a variety of plastic materials. FDM process also induces intrinsic shape memory programming enabling controlled actuation. To make optical fibers made of polymers that could remember their shapes, scientists used a scalable preform-to- fiber sketching technique. Using the shape-memory properties of the waveguide, the study explores light-guiding elongated structures and creates an adaptive sensor based on evanescent wave absorption spectroscopy [29]. FDM technique has been used in various studies to fabricate 3D/4D printed optics and photonics devices, including terahertz gratings [30], flat optics [31], prisms [32], optical fibers [33], including step-index and micro-structured shape memory polymer optical fibers [34], special-shaped cores, hollow-core waveguides [35], and lenses. However, poor resolution and anisotropic mechanical properties are the main drawbacks. Direct ink writing (DIW) is another advanced additive manufacturing process closely related to AM techniques. In DIW, the polymer material, typically in a liquid state, is extruded from a nozzle. This extrusion can occur either in a molten form, with or without heating. DIW allows for precise control over the deposition of material, making it a versatile method for creating intricate structures and functional prototypes. Researchers have used this technique and silk fibroin ink to create optical waveguides with good optical properties, biocompatibility, and biodegradability, making them interesting for bio photonic sensing devices [36]. We present a hybrid ink that allows for the direct injection welding (3D printing) of stretchy semi-IPN elastomer. In addition to having high-strain SM and shape memory-assisted self-healing capabilities, the ink comprising urethane diacrylate and linear semicrystalline PCL may stretch up to 600%. 4D printing, biological components with shape memory and self-healing capabilities, photonics devices, soft robots, and other fields might benefit from this study [37].

9.3.1.2 Polymerization

Polymerization is a crucial method in additive manufacturing for 4D printing, where the initial material is in liquid form. This process involves curing, where the liquid polymer solidifies to form the desired part. Polymerization can take various forms, each offering unique advantages and applications in 4D printing.

Stereolithography: This process includes the curing of a liquid photopolymer (or monomer) by a laser beam where the vat contains a liquid photopolymer. There is an elevator platform, which is driven vertically by numerical control (NC) drive system from top of the polymer. A suitable laser beam is an incident onto the liquid surface with the help of Galvano mirror, which is con- trolled by the computer. The movement of the reflecting mirror helps in tracing the required paths on the surface to generate the cross-section of the first layer of the object. The platform moves down after the initial layer is done. Based on the value of each layer's thickness provided by the system. Thus, this process continues to form the 3D part layer by layer. Stereolithography, microlithography, and micro stereolithography is used to disperse amorphous silica nanoparticles in a monomeric matrix, primarily hydroxyethyl methacrylate (HE-MA). The nanocomposite is shaped using stereolithography, resulting in a green part and a brown part. The green component undergoes further thermal or solvent polymerization, while the brown component is thermally debonded and sintered at 1,300°C to create optics-grade fused silica glass free of residual porosity and fractures [38]. A hybrid fabrication approach using 3D printing (SLA), metal coating, and wet etching to create complex 2D and 3D metamaterials with novel functionalities. It enables 3D pattering of dielectric layers, enabling unique electromagnetic functionality, and fuses multiple functions into a single metamaterial-embedded geometrical optical device [39]. An optimized process for creating plano-convex lenses using an SLA printer is presented. The lenses were compared to glass lenses and compared using white light interferometry. Results showed low form RMS values, while spin coating resulted in poorer quality. The process allowed for freeform elements production [40].

Polyjet: Polyjet printers utilize different print heads to splash/spray the polymer fluid through a small micro-opening inside the print head like a regular inkjet printer. When the sprayed photopolymers get deposited on to the build platform in a specified design as required, then quick solidification by ultraviolet (UV) light takes place. When the first layer is done, the assembled stage drops along the Z-axis. This process is continued repeatedly until the final part is completely built. A new direction in 4D printing design for programmable actuators, demonstrating thermal actuation in rotational bistable structures and extending them to multistable ones has been demonstrated [41]. The multistable structures enable room-temperature programming and thermal actuation at higher temperatures, simplifying the programming process and paving the way for potential applications in smart and programmable actuators. In another work, researchers have demonstrated the possibility of combining polyjet printing with shape memory material embedding to create actuated structures [42], which can be potentially utilized for creating thermos responsive substrate for various photonic devices like photonic crystals. In another study, 3D printed THz waveguides using Ka-gome photonic crystal structure, inspired by low-loss hollow-core terahertz inhibit coupling fibers is demonstrated. The micro-structured THz fiber is more attractive due to its customizable core and cladding structure. The study fabricated various waveguides with identical cross-sections and lengths, using VeroWhitePlus polymer in Polyjet system [43].

Digital Light Processing (DLP): DLP 3D printer uses light processing the same as that of a projector. It consists of many micromirrors and other micro electrical mechanical system (MEMS) devices that reflect the required light in a certain direction and in a controlled way, due to which the pattern is developed onto the vat of polymer where the curing is done using a UV source. A hydrophilic UV-curable SMP has been developed for dual information encryption using gray-scale- g-DLP 3D printing [44]. This system allows for erasable, cyclically displayed, and permanently encrypted or decryption performances. The grayscale images create differences in crosslinking degree, and the encryption/decryption process can be controlled through grayscale and 3D geometries. The material also allows for permanent storage or destruction through UV irradiation. This innovative method offers a reliable, pollution-free method for dual information encryption. The use of polymerization techniques is explored, which provides a finely detailed and fine surface finish, in 3D-printed optical components. Its objectives are to evaluate these components' performance and quality and investigate potential new optical features. The production of diffraction gratings by AM, by combining several optical components into a single printed element, is a future direction of this kind of research. The study also discusses additive manufacturing techniques, particularly the DLP method, and the effects of layer thickness on resulting gratings. The findings could lead to the production of tunable and repeatable diffraction gratings [45]. Another innovative new 3D printing method is called oscillation-assisted DLP-based 3D printing, which allows for the creation of optically smooth microlens arrays with a single exposure of grayscale UV light. This method eliminates the traditional layer-by-layer effect and ensures a 1 nm surface roughness. The method also allows for the fabrication of various profiles and curvatures, enhancing the quality and speed of 3D printing [46]. A digital 4D printing strategy with fast, simple setups, designable 3D shapes, tunable shape memory, and versatility for combining multi-SMP within a single material construct. This approach offers unique benefits for nano-photonics and shrinkable electronics and can be used for other multifunctional devices. The chemistry principle can be expanded to various materials [47].

Multi-Jet Printing: This is a printing technique that is the same as that of an inkjet printer. It uses a piezoelectric-based print head to deposit photocurable resin layer by layer. It can also be used to deposit wax onto the built platform. The feature resolution is very high. It is easy to remove support materials which are soluble in water and hence require very less post-processing. Although this technique of AM was utilized first for various 4D printing applications it has not been explored much for the photonics applications.

Two Photon Polymerization: Two-photon polymerization (TPP) is an advanced additive manufacturing technique that utilizes a highly focused laser to polymerize a photosensitive resin with exceptional precision. This process enables the creation of complex three-dimensional structures at the nanoscale level, making it invaluable in fields such as microelectronics, photonics, and biomedical engineering. This is also known as direct laser writing (DLW). By exploiting the nonlinear absorption of photons, two-photon polymerization achieves resolution beyond the diffraction limit, resulting in intricate and high-resolution structures with applications ranging from micro-optics to tissue engineering. This technique is the most utilized for creating photonics devices, and recent advances in research have also provided many novel applications for 4D micro and nano printing [48]. A new approach for fabricating photo-responsive 4D microstructures that can be actuated at different wavelengths in the visible range is presented [49]. The process involves TPP 3D microstructures without dye, replacing them with a dye or mixture of dyes, and curing using UV light. The method has been successfully used to incorporate five different dyes into liquid crystal microstructures, enabling actuation at wavelengths from 400 to 650 nm. Multi-responsive liquid crystal microactuators can also be created using combinations of dyes absorbing at different wave- lengths. Another study presents a 4D micro-printing scheme for creating reconfigurable compound micromachines using a single-material, one-step DLW process [50]. The technique uses two-photon printable smart hydrogel materials to create 3D microstructure with a wide range of gradient deformations. This approach enables the creation of mechanically enhanced micromachines like micro stents, micro cages, and micro-umbrellas. The technology behind TPP uses a photoresist that consists of MBAm ($CH_2[NHCOCH=CH_2]_2$), and a water-soluble photoinitiator. It enables localized property tuning in 3D microstructures by exposure dose tuning in 3D laser lithography, leading to substantial property variations [25]. The photoresist also induces local temperature increases and actuation through two-photon absorption, potentially initiating local responses in 3D structures.

9.3.2 Solid Based

9.3.2.1 Laminated Object Manufacturing (LOM)

This additive manufacturing method uses a combined cutting and adding of material to build the part layer by layer. The profile of the part so received from the slicing data is cut using a laser source. Then each of the layers after being cut based on the part to be built is added to each order. All the remaining excess material has been removed. This is a fast process that produces large-sized parts or prototypes. The material used in this method of additive manufacturing is in the form of powders. As these processes are utilized for large scale AM, these are few previous literatures related to 4D printing in photonics applications.

9.3.2.2 Powder Bed Fusion

This process employs either a laser or an electron beam to melt or fuse metal powder particles layer by layer according to the supplied 3D data, thereby creating the component. Nearly all powder bed fusion technologies utilize a mechanism to evenly spread the powder after each previously fused layer. Once a layer is completed, the build platform descends in the Z direction. After the entire part is printed, the platform rises, and the component is removed from the powder bed, which requires further post-processing. This study examines the use of selective laser melting (SLM) for producing optical components, highlighting its efficiency and carbon sustainability. However, achieving a high reflection grade presents a challenge, as the typical SLM aluminum alloy AlSi10Mg yields a reflection grade of only 50–70% after polishing [51]. The production of miniaturized reflectors and a compact metrology system has been demonstrated for industrial applications. The process chain encompasses roughness measurements, mirror fabrication, and the application of a sensor system. This system creates a point cloud of internal surfaces utilizing a 3D acquisition algorithm.

9.3.2.3 Direct Energy Deposition

This method uses small nozzles through which the metal powders are sprayed and deposited at the same time. In this case, a laser or electron beam fuses it by creating a melt pool and finally solidifying it. The speed is very high as the laser only scans those areas when the power is jetted through the nozzle. It provides good functionality for the deposition of material at only those places where it is required, and hence, it is utilized to build parts above already existing parts [52].

9.3.2.4 Binder Jetting

This process utilizes two types of materials. First, a powder is deposited on the build platform, while the second material is a binder or adhesive supplied by the print head. The print head moves in the X and Y directions, applying the binder only to the areas that need to be solidified. After completing the first layer, the platform lowers in the Z direction by an amount equal to the width or thickness of the part being constructed [53].

9.4 4D Printed Optical Components and Their Applications in Optics and Photonics

Extensive research is being done in 3D printing to fabricate complex 3D optical and photonic devices that allows the light matter interactions beyond 2D planar structures, which were hard to fabricate using conventional lithographic techniques. Using smart materials to fabricate the nanostructures effectively translates external stimuli into optical signals throughout shape memory cycles. The 4D printing of complex optical components have opened new avenues in researching reconfigurable and dynamic optics/photonics devices. Since 4D printing is a new and developing research area, much work still needs to be done, and many applications are still to be explored, especially in optics and photonics. This section discusses various 4D-printed optical components such as fibers/waveguides, optical resonators, photonic crystals, and metamaterials and their applications in color printing, anticounterfeiting, wavefront manipulation, sensors, and actuators.

9.4.1 4D Printed Fibers

Fiber, a technology that uses strands of glass to transmit data, is among the most famous. The capacity to see and react to one's surroundings is provided by the integration of various visual, electrical, and mechanical capabilities. The production of adaptive devices using SMPs has a lot of impact in technological fields. The optical actuation provides flexibility and enhanced capabilities for numerous applications in biomedicine to oceanography. The requirement of controlled and consistent illumination of photo actuator over a long distance is possible by using a waveguide. The soft photo actuator based on liquid crystal elastomer has been developed to study the bending angle more than 90°, which is further used as the micro-hook. This wavelength-dependent fiber photo actuator opens the way to remotely control the cases where direct illumination is not possible [54]. Photonic crystals (PCs) possess a distinctive capability to regulate and manipulate the movement of photons. The fiber-shaped PCs, also called photonic crystal fibers (PCF), give the control for light manipulation. These are characterized by the coaxially grown photonic structures onto the fibers, resulting in optical and electrical functionalities. Combining the shape memory polymers onto PCF helps in achieving high-performance smart wearable devices [16]. Numerous efforts have been made to fabricate the fibers with low levels of attenuation and defect-free structures. Additionally, the choice of materials has also been expanded including doped polymer matrices and even inorganic substances like silica and soft glasses. This leads to achieving intricate configurations by exploiting the shape memory polymers. The semi-crystalline property of polylactic acid (PLA) makes it difficult to choose for drawing fibers. It's possible to regulate the flow of PLA molten during thermal elongation by cladding it with an amorphous substance like polyethylene terephthalate glycol (PETG). Scalability of ultra-compact devices with optical sensing and delivery is made possible by the study of rheological and optical compatibility made possible by the variety of fiber designs [29].

9.4.2 4D Printed Optical Resonators

Optical resonators are structures that confine light through multiple reflections between its surfaces and create standing wave patterns known as resonance at certain wavelength. The characteristic feature of the resonator is its quality (Q) factor, which measures the strength of the light confinement and loss of energy. The optical resonators are generally passive but they can be made tunable by utilizing the liquid crystals [55]. Achieving controllable reversibility of tuning is challenging and cannot be integrated into a single on-chip device. However, employing photo responsive smart materials such as nematic liquid crystal elastomers with azo-dye enables reversible tunability of micro/nanostructures as shown in Figure 9.5. The 3D printing of the nematic LCE using the DLW technique photopolymerize the LC monomer mixture resulting in the photo tunable microstructure. The photo responsivity is due to the azo-dye. The LCE with DLW can be used for fabricating the tunable whispering gallery mode (WGM) by employing two strategies: firstly, use LCE as an actuator or make a resonator cavity of LCE itself. In fully 3D geometry, with the help of DLW it is possible to integrate different photopolymerizable materials such as elastic WGM cavities with the polymeric photonic circuits on the same chip. WGM resonators are a type of optical resonators that have small modal volume and ultrahigh Q factor [56]. As part of the first strategy, two-photon polymerized

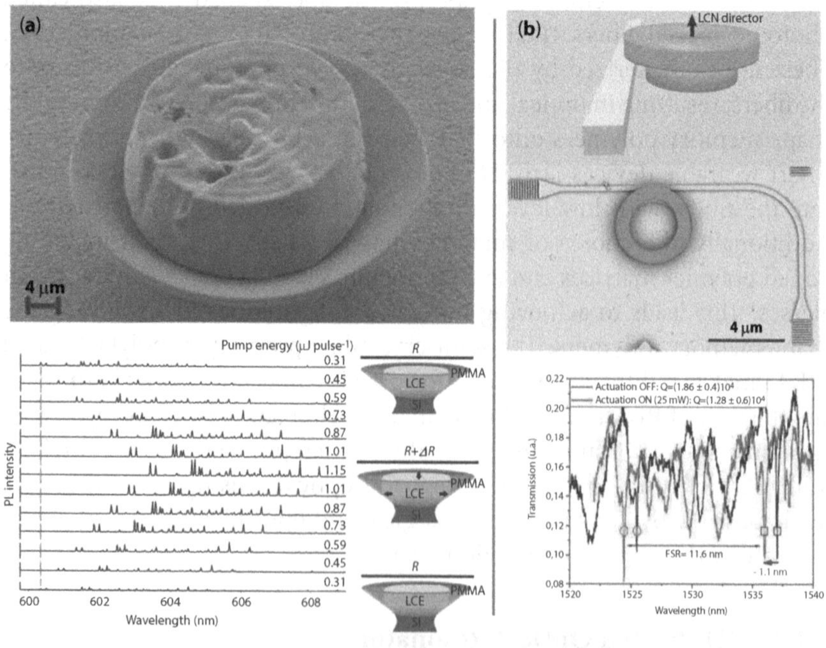

Figure 9.5 4D printed optical resonator (a) From top to bottom: SEM image of the fused LCE micro-cylinder acting as actuator and dye-doped PMMA micro resonator, Photo-induced tuning of laser mode for different pump energies in fused microstructure. Adapted with permission from [57]. Copyright 2024, Springer Nature. (b) From top to bottom: Rendering and SEM image of the LCN resonator (false color for the LCN), transmission spectrum of the LCN ring resonator coupled to the polymeric waveguide with and without actuation. Adapted with permission from [58]. Copyright 2024, American Chemical Society.

LCE microcylinder is placed on the top of the laser dye (PM 597) doped PMMA microcavity, combined through the thermal reflow process. The combined structures have almost same Q factor as when there was no LCE microstructure. Hence, LCE microstructure does not deteriorate the Q factor of the polymeric cavity as the modes are confined near the rim rather than the center. However, when the combined structure is optically pumped with a nanosecond laser pulse, the polymeric structure expands and the lasing modes redshift as a result of light-induced mechanical deformation of the LCE microstructure. Both conditions are entirely reversible [57]. In the second case, where the LCE constitutes the cavity itself, it aims at maximizing the light-matter interaction to exploit the refractive index variation of LCE for tunability of resonant wavelength. When the LCE cavity is illuminated, the extraordinary refractive index, parallel to the cavity's axis of symmetry decreases in relation to light intensity. Conversely, the ordinary refractive index in the perpendicular plane increases [58]. As the LCE cavity expands the resonant modes exhibits blue shift. Here, the blue shift in the resonant modes is observed as the LCE cavity expands. These properties of 4D printed LCE based optical resonator can be used for various tunable integrated photonics applications.

9.4.3 Structural Color Generation and Anticounterfeiting

Structural color in photonics is generated from the micro-and nanostructures arranged in a pattern that manipulates light through interference, diffraction, and scattering, rather than from pigments. These colors are generally angle-dependent. Since the dimensions and pattering of the micro-nanostructures are fixed, the generated colors are passive, which limits their applicability in dynamic displays and anticounterfeiting, and rewritable papers. Reconfigurable structural colors have been achieved by using tunable materials, changing the refractive index of the surrounding environment, or the geometrical parameter of the micro-nanostructures. All the existing work on color optics is based on 2D planar designs that are fabricated by a conventional lithographic process. However, the introduction of 3D printing in optics and photonics has opened the doorway to harnessing the potential of three-dimensional complex structures made of low refractive index materials in a single lithographic process. The structures such as photonic crystals, diffraction grating, nanopillars are generally used to achieve the full color or grayscale painting. The static color generation can be made active by incorporating the SRMs with 3D printing.

The 3D printing of the cholesteric liquid crystals (CLC) to make different shapes of photonic structure such as flower, a butterfly and a wing pattern leads to the generation of colors due to the reflection from the surface of the structure. The varying humidity and temperature change the pitch of CLC network that modulates the shape and color of the structure due to the hygroscopic nature of the polymer. When nanostructures of optically responsive ionogel is fabricated using DLW they display varying structural colors. The redshift in structural colors comes from the hydrophilic nature of the ionogel that swells in presence of polar solvents such as water, ethanol, isopropanol either in liquid or vapor state indicating the change in geometrical parameters.

9.4.4 Infrared Detector

Infrared (IR) imaging has become a crucial technology in various field ranging from thermography and medical diagnostics to firefighting and security. Conventionally there are two types of IR detectors technologies: first, cooled photon detectors and second, uncooled thermal detectors. Photon detectors operate at cryogenic temperature to minimize the thermal heat for its fast and sensitive IR detection. However, this increases its size and cost [59]. Whereas thermal detectors are small and power efficient but less sensitive and slow response time as compared to the cooled detectors as they include bolometers, pyroelectric detectors, thermopiles and other such devices to measure the temperature dependent physical properties [60–64]. Resonant IR detectors can be a way out for thermal detector in terms of sensitivity and response time.

The figure of merit for an infrared (IR) detector is the noise equivalent temperature difference (NETD), which quantifies the ratio of the temperature change at the R source to the signal-to-noise ratio (SNR) of the IR detector's output. To enhance the sensitivity of a resonant IR sensor, it is essential to either maximize absorbance, reduce thermal losses in the detector, or increase the temperature coefficient of the resonant frequency (TCF). It is possible to boost the TCF by using shape memory polymers, which have a high Young's modulus that depends on temperature. Shape memory polymers, which are programmable materials, have the ability to remember and revert to their original shape when subjected to environmental influences such as heat.

The shape memory polymer such as MM4520 membrane is patterned using laser cutter into the circular and square shaped resonator that is assembled on the PMMA substrate to form a single resonator unit (Figure 9.6(a) and (c)). The SMP resonator is heated due to IR radiation form the source that modifies its mechanical properties. The phase of SMP

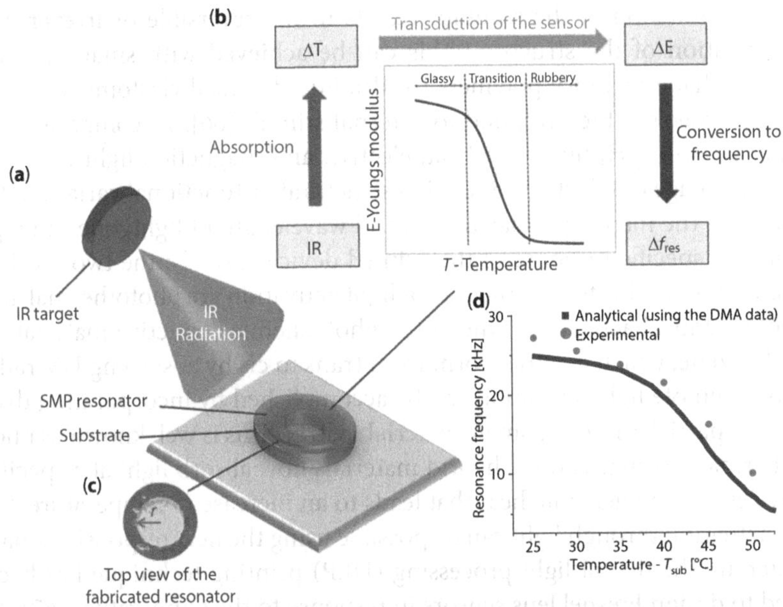

Figure 9.6 Illustration of SMP IR detector and its working principles. (a) Schematic of SMP resonator. (b) The working principle of the IR detector. (c) Top view. (d) Resonance frequency with respect to temperature compared with the analytical values.

transitions from a rigid state to a rubbery state as it crosses the glass transition temperature, both below and above this threshold (Figure 9.6(b)). As a result, this modifies the resonator's mechanical resonance frequency, which can be determined using a frequency reading device and translated back into fluctuations in temperature (Figure 9.6(d)). That is used to calculate the TCF. The larger the TCF, the larger the temperature change, and the larger the frequency shift. Higher TCH will lead to higher sensitivity of the IR detector [65]. Such SMP based detectors have high TCF; hence, are very sensitive. The NETD and Q factor can be further enhanced by using SiN membrane as substrate. The 4D printed SMP based IR detectors are particularly promising in terms of sensitivity at room temperature, flexibility and reduced costs.

9.4.5 Sensors and Actuators

3D-printed structures give us a fixed shape which cannot be changed further. If we want to add responsive behavior in our structure, then 4D printing is the way to achieve this. One of the factors involved in 4D printing is the type of stimulus, which can be biological, chemical, or physical.

The introduction of the stimulus results in the reversible or irreversible deformation of the structure. This can be achieved with smart materials like shape memory polymers (SMP), liquid crystal elastomers (LCE), and hydrogels, which respond to external stimuli [66]. In comparison to other stimuli (pH, humidity, heat, electric, and magnetic), light provides the advantage of selectively inducing structural or functional variation. By adjusting the intensity, polarization, and wavelength of light, one can regulate the specific response of the printed device. Usually, the two mechanisms that are able to play the role of light activation are photothermal and photochemical activation. One of the photochemical reactive materials is azobenzene, which can transform from trans to cis by absorbing UV radiation. Remote light activation can be accomplished by incorporating dyes or nanoparticles into the smart material matrix. This is well known as photothermal heating, as these hybrid materials now absorb light at a specific wavelength and generate heat that leads to an increase in temperature. So, the actuation through light can be possible using the heat responsive smart material. The digital light processing (DLP) printing technique has been used to design Fresnel lens sensors in response to the temperature [67]. In this work, they have discussed the fabrication of Fresnel lens sensors using thermochromic materials. This type of material undergoes coloration or decoloration at the transition temperature. The thermochromic pigment powders are used as adding to the resin and are magnetically stirred, which do not provide any impact on the optical phenomena through the lens as shown in Figure 9.7. This 4D printed fresnel lens is used for temperature monitoring as it exhibits color variation due to the addition of thermochromic materials in the transparent resin, which is used for the printing of the lens. Liquid crystal elastomers (LCE) have the ability to make reversible morphing, which can be used in soft robotics and biomedical purposes [68]. In this work Azobenzene functionalized LCEs (ALCEs) are used that come in the category of photo-responsive material. On exposure to light, these can undergo a stable transformation from one shape to another by instigating supramolecular interactions. On exposure to 365 nm ultraviolet (UV) light, the absorption of photons in the surface layer itself takes place, resulting in uneven distribution, and hence, bending is observed. Illuminating visible light leads to a decrease in the absorption band of the shape-switching LCE, further speeding up the process of reversibility that indicates the cis-trans reverse isomerization [69]. In this work, PUCB is taken as the photo-responsive material and further FDM is done. Here, it is shown that the intensity $76mW/cm^2$ of sunlight is able to transform the 3D printed shape. They have printed the sunflower using PUCB (petals) and PLA (vase). The shape memory behavior of PUCB, which is triggered due

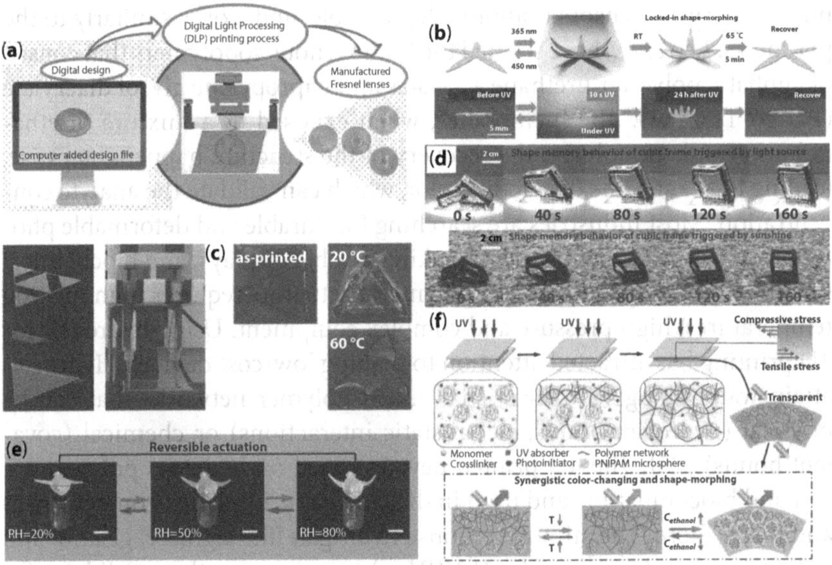

Figure 9.7 (a) Illustration of 3D printing through direct laser printing process [67] (b) The photo-switchable deformations are shown in flower-like actuator [68]. (c) Bilayers formed from the temperature-sensitive polymer and inactive material show the thermal transition above 32°C [70]. (d) The recovery of the cubic frame triggered by external illumination [69] (e) The reversible actuation of the 3D aircraft model in response to relative humidity change from 20% to 80% [71] (f) The fabrication principle of 3D photonic (color-changing) hydrogel actuator is illustrated [72].

to the illumination of light, replicates the phenomenon of heliotropism of plants. The development of shape and color morphing technologies used in soft robotics, data encryption, and displays has been inspired by color and shape- changing organisms in nature, like chameleons, that respond to external stimuli. The integration of a color-changing layer with an active polymer helps in the simultaneous morphing of color and shape [73]. The well-known phenomenon of creating a forbidden gap by periodically modulating the refractive index gives rise to the photonic crystals. The color-changing actuators can be obtained by integrating hydrogel actuators (that can show shape deformation) with photonic crystal (variation of color due to the change in lattice constant). Researchers have developed a method to create a photonic hydrogel actuator based on soft permeable photonic crystals with switchable structural colors [72]. In the presence of external stimuli like ethanol concentration, the color and shape of the structure can be reversed. This brings significant advancements to biomimetic color-changing soft robotics. The ongoing research on chromogenic photonic crystal sensors is affordable, reusable, and works similarly to the pH sheets. Leo *et al.*, have studied the homogenous copolymers that consist of aliphatic polyester urethane acrylates and tripropylene glycol diacrylate (PU-co-TPGDA). The membranes, when exposed to a mixture of ethanol or any alkane solvent, the recovery of the structure happens stepwise, leading to the specific diffractive color, which can validate the analyte concentration. Most industries are searching for durable and deformable photonic devices that can be used in making flexible displays, and biochemical sensors [74]. The production of polymeric actuators requires high melting temperatures, high pressure and complex equipment. Using hydrogels for 4D printing has attracted attention to making low-cost devices. Hydrogels attain water along with their crosslinked polymer networks that can be physical (hydrogen bonds, electrostatic interactions) or chemical (covalent bonds) [75]. The 4D printed devices based on hydrogel polymer are soft and biocompatible and used in drug delivery, implants, contact lenses and many more. The stimuli response hydrogels like alginate, gelatin, and poly(n-isopropyl acrylamide) (PNIPAm) have become the candidates for the application of actuators, grippers, and drug delivery. The researchers have presented the bio polyester that has an interpenetrating network of hydrogen and covalent bonding, which also exhibits a self-healing ability. The bilayer actuator consists of hygroscopically active at one layer and inactive at another, showing the bending towards the inactive layer in the presence of humidity [71]. It also maintains its performance due to its self-healing property. Hydrogels are also sensitive to temperature. The inks containing linear polyurethane polymers and UV-curable monomers have the ability to show tunable mechanical and swelling properties [70]. The well-known technology of infrared (IR) imaging has been adopted in various fields like security and food inspection. The IR detectors generally used are photon detectors (work at cryogenic temperature) and thermal detectors (work at uncooled temperature but have low sensitivity and response time). Adiyan *et al.*, have reported the IR sensor using shape memory polymers that show thermal response.

Despite operating in a vacuum, this uncooled infrared detector maintains high sensitivity even when exposed to air pressure. The MM4520 SMP pellets are used in the fabrication of the resonators that sit above the polymethyl methacrylate (PMMA). The piezoelectric actuator, when put on the heater, is able to adjust the temperature further. In this case, the phase change polymer serves as both an absorber of long-wavelength (7-14 μm) infrared light and an inducer of the transduction process [65]. Incorporating both sensing and actuation in a single device eradicates the assembly and compatibility issue, thus enhancing the efficiency of the device. Embedding the piezoelectric nanoparticles into the polymer matrix for realizing the multifunctional structures has been explored for integrating both sensing and actuation with the single material [76]. Two-photon polymerization direct laser writing (TPP-DLW) is utilized to create micro actuators using hydrogels with a low cross-linking density. However, the resolution is negatively affected by the swelling of the surrounding photoresist, which disrupts the low cross-link density networks. Researchers have developed a photonic photoresist for manufacturing stimuli-responsive photonic micro actuators based on supramolecular cholesteric liquid crystals (CLC) using TPP-DLW. The ability of cholesteric liquid crystals (CLC) to reflect light has been employed to illustrate structural color and shape. The structures developed from the supramolecular photonic photoresist show the dual response (by changing the shape due to humidity and temperature) [77]. The inspiration taken from the cuttlefish has evolved the idea of pigmented structural color actuators. A cholesteric liquid crystal elastomer has been utilized with a reduced actuation temperature to efficiently create a pigmented structural color actuator. Both the structural and pigmented coloring are controllable using infrared light and temperature. The NIR-absorbing dye is used that can cause a photothermal effect and drive the actuation process. In soft robotics, the pigmented structural color actuators could be used for signaling and camouflage [78]. Structural colors produced through the interaction of light with microstructures and nanostructures are essential for blending, signaling, and recognition. The domain of optical information encryption leverages the resistance to photochemical degradation and the programmability of flexible colors [79]. The dynamically reconfigurable colors can be created using SMPs in optical devices that offer submicron scale resolution, rapid color changes, and the ability to recover shape upon heating. The multicolor invisible inks created through high-resolution 4D printing techniques by using the optically transparent thermosetting polymer resin have potential applications in information encryption, anti-counterfeiting, and sensors [80]. Light has the ability to induce stimuli-responsive structural changes remotely. The multicolor 4D printing approach for creating SMP structures is used to remotely actuate the structures by light through selective heating. The SMP sample is first deformed and then heated to a temperature at which the structure is locked in place after cooling. The thermomechanical programming is needed for SMPs to recover their original shapes. This leads to the use of SMP-based color-dependent structural changes, which can be used multiple times and offer remote actuation [81].

9.5 Future of 4D Printed Photonics and Emerging Novel Applications

To develop and enhance photonics using 4D printing technology, researchers are exploring a new paradigm. Many industries stand to benefit greatly from this confluence of technology, including aviation, transportation, healthcare, and telecommunications, among many others. The early advancements in 3D-printed products have set the stage for competition with conventional manufacturing methods. However, one significant limitation of 3D-printed structures is the post-processing requirements and the overall product quality. To address these issues and facilitate large-scale rapid prototyping and manufacturing, the 4D printing process was introduced, incorporating the element of time. This innovative printing category provides various capabilities for altering structure and function over time. In addition to materials that can change shape, 4D printing also employs materials that can change color when exposed to visible and ultraviolet light. The color and shape-shifting properties can be modified based on the printing parameters and stimulus conditions. With its overall merits, the domain of 4D printing has become one of the fast-growing researches which has the ability to revolutionize also the fashion, and robotics sectors. In the future, 4D printing can bring endless contributions by integrating machine learning to tackle and optimize the precise problem. Research has steadily progressed over the years toward developing materials suitable for 4D printing that also support nano printing. Significant efforts have been made to incorporate active materials into various engineering sectors, including soft robotics and the aerospace industry. These materials can also be configured to take on a variety of shapes to fulfil the necessity accordingly. Developing the materials with several functionalities has the ability to go for significant adaptations that can be used in various fields of research. This technology allows us to go for complicated and customizable photonic devices that enable tailoring to specific applications. Controlling the material properties helps in creating gradient profiles possible through 4D printing technology. This technology also streamlines the fabrication process and reduces production time, which helps in designing customizable products. It also paves the way for creating reconfigurable optical components used in the telecommunication industry. Adaptive imaging systems for virtual and augmented reality applications can also be built using the 4D printed photonic components. Incorporating responsive materials into photonics can help in developing dynamic optical systems. There is a possibility of programmable light manipulation, which can be achieved using 4D printed devices. This flipping of optical states would help in creating reconfigurable optical switches and adaptive filters. One can draw inspiration from nature and think of creating a biomimetic photonic structure that can show adaptable characteristics according to environmental stimuli. Signaling and camouflaging are possible in the presence of external stimuli. Using this, the era of smart textiles can also flourish. Although 4D printing technology has emerged in the last few years, there is still scope for refining the printing techniques and developing feasible materials. More reliable commercial products are yet to be realized that would become cutting-edge developments in the near future.

9.6 Conclusion

This review highlights a new wave of innovation and application possibilities in various fields through the integration of 4D printing technology. The study focuses on the potential of 4D printing within the field of photonics, showcasing its capability to create flexible structures with enhanced optical properties. The main findings and future directions are summarized as follows:

- By employing photo-responsive smart materials and the innovative design of micro-nanostructures, it becomes possible to produce vibrant colors that may change dynamically, thus surpassing the constraints of fixed color generation. Subsequent research should prioritize improving the extent and durability of these alterations in color to expand their utilization in dynamic exhibits and anti-counterfeiting systems.
- By incorporating thermo-responsive smart materials, 4D printing processes can create objects capable of changing their shape and optical characteristics in reaction to temperature variations. Additional research should focus on enhancing the responsiveness and durability of these materials to broaden their utilization in practical applications.
- SMP-based IR detectors have shown substantial enhancements in sensitivity and re-action time at room temperature, making them an attractive alternative to conventional detectors. Further research should investigate the incorporation of these sensors into more intricate systems and improve their efficiency in different conditions.

- Materials produced through 4D printing possess the beneficial ability to alter their shapes when exposed to external stimuli, which makes them particularly well-suited for applications in soft robotics and biomedical devices. Further research should be conducted to further the development of biocompatible materials and investigate their durability and safety in medical applications over an extended period of time.
- 4D printing allows for the production of adaptable optical components, including imaging systems that may be adjusted for use in virtual and augmented reality. Subsequent investigations should prioritize enhancing the accuracy of these elements and investigating their incorporation into marketable goods.
- Stimuli-responsive hydrogels, which possess features that may be altered in response to external factors, provide cost-effective solutions for several applications, including photo-responsive drug delivery and intelligent actuators. Subsequent research should focus on enhancing the mechanical strength and responsiveness of these hydrogels to improve their practical usefulness.

Photonics devices can take advantage of 4D printing to provide dynamic functionality. This allows for the creation of tunable photonic crystals, reconfigurable diffraction gratings, and adaptive lenses, expanding the range of possibilities in this field. These devices can be used in telecommunications, environmental sensing, and smart textiles. Subsequent investigations should prioritize the integration of these devices into more extensive systems to optimize their influence. Combining machine learning with 4D printing can improve the design and performance of printed structures. Further investigation is warranted to examine the capacity of machine learning algorithms in forecasting and improving the efficiency of 4D printed materials. The prospects for 4D printed photonics are promising, offering limitless opportunities for creativity and practical use. With the continuous advancement of research, we expect the emergence of increasingly reliable and economically feasible solutions that will profoundly transform various industries.

Acknowledgement

This research received support from the Ministry of Human Resource Development (MHRD), Government of India.

References

1. Yeh, C., *Applied photonics*, Academic Press, San Diego, California, USA, 1994.
2. Saleh, B.E. and Teich, M.C., *Fundamentals of photonics*, John Wiley & Sons, New York, USA, 2019.
3. Winzer, P.J., Neilson, D.T., Chraplyvy, A.R., Fiber-optic transmission and networking: the previous 20 and the next 20 years. *Opt. Express*, 26, 18, 24190–24239, 2018.
4. Chuang, S.L., *Physics of photonic devices*, John Wiley & Sons, Weinheim, Germany, 2012.
5. Gibson, I., Rosen, D.W., Stucker, B., *et al.*, *Additive manufacturing technologies*, p. 17, Springer, Singapore, 2021.
6. Balashanmugam, N., *Perspectives on additive manufacturing in Industry 4.0*, pp. 127–150, Elsevier, Amsterdam, Netherlands, 2021.
7. Gao, H., An, J., Chua, C.K., Bourell, D., Kuo, C.N., Tan, D.T., 3D printed optics and photonics: processes, materials and applications. *Mater. Today*, 69, 107–132, 2023.
8. Baev, A., Prasad, P.N., Ågren, H., Samóc, M., Wegener, M., Metaphotonics: An emerging field with opportunities and challenges. *Phys. Rep.*, 594, 1–60, 2015.
9. Camposeo, A., Persano, L., Farsari, M., Pisignano, D., Additive manufacturing: applications and directions in photonics and optoelectronics. *Adv. Opt. Mater.*, 7, 1, 1800419, 2019.
10. Tibbits, S., *The emergence of "4D printing"*, TED Conferences, LLC (Technology, Entertainment, Design), New York, USA, 2013.
11. Samal, B.B., Jena, A., Varshney, S.K., Kumar, C.S., 4D printing of shape memory polymers: a comparative study of programming methodologies on various material properties. *Smart Mater. Struct.*, 32, 7, 074003, 2023.
12. Jeong, H.Y., Lee, E., An, S.C., Lim, Y., Jun, Y.C., 3D and 4D printing for optics and metaphotonics. *Nanophotonics*, 9, 5, 1139–1160, 2020.
13. Vainsencher, A., Satzinger, K., Peairs, G., Cleland, A., Bi-directional conversion between microwave and optical frequencies in a piezoelectric optomechanical device. *Appl. Phys. Lett.*, 109, 3, 33107, 2016.
14. Mirhosseini, M., Sipahigil, A., Kalaee, M., Painter, O., Superconducting qubit to optical photon transduction. *Nature*, 588, 7839, 599–603, 2020.
15. Liu, J., Tian, H., Lucas, E., *et al.*, Monolithic piezoelectric control of soliton microcombs. *Nature*, 583, 7816, 385–390, 2020.

16. Cheng, J., Zhang, L., Zhao, K., *et al.*, Flexible multifunctional photonic crystal fibers with shape memory capability for optical waveguides and electrical sensors. *Ind. Eng. Chem. Res.*, 60, 23, 8442–8450, 2021.
17. Liu, X., Bruch, A.W., Tang, H.X., Aluminum nitride photonic integrated circuits: from piezo- optomechanics to nonlinear optics. *Adv. Opt. Photonics*, 15, 1, 236–317, 2023.
18. Fan, L., Sun, X., Xiong, C., Schuck, C., Tang, H.X., Aluminum nitride piezo-acousto-photonic crystal nanocavity with high quality factors. *Appl. Phys. Lett.*, 102, 15, 153507, 2013.
19. Bochmann, J., Vainsencher, A., Awschalom, D.D., Cleland, A.N., Nanomechanical coupling between microwave and optical photons. *Nat. Phys.*, 9, 11, 712–716, 2013.
20. Wang, Y., Li, M., Chang, J.K., *et al.*, Light-activated shape morphing and light-tracking materials using biopolymer-based programmable photonic nanostructures. *Nat. Commun.*, 12, 1, 1651, 2021.
21. Li, H., Li, C., Sun, W., *et al.*, Single-Stimulus-Induced Modulation of Multiple Optical Properties. *Adv. Mater.*, 31, 23, 1900388, 2019.
22. Wei, H., Han, L., Yin, R., *et al.*, Micro-3D printed Concanavalin A hydrogel based photonic devices for high-sensitivity glucose sensing. *Sens. Actuators, B*, 386, 133707, 2023.
23. Ge, J., Hu, Y., Yin, Y., Highly tunable superparamagnetic colloidal photonic crystals. *Angew. Chem.*, 119, 39, 7572–7575, 2007.
24. Nocentini, S., Martella, D., Parmeggiani, C., Zanotto, S., Wiersma, D.S., Structured optical materials controlled by light. *Adv. Opt. Mater.*, 6, 15, 1800167, 2018.
25. Hippler, M., Blasco, E., Qu, J., *et al.*, Controlling the shape of 3D microstructures by temperature and light. *Nat. Commun.*, 10, 1, 232, 2019.
26. Zhou, Y., Layani, M., Wang, S., *et al.*, Fully printed flexible smart hybrid hydrogels. *Adv. Funct. Mater.*, 28, 9, 1705365, 2018.
27. Zhuo, X., Shen, H., Bian, Y., Xu, A., Zhu, R., Projection-suspended stereolithography 3D printing for low-loss optical hydrogel fiber fabrication. *APL Photonics*, 6, 12, 121302, 2021.
28. Zhao, P., Chen, H., Li, B., Tian, H., Lai, D., Gao, Y., Stretchable electrochromic devices enabled via shape memory alloy composites (SMAC) for dynamic camouflage. *Opt. Mater.*, 94, 378–386, 2019.
29. Strutynski, C., Evrard, M., Désévédavy, F., *et al.*, 4D Optical fibers based on shape-memory polymers. *Nat. Commun.*, 14, 1, 6561, 2023.
30. Sun, J. and Hu, F., Three-dimensional printing technologies for terahertz applications: a review. *Int. J. RF Microwave Comput. Aided Eng.*, 30, 1, e21983, 2020.
31. Squires, A., Constable, E., Lewis, R., 3D printed terahertz diffraction gratings and lenses. *J. Infrared, Millimeter, Terahertz Waves*, 36, 72–80, 2015.

32. Hernandez-Serrano, A. and Castro-Camus, E., Quasi-Wollaston-prism for terahertz frequencies fabri- cated by 3D printing. *J. Infrared, Millimeter, Terahertz Waves*, 38, 567–573, 2017.
33. Canning, J., Hossain, M.A., Han, C., Chartier, L., Cook, K., Athanaze, T., Drawing optical fibers from three-dimensional printers. *Opt. Lett.*, 41, 23, 5551–5554, 2016.
34. Cook, K., Canning, J., Leon-Saval, S., *et al.*, Air-structured optical fiber drawn from a 3D-printed preform. *Opt. Lett.*, 40, 17, 3966–3969, 2015.
35. Zhao, Q., Tian, F., Yang, X., *et al.*, Optical fibers with special shaped cores drawn from 3D printed preforms. *Optik*, 133, 60–65, 2017.
36. Parker, S.T., Domachuk, P., Amsden, J., *et al.*, Biocompatible silk printed optical waveguides. *Adv. Mater.*, 21, 23, 2411–2415, 2009.
37. Kuang, X., Chen, K., Dunn, C.K., Wu, J., Li, V.C., Qi, H.J., 3D printing of highly stretchable, shape- memory, and self-healing elastomer toward novel 4D printing. *ACS Appl. Mater. Interfaces*, 10, 8, 7381–7388, 2018.
38. Kotz, F., Arnold, K., Bauer, W., *et al.*, Three-dimensional printing of transparent fused silica glass. *Nat.*, 544, 7650, 337–339, 2017.
39. Sadeqi, A., Rezaei Nejad, H., Owyeung, R.E., Sonkusale, S., Three dimensional printing of meta- material embedded geometrical optics (MEGO). *Microsyst. Nanoeng.*, 5, 1, 16, 2019.
40. Berglund, G.D. and Tkaczyk, T.S., Fabrication of optical components using a consumer-grade lithographic printer. *Opt. Express*, 27, 21, 30405–30420, 2019.
41. Jeong, H.Y., Lee, E., Ha, S., Kim, N., Jun, Y.C., Multistable thermal actuators via multimaterial 4D printing. *Adv. Mater. Technol.*, 4, 3, 1800495, 2019.
42. Samal, B.B., Jena, A., Mishra, D., *Controlled shape changing components by using 4D printing technology*, pp. 78–81, 10th International Conference on Precision, Meso, Micro and Nano Engineering, IIT Madras, 2017.
43. Yang, J., Zhao, J., Gong, C., *et al.*, 3D printed low-loss THz waveguide based on Kagome photonic crystal structure. *Opt. Express*, 24, 20, 22454–22460, 2016.
44. Liu, S., Zhao, H., Lyu, Y., *et al.*, Grayscale stereolithography 3D printing of shape memory polymers for dual information encryption based on reconfigurable geometry and tunable optics. *Chem. Eng. J.*, 487, 150552, 2024.
45. Vallejo-Melgarejo, L.D., Reifenberger, R.G., Newell, B.A., Narváez-Tovar, Garcia-Bravo, J.M., Characterization of 3D-printed lenses and diffraction gratings made by DLP additive manufacturing. *Rapid Prototyping J.*, 25, 10, 1684–1694, 2019.
46. Yuan, C., Kowsari, K., Panjwani, S., *et al.*, Ultrafast three-dimensional printing of optically smooth microlens arrays by oscillation-assisted digital light processing. *ACS Appl. Mater. Interfaces*, 11, 43, 40662–40668, 2019.
47. Zhang, Y., Huang, L., Song, H., *et al.*, 4D printing of a digital shape memory polymer with tunable high performance. *ACS Appl. Mater. Interfaces*, 11, 35, 32408–32413, 2019.

48. Korrayi, R.R., Jena, A., Marupalli, B.C., Samal, B.B., Varshney, S.K., Kumar, C.S., Four-Dimensional (4D) Microprinting: Materials, Processes, Challenges and Applications, in: *Additive Manufacturing with Novel Materials: Processes, Properties and Applications*, vol. 2024, pp. 429–457, River Street, Hoboken, USA & Beverly, MA, USA.
49. Hsu, L.Y., Mainik, P., Münchinger, A., *et al.*, A Facile Approach for 4D Microprinting of MultiPhotoresponsive Actuators. *Adv. Mater. Technol.*, 8, 1, 2200801, 2023.
50. Jin, D., Chen, Q., Huang, T.Y., Huang, J., Zhang, L., Duan, H., Four-dimensional direct laser writing of reconfigurable compound micromachines. *Mater. Today*, 32, 19–25, 2020.
51. Heinrich, A., Rank, M., Maillard, P., *et al.*, Additive manufacturing of optical components. *Adv. Opt. Technol.*, 5, 4, 293–301, 2016.
52. Alojaly, H.M., Hammouda, A., Benyounis, K.Y., *Review of recent developments on metal matrix composites with particulate reinforcement*, Elsevier, Amsterdam, Netherlands, 2023.
53. Feng, K., Hu, S., Zhao, W., *et al.*, Bimodal powder optimization in SiC binder jetting for mechanical performance. *Int. J. Mech. Sci.*, 274, 109278, 2024.
54. He, Y., Liang, H., Chen, M., *et al.*, Optical fiber waveguiding soft photo-actuators exhibiting giant reversible shape change. *Adv. Opt. Mater.*, 9, 21, 2101132, 2021.
55. Humar, M., Ravnik, M., Pajk, S., Mŭsevič, I., Electrically tunable liquid crystal optical microresonators. *Nat. Photonics*, 3, 10, 595–600, 2009.
56. Armani, D., Kippenberg, T., Spillane, S., Vahala, K., Ultra-high-Q toroid microcavity on a chip. *Nat.*, 421, 6926, 925–928, 2003.
57. Flatae, A.M., Burresi, M., Zeng, H., *et al.*, Optically controlled elastic microcavities. *Light Sci. Appl.*, 4, 4, e282–e282, 2015.
58. Nocentini, S., Riboli, F., Burresi, M., Martella, D., Parmeggiani, C., Wiersma, D.S., Three-Dimensional Photonic Circuits in Rigid and Soft Polymers Tunable by Light. *ACS Photonics*, 5, 8, 3222–3230, 2018. doi: 10.1021/acsphotonics.8b00461.
59. Rogalski, A., Infrared detectors: status and trends. *Prog. Quantum Electron.*, 27, 2, 59–210, 2003. doi: https://doi.org/10.1016/S0079-6727(02)00024-1.
60. Rogalski, A., Novel uncooled infrared detectors. *Opto-Electron. Rev.*, 18, 4, 478–492, 2010.
61. Chen, C., Yi, X., Zhao, X., Xiong, B., Characterizations of VO2-based uncooled microbolometer linear array. *Sens. Actuators, A*, 90, 3, 212–214, 2001.
62. Kang, D.H., Kim, K.W., Lee, S.Y., Kim, Y.H., Gil, S.K., Influencing factors on the pyroelectric properties of Pb (Zr, Ti)O3 thin film for uncooled infrared detector. *Mater. Chem. Phys.*, 90, 2–3, 411–416, 2005.
63. Schaufelbuhl, A., Schneeberger, N., Munch, U., *et al.*, Uncooled low-cost thermal imager based on micromachined CMOS integrated sensor array. *J. Microelectromech. Syst.*, 10, 4, 503–510, 2001. doi: 10.1109/84.967372.

64. Hui, Y., Gomez-Diaz, J.S., Qian, Z., Alu, A., Rinaldi, M., Plasmonic piezoelectric nanomechanical resonator for spectrally selective infrared sensing. *Nat. Commun.*, 7, 1, 11249, 2016.
65. Adiyan, U., Larsen, T., Zárate, J.J., Villanueva, L.G., Shea, H., Shape memory polymer resonators as highly sensitive uncooled infrared detectors. *Nat. Commun.*, 10, 1, 4518, 2019.
66. Ahmed, A., Arya, S., Gupta, V., Furukawa, H., Khosla, A., 4D printing: Fundamentals, materials, applications and challenges. *Polymer*, 228, 123926, 2021.
67. Ali, M., Alam, F., Fah, Y.F., Shiryayev, O., Vahdati, N., Butt, H., 4D printed thermochromic Fresnel lenses for sensing applications. *Composites, Part B*, 230, 109514, 2022.
68. Lu, X., Ambulo, C.P., Wang, S., *et al.*, 4D-printing of photoswitchable actuators. *Angew. Chem. Int. Ed.*, 60, 10, 5536–5543, 2021.
69. Yang, H., Leow, W.R., Wang, T., *et al.*, 3D printed photoresponsive devices based on shape memory composites. *Adv. Mater.*, 29, 33, 1701627, 2017.
70. Naficy, S., Gately, R., Gorkin, I.I.I.R., Xin, H., Spinks, G.M., 4D printing of reversible shape morphing hydrogel structures. *Macromol. Mater. Eng.*, 302, 1, 1600212, 2017.
71. Cao, J., Zhou, C., Su, G., *et al.*, Arbitrarily 3D configurable hygroscopic robots with a covalent–noncovalent interpenetrating network and self-healing ability. *Adv. Mater.*, 31, 18, 1900042, 2019.
72. Duan, J., Cui, L., Li, M., Fan, W., Sui, K., Biomimetic 3D Color-Changing Hydrogel Actuators Constructed Based on Soft Permeable Photonic Crystals. *ACS Appl. Mater. Interfaces*, 15, 46, 54018–54026, 2023.
73. Jeong, H.Y., An, S.C., Jun, Y.C., Light activation of 3D-printed structures: from millimeter to submicrometer scale. *Nanophotonics*, 11, 3, 461–486, 2022.
74. Umar, M., Min, K., Kim, S., Advances in hydrogel photonics and their applications. *APL Photonics*, 4, 12, 120901, 2019.
75. Dong, Y., Wang, S., Ke, Y., *et al.*, 4D printed hydrogels: fabrication, materials, and applications. *Adv. Mater. Technol.*, 5, 6, 2000034, 2020.
76. Bodkhe, S. and Ermanni, P., 3D printing of multifunctional materials for sensing and actuation: Merging piezoelectricity with shape memory. *Eur. Polym. J.*, 132, 109738, 2020.
77. Del Pozo, M., Delaney, C., Bastiaansen, C.W., Diamond, D., Schenning, A.P., Florea, L., Direct laser writing of four-dimensional structural color microactuators using a photonic photoresist. *ACS Nano*, 14, 8, 9832–9839, 2020.
78. Zhang, P., Debije, M.G., dLT, H., Schenning, A.P., Pigmented structural color actuators fueled by near-infrared light. *ACS Appl. Mater. Interfaces*, 14, 17, 20093–20100, 2022.
79. Huang, F., Weng, Y., Lin, Y., Zhang, X., Wang, Y., Chen, S., Wetting-enhanced structural color for convenient and reversible encryption of optical information. *ACS Appl. Mater. Interfaces*, 13, 35, 42276–42286, 2021.

80. Zhang, W., Wang, H., Wang, H., *et al.*, Structural multi-colour invisible inks with submicron 4D printing of shape memory polymers. *Nat. Commun.*, 12, 1, 112, 2021.
81. Jeong, H.Y., Woo, B.H., Kim, N., Jun, Y.C., Multicolor 4D printing of shape-memory polymers for light-induced selective heating and remote actuation. *Sci. Rep.*, 10, 1, 6258, 2020.

10

Methods, Materials, Shape Programming, and Applications of 4D Printing

Jian Ming Lee[†], Jie Wei Chee[†], Joel Zi Xu Wong[†], Li Yang Foong[†] and Marwan Nafea*

Department of Electrical and Electronic Engineering, Faculty of Science and Engineering, University of Nottingham Malaysia, Semenyih, Selangor, Malaysia

Abstract

Four-dimensional (4D) printing has emerged as a modern technology that allows the development of dynamic structures that are printed from smart materials. The 4D printing process offers several advantages compared to traditional fabrication methods of smart materials in terms of the speed, complexity, and cost of the fabrication process. Therefore, this technique has caught the attention of researchers around the world within a short period of time after it was proposed. Motivated by these attractive features, this chapter discusses the concept of 4D printing, as well as its main methods and materials. The main characteristics, advantages, and disadvantages of these methods and materials are also discussed. The chapter also discusses the main shape programming and shape change approaches in this area. In addition, the chapter discusses some of the main applications of 4D printing in the areas of electronics, biomedicine, origami, and robotics.

Keywords: 4D printing, shape memory polymer, shape memory effect, shape programming, smart material

**Corresponding author*: marwan.nafea@nottingham.edu.my
[†]*Equal contribution*

Bijaya Bikram Samal, Cheruvu Siva Kumar and Shailendra Kumar Varshney (eds.) 4D Printing Technology: Principles, Materials and Applications, (315–344) © 2025 Scrivener Publishing LLC

10.1 Introduction

Over the years, design and manufacturing have been done through standardized and inflexible ways. Traditional manufacturing processes include casting and molding, machining, joining, and shearing and forming. These methods require high manufacturing time, cost, and complexity [1]. In various industries, the rising demand of consumers in terms of diversity of needs and fast-changing demand have been causing a bottleneck among manufacturers. Therefore, the emergence of additive manufacturing (AM) has a great impact on the efficiency of manufacturing, prototyping complex structures, mass production, and customizability to fulfill the great demand of the stakeholders, especially the customers and clients [2]. Additive manufacturing, or three-dimensional (3D) printing, provides low-cost, fast, ease of use, and direct digital fabrication approaches that facilitate computer-aided design (CAD) to computer-aided manufacturing (CAM), providing endless opportunities in a wide range of applications [3].

Due to the attractive features of 3D printing, this technology gained a lot of attention in the past decade. This led to extending the technology by utilizing smart materials, which opened doors for a new technology that is known as four-dimensional (4D) printing [4]. The concept of 4D printing was first introduced to the world by Skylar Tibbits at the Ted conference in 2012, where he demonstrated the transformation of a 3D printed structure into a more complex shape when placed under a certain stimulus. This technology gained interest due to its potential to reduce the inefficiencies within the manufacturing, construction, and infrastructure industry with the newfound self-assembly aspect of 4D printing [5]. 3D-printed structures alter their shapes to their programmed shape over time under the influence of external stimuli. Thus, 4D printing is 3D printing but with an additional dimension of time [6]. Therefore, with the appropriate printing and design configuration along with the utilization of the smart properties of 4D printing materials, products that are capable of self-construction can be fabricated from a single printing session. This approach also simplifies the post-production process, where the assembly and integration processes can be achieved in more efficient approaches compared to standard fabrication methods [7].

There are six factors that act as a base for all 4D printing designs, as illustrated in Figure 10.1. These factors include the printing method, printing material, activation stimulus, shape programming, shape memory effect (SME) type, and the applications of the 4D-printed structure.

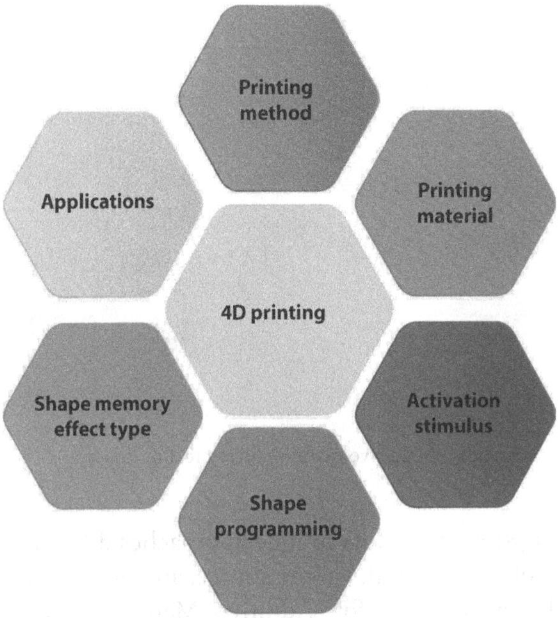

Figure 10.1 Important factors in 4D printing.

This chapter[1] focuses on discussing these important factors and the concepts related to them. In addition, this chapter aims to link these factors in a way that allows researchers to understand the basics of 4D printing and select suitable printing methods and printing materials that suit their desired actuation performance and application. The following sections discuss the main printing methods used in this area, 4D printing materials and their stimuli, shape programming methods, the main shape changes in 4D printed structures, the main application of 4D printing, and a future perspective and conclusion.

10.2 4D Printing Methods

The trend of AM, or 3D Printing technology is rising rapidly with the active promotion of global leading industrial countries [7]. The technology has been developed into various distinct branches, with varying

[1]Parts of this chapter have appeared in a group project's thesis entitled: "Jie Wei Chee, Li Yang Foong, Joel Zi Xu Wong, Jian Ming Lee, Development of 4D Printed Origami Structures, Group Project Thesis, University of Nottingham Malaysia, 2022." The project was supervised by Marwan Nafea. The authors declare that they have no conflict of interest related to this publication.

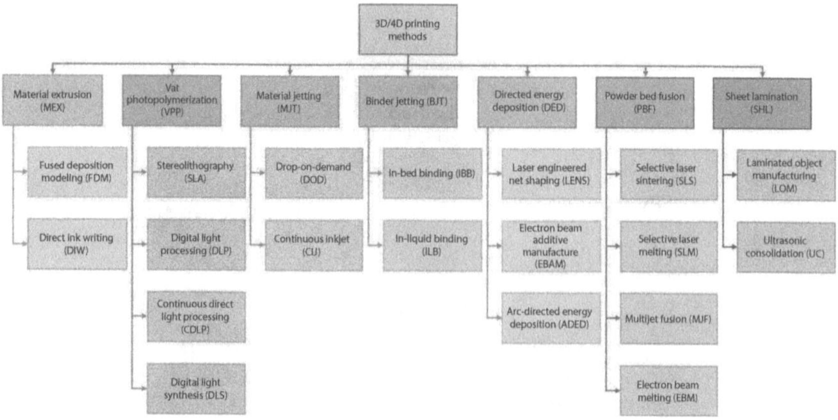

Figure 10.2 Classification of additive manufacturing technologies.

characteristics [8] each with its unique approaches differentiated by layer creation technology, material, energy source, and build volume [7]. Based on Standard Terminology for Additive Manufacturing Technologies F2792-12, a designation of the American Society for Testing and Materials (ASTM) International Committee F42 on Additive Manufacturing, AM methods have been classified and standardized into seven different categories [7–9]. The seven main categories of AM technologies are material extrusion (MEX), vat photopolymerization (VPP), material jetting (MJT), binder jetting (BJT), directed energy deposition (DED), powder bed fusion (BPF), and sheet lamination (SHL) [9], as shown in Figure 10.2. Along with the introduction of smart materials and multi-material structures, additive manufacturing could expand its endless possibilities in the 4D printing field. In this chapter, the emphasized discussion on AM technologies is based on the widely used printing techniques, where some of the main methods will be selected from Figure 10.2.

10.2.1 Material Extrusion (MEX)

MEX is a printing technique in which material that is in filament, paste, or semi-liquid form is programmed to be extruded through a nozzle or orifice and be printed layer-by-layer [8, 9]. It can be classified into two main sub-groups, which include fused deposition modeling (FDM), which extrudes melted thermoplastic polymers, and direct ink writing (DIW) which extrudes ink directly without heat [7].

Fused deposition modeling (FDM), also known as fused filament fabrication (FFF) is the most popular AM method [10, 11]. FDM uses thermoplastic polymer filaments that are heated above their melting temperature and extruded in line-by-line and layer-by-layer form using a printing head [12]. This method of printing is less costly, provides an easy printing process, utilizes a wide range of filaments, and is multi-material applicable. Multi-material printing commonly appears in the market as multi or dual-extruder printers [7], that can be used to create ornamental two-color structures [13], functional two materials structures such as self-folding hinges [14], or even conductive circuits [15]. Furthermore, FDM technology also provides a manufacturing solution, a bi-extruder that melts two kinds of filaments into a single nozzle. By adjusting the ratio of the filaments, parameters such as the conductivity of the structure could be controlled [16, 17]. Another reason that FDM is widely accepted is the wide range of filaments mentioned previously [18]. Common materials such as pure polylactic acid (PLA), a mixture of other materials with PLA, and acrylonitrile butadiene styrene (ABS), and other materials, such as nylon, epoxy resin, carbon, glass, and Kevlar fibers could be used as FDM filaments [7]. However, the drawbacks of FDM are the low resolution and roughness issues [13].

Direct ink writing (DIW) is an extrusion printing process without heating, also known as syringe printing [18]. DIW extrudes semi-liquid, paste-form material or polymetric ink through the syringe onto the printing bed. The material is dispensed through pressured air, piston, or screw [7]. DIW holds the advantage of having diverse materials, versatility, and higher resolution than FDM [13]. This method of printing allows lots of freedom since heat is not necessary for printing, hence being suitable for biomedical applications [18]. However, DIW requires post-curing after the printing of every layer and has a small build volume [7, 13]. Table 10.1 presents a comparison between several features associated with FDM and DIW.

10.2.2 Vat Photopolymerization (VPP)

VPP is an additive fabrication method that uses photopolymer liquids, or radiation-curable resins as its stock material [10]. During the process, liquid photopolymer in a vat is programmed to be cured by radiation in the ultraviolet (UV) light or some specific range of visible light to transform into a solid through a chemical reaction [9]. Photopolymerization techniques can be classified into three main categories, which are stereolithography (SLA), digital light processing (DLP), continuous digital light processing (CDLP), and digital light synthesis (DLS) [7].

Table 10.1 Comparison between FDM and DIW.

Feature	FDM	DIW
Material Type	Thermoplastic filaments	Ceramics, pastes, bio-inks, metals
Material State	Solid filament	Liquid suspension
Post-processing	Minimal	Sintering, de-binding, finishing
Applications	Rapid prototyping, functional parts	Bioprinting, microfluidics, electronic circuits
Advantages	Affordable, easy to use, wide material variety	Advanced materials, diverse functionalities
Disadvantages	Limited to thermoplastics	More complex process, post-processing required

SLA is a process that cures the resin mixture with a laser emitting in UV range to a solidified polymer. SLA has high spatial resolution and fine finish as the area triggered by the polymerization reaction is targeted precisely by the continuous emitting laser beam [19]. The laser movement ensures the smoothness of the surface which is preferred by high precision applications [18]. On the other hand, the mechanical properties of photopolymers cause instability of printed objects [7].

While SLA is cured by laser point-by-point, DLP is cured by a projector all at once [19]. Therefore, DLP has a faster printing speed but with rougher finishing of the object, while SLA requires a lower cost of printers but a higher cost of materials [7]. Printable ink of DLP is also reported with very high mechanical strength and good shape memory properties which are applicable in industrial use [20]. CDLP is similar to DLP but uses a powerful projector to cure a layer with a single, continuous exposure. This method offers even faster printing speeds compared to standard DLP due to the continuous curing. However, this approach requires specialized resins for CDLP's high-intensity light source.

On the other hand, DLS is similar to DLP but uses a permeable membrane at the bottom of the resin vat. This allows the uncured resin to be continuously removed through the membrane while printing, which creates a cleaner printing process with reduced support structures. Moreover, this method offers high-quality prints with improved surface finish and

Table 10.2 Comparison between SLA, DLP, CDLP, and DLS.

Feature	SLA	DLP	CDLP	DLS
Light Source	Focused UV laser	Digital projector	High-intensity digital projector	Digital projector
Layer Curing	Point-by-point scanning	Entire layer at once	Continuous exposure	Entire layer at once
Printing Speed	Low	Medium	High	Medium
Resolution	Higher	Medium	Medium	High
Applications	Detailed parts, high accuracy	Faster prototyping, larger prints	Very fast prototyping	High-quality parts, reduced supports

faster printing speeds. A comparison between the main VPP-based printing methods is presented in Table 10.2.

10.2.3 Material Jetting (MJT)

MJT is an additive technique in which droplets of build material such as photopolymer or thermoplastic materials are selectively deposited layer by layer [19]. MJT is the second most widely used 3D printing technology that secures its position with high accuracy, high resolution, and high printing freedom and allows multi-material printing [18]. The liquid material that is jetted solidifies after deposition via chemical changes, cooling, or solvent evaporation [21]. In material jetting, inkjet, or polyjet, material droplet creation is categorized into two common yet different methods, termed drop-on-demand (DOD) and continuous inkjet (CIJ) [21, 22]. DOD has advantages in higher accuracy, smaller drop size, and fewer restrictions on ink properties, while CIJ is used where higher printing speed, higher drop velocity, or fluid with lower viscosity is required [8].

Multijet modeling (MJM) and polyjet modeling (PJM) use the concept of DOD to 3D print objects. MJM utilizes inkjet technology to simultaneously deposit multiple materials, such as thermoplastic polymers and waxes onto a build platform. Meanwhile, PJM uses a similar concept, but

Table 10.3 Comparison between DOD and CJ.

Feature	DOD	CJ
Jetting Process	Deposits droplets on demand	Continuous jet with deflection
Printing Speed	Medium	High
Resolution	High	Medium
Applications	Detailed parts, high accuracy	Faster prototyping, larger prints

it is focused on photopolymer materials. Multi-material or multi-color structures can be achieved by jetting different materials together, allowing active and passive layers to form single bending or sequentially triggered bending of objects [7]. Although it offers multiple advantages in terms of printing results, this manufacturing process is relatively expensive. DOD and CJ are briefly compared in Table 10.3.

10.2.4 Binder Jetting (BJT)

BJT uses ceramic powder, metal powder, or polymer powder as materials, where a liquid bonding agent is jetted to bond them together to form a structure layer-by-layer [7]. This method offers a wide range of materials that could fabricate objects in full color but exhibit poor strength and rough finish. This printing method can be classified into two subcategories, including in-bed binding (IBB) and In-liquid binding (ILB). When using IBB, a layer of powder material is spread across a platform. Then, the printhead selectively jets a binding agent onto the powder within the powder bed itself. This causes the binder to adhere the targeted powder particles together according to the design. The process is repeated layer-by-layer to build the entire object. On the other hand, when using ILB, the powder material is first suspended in a liquid bath containing the binding agent. Then, the printhead disrupts the liquid bath selectively, causing the binder to adhere to the desired powder particles in the liquid. Later, the binder-coated particles accumulate and solidify to form the 3D object layer-by-layer. A brief comparison between these methods is presented in Table 10.4.

Table 10.4 Comparison between IBB and ILB.

Feature	IBB	ILB
Binding Location	Within the powder bed	In a liquid bath
Printhead Action	Jets binder onto a powder	Disrupts liquid, promoting selective binding
Support Structure	Unbound powder	Requires additional support or specific bath properties

10.2.5 Directed Energy Deposition (DED)

DED uses focused thermal energy through laser or plasma to fuse or melt molten metal powder together and deposit it on the printing bed layer-by-layer [8, 19]. This method can be used for the fabrication of new objects or to repair damaged structures. However, the materials are limited to metals and metal alloys, and the process generates poor dimension accuracy and poor resolution [23]. The printing technologies that are branched under DED are laser engineered net shaping (LENS), electron beam additive manufacture (EBAM), and arc-directed energy deposition (ADED) [7]. LENS utilizes a high-powered laser beam as the energy source that melts either a powder feedstock or a wire feedstock to deposit the molten material onto the build platform. Meanwhile, EBAM uses a high-energy electron beam to melt the material feedstock. On the other hand, ADED utilizes an electric arc created between the print head and a wire electrode to melt the wire material, which is then deposited onto the build platform. These three methods are compared in Table 10.5.

10.2.6 Powder Bed Fusion (PBF)

PBF is an AM technology in which regions of a powder bed are selectively fused by thermal energy [9]. Materials such as ceramic powder, polymers, and atomized metal powder are used while heat sources are applied to melt the powder into desired patterns. Among the branches of powder bed fusion, there are selective laser sintering (SLS), selective laser melting (SLM), multijet fusion (MJF), and electron beam melting (EBM) [7].

One of the better-known PBF fabrication methods is SLS, in which a laser is used to selectively melt thin layers of plastic powder during the process and build up layer-by-layer in the powder bed [7]. Among the advantages, it is worth mentioning that SLS is able to manufacture parts with

Table 10.5 Comparison between ADED, LENS, and BEAM.

Feature	ADED	LENS	EBEAM
Energy Source	Electric arc	High-powered Laser Beam	High-energy electron beam
Advantages	High deposition rates, affordable	Higher accuracy, resolution	Highest power density, wider material compatibility
Disadvantages	Lower resolution, higher heat input	Slower printing speeds, higher equipment cost	Most complex and expensive, requires a high-vacuum environment

good mechanical properties in comparison with standard manufacturing technology, such as injection molding although it does not necessarily have the same properties [7]. Furthermore, SLS allows a great collection of materials that is still constantly expanding, which includes metals, polymers, and a mixture of metals and ceramic or polymers [19]. SLM relies on a similar concept, but it is used for metal powders. Meanwhile, EBM uses an electron beam in a high-vacuum environment to perform the AM process of metal powders. On the other hand, MJF uses droplets of binding agents heated to a high temperature by an infrared lamp to print objects using polymer powders. Table 10.6 presents a comparison between PBF-based printing methods.

10.2.7 Sheet Lamination (SHL)

SHL method bonds sheets of materials such as paper, ceramic tape, metallic sheet, or plastic film to create an object by lamination, cutting, and pasting [19]. This method of printing is fast and inexpensive but requires much labor for de-cubing [7]. Development of technology has generated laminated objected manufacturing, (LOM) and ultrasonic consolidation (UC) [8].

LOM uses paper sheets with a heat-activated adhesive coating. A laser cutter forms the desired shapes from each sheet. Then, the sheets are bonded together layer-by-layer using heat and pressure to adhere to the adhesive coating. On the other hand, UC uses metal sheets as the material, where ultrasonic welding is used to create bonds between the metal layers. Table 10.7 presents a comparison between LOM and UC.

Table 10.6 Comparison between SLS, SLM, MJF, and EBM.

Feature	SLS	SLM	MJF	EBM
Heat Source	High-powered laser beam	High-powered laser beam	Infrared lamp (heated binding agent droplets)	Electron beam (high vacuum)
Material	Polymers (nylon, PA, thermoplastics)	Metals (stainless steel, titanium, aluminum, alloys)	Polymer powders with functional additives	Metals (including reactive metals like titanium)
Process	Selective sintering of powder particles	Complete melting of powder particles to form a dense metal structure	Selective binding of powder particles with heated droplets	Melting powder particles within a vacuum chamber

Table 10.7 Comparison between LOM and UC.

Feature	LOM	UC
Material	Paper sheets with adhesive coating	Thin metal sheets or ribbons
Process Mechanism	Laser cutting and heat/ pressure bonding	Ultrasonic welding
Layer Bonding	Adhesive activation	Solid-state metal bonding
Advantages	Affordable, easy to use, rapid prototyping	Functional metal parts, good conductivity
Disadvantages	Limited material options (mainly paper)	Complex process

10.3 4D Printing Materials

The materials used in 4D printing play an important role in achieving the required actuation. These materials should possess the characteristics of responding to stimuli by altering their properties, such as shape and color. These materials are known as programmable materials, stimulus-responsive materials, or smart materials [5]. Using these materials grants 3D printed structures the ability to be programmed so that they can be activated to predicted designs under the influence of external stimuli. This is the result of the SME that smart materials possess [24, 25]. Smart materials can be classified into several categories, where shape-memory materials (SMMs) are the most commonly used category in 4D printing [26]. SMMs can be further classified into shape memory polymers (SMPs) [27], shape memory alloys (SMAs) [28], magnetic shape memory alloys (MSMAs) [29], and shape memory ceramics (SMCs) [30], where SMPs are the most commonly used SMMs in 4D printing. SMPs are usually preferred due to their wider range of glass transition temperatures, which range from -70 °C up to 100 °C. This allows their elastic properties to be adjusted based on various applications. SMPs also have a higher recovery rate as compared to SMAs, while SMAs commonly require a higher cost and rather complex manufacturing procedures as compared to SMPs [7, 31]. Polymer-derived

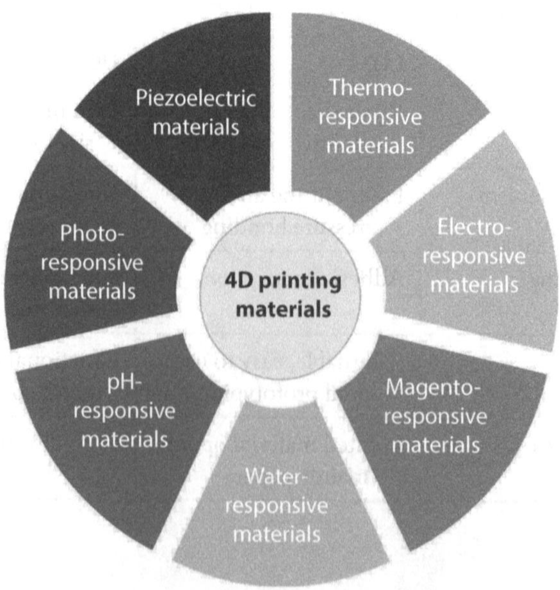

Figure 10.3 Classification of the main 4D printing materials.

ceramics have also been used in 4D printing, as they possess remarkable properties, such as corrosion and oxidation resistance and high thermal stability [32]. SMMs can be used in different types of applications, such as in miniature sensors, actuator systems, space systems [33], self-healing applications, and many more [26].

4D printing materials can be classified based on the types of stimuli that result in changes in their properties. Generally, the most commonly used 4D printing materials can be classified as thermo-responsive, electro-responsive, magneto-responsive, water-responsive, pH-responsive, photo-responsive, and piezoelectric materials, as illustrated in Figure 10.3. These different categories allow greater flexibility for the user to decide the type of stimuli that would be suitable for the application of the design [5].

10.3.1 Thermo-Responsive Materials

Thermo-responsive materials are SMMs that respond to heat stimuli, which have a property called the glass transition temperature (T_g) [34]. When a structure is programmed and cooled to take the form of a temporary shape, that structure returns to its original shape once its temperature is raised above T_g. The reason for this behavior can be explained when observing the internal energy of the polymer chains when the temperature is below T_g [35]. At that range, the internal energy is low, resulting in a lack of freedom for the polymer chains that cause the material to be rigid. When the temperature is raised beyond T_g, there is sufficient internal energy in the polymer chains, which increases the freedom of movement and results in the material being prone to deformation, causing it to return to its original shape [5]. Another way thermo-responsive materials respond is by the shape-changing effect, which occurs in bilayer structures that have a significant difference in the coefficient of thermal expansion. In this manner, one layer will expand or contract more than the other and result in shape deformation to occur [36].

The most common thermos-responsive material that is used in 4D printing is polylactic acid (PLA) [19, 37]. PLA is made up of renewable resources like corn starch and is biodegradable, which presents itself as a bioplastic and is used in many applications, such as the biomedical field [38]. Moreover, PLA has a low melting point, which results in the energy needed to print being lower as compared to other materials. Furthermore, the low glass transition temperature of PLA also plays a huge role that contributes to the popularity of this material in the field of 4D printing [38].

10.3.2 Electro-Responsive Materials

SMMs that respond to electric current are known as electro-responsive materials. Electricity itself can be referred to as an indirect stimulus, as it is mostly used to apply a Joule heating effect on the structure using the resistance of the material itself when the electric current passes through [5], [39]. Such SMPs can be electrically activated by incorporating conductive fillers within the polymer matrix, such as copper [40], carbon nanotubes [41], carbon fibers [42], and graphene [43]. Such an approach increases the electrical conductivity of SMPs, which allows the use of Joule heat for the activation process [44]. The use of this approach allows precise control of the temperature of the conductive SMP by controlling the supplied power. In addition, using conductive fillers increases the strength of the printed structure, which broadens the range of potential applications for such materials [45].

10.3.3 Magneto-Responsive Materials

SMMs affected by the activation of a magnetic field are referred to as magneto-responsive materials. These materials are usually polymers that are embedded with magnetic particles, such as magnetite iron oxide nanoparticles [46]. These materials have potential as they can allow the printed structure to be activated wirelessly. However, if the structure is to be affected by an activation magnetic field, it must be lightweight, therefore creating limitations on the overall size of the printed structure [18, 36].

10.3.4 Water-Responsive Materials

A functional water-responsive SMM system consists of two components, which are the hydrophilic component and the hydrophobic component [47]. The overall structure is immersed in water, which allows deformation to happen. After deformation, the new shape is fixed by drying the structure [48]. Hydrogel is a great example of water-responsive material as its hydrophily allows it to expand up to twice its original volume, and it is also highly printable using DIW [49]. However, the drying process of a water-responsive material requires a long time that may even lead up to hours [36].

10.3.5 pH-Responsive Materials

These SMMs can respond to pH values and transform accordingly. Polyelectrolytes are common examples of such materials as they are able to transform at different pH values [50]. These materials are polymers that contain ionizable groups, which provide them with unique properties such as high solubility in water and responsiveness to changes in pH and ionic strength [5, 51]. This phenomenon is caused by the polymer chain that stretches and deforms as a result of the protons being released. When the protons are accepted, the structure neutralizes [5].

10.3.6 Photo-Responsive Materials

Photo-responsive materials refer to SMMs that are responsive to light, which is also another indirect stimulus similar to electricity [52]. When a photo-sensitive area of a structure is exposed to light, it absorbs the heat which results in the heating of the structure [53]. This heat will act as a trigger for the transformation of the structure to the programmed shape [5]. An example of such materials is photoreactive polymers, which contain photoreactive groups that undergo crosslinking under light irradiation [54].

10.3.7 Piezoelectric Materials

Piezoelectric materials possess a crystal structure that enables them to convert mechanical energy into electrical energy and vice versa through the piezoelectric effect. By applying controlled electrical signals, the piezoelectric elements can induce mechanical deformation, shape changes, or actuation in the printed structures [55]. Therefore, this process is referred to as piezoelectric stimuli [5, 56]. These materials can be in the form of ceramics or polymers, such as polyvinylidene fluoride (PVDF) [57]. Such materials are frequently used in smart scaffolds and bone tissue engineering [58].

10.4 Shape Programming

SMMs rely on the SME, which can be defined as the transformation of a material into another shape by applying external stimuli to it [5, 26]. The cycle of the SME can be introduced in two stages, which are the programming stage and the recovery stage. In the programming stage, the printed structure is deformed to a temporary shape, and it returns to its original printed shape during the recovery stage when an appropriate stimulus is applied to it externally. It is worth noting that shape programming can be achieved using manual and non-manual shape programming approaches [59].

Manual shape programming relies mainly on deforming the structure under an external force while subjecting it to a stimulus, and then gradually removing the stimulus while maintaining the force. On the other hand, non-manual shape programming can be achieved during the printing process using the printing parameters, which allow controlling the induced strain during printing. Some of these parameters include the printing speed, printing temperature, layer height, and passive-to-active layers ratio [60].

The SME can be categorized into one-way, multi-way, and two-way SMEs [18, 61]. A 4D-printed actuator with a one-way SME can be programmed to transform into a specific shape that is activated using an external stimulus and remains at that shape even after the stimulus is removed. In this case, the structure will not return to its initial shape after removing the stimulus. Meanwhile, multi-way SME allows 4D-printed actuators to be programmed with one or two temporary shapes and a final permanent shape [53, 62]. On the other hand, two-way SME allows 4D-printed actuators to transform back and forth between their programmed shape and initial shape by activating and de-activating the external stimulus [61, 63, 64]. Among these types, one-way SME is the most used actuation principle in 4D printing due to its simplicity [18, 65].

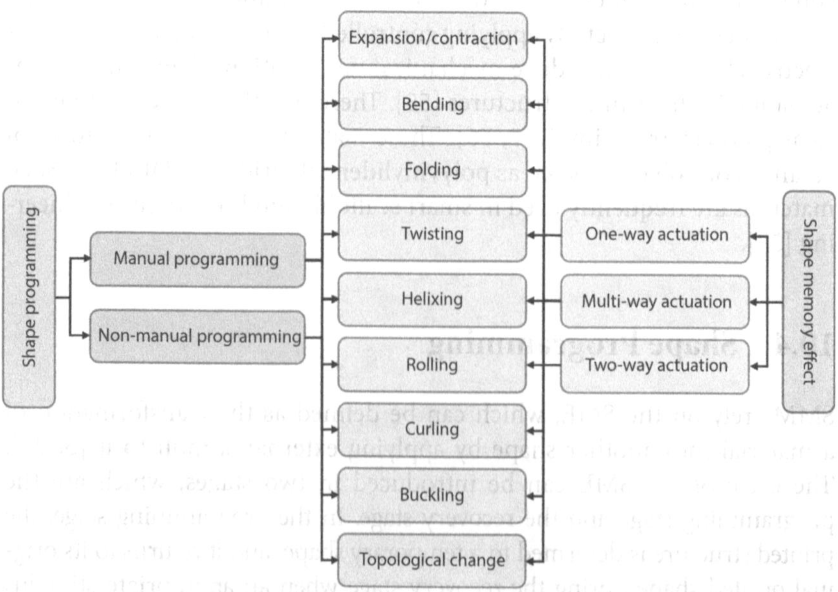

Figure 10.4 Illustration of the relationships between shape programming, shape memory effect, and shape change in 4D printed structures.

The aforementioned types of shape programming and shape memory effect cause 4D printed structures to change their shape in different ways. This shape change is controlled during the shape-programming process [66]. 4D printed structures can change their shape based on different principles, including expansion/contraction, bending, folding, twisting, helixing, rolling, curling, buckling, and topographical change [67]. In addition, combinations of the shape change principles can be used simultaneously or sequentially to achieve complex shape changes. The general principles of shape programming, shape memory effect, and shape change in 4D printing are summarized in Figure 10.4.

10.5 Applications of 4D Printing

4D printing has significant potential in various fields such as electronics, biomedical, origami, and robotics. For instance, in electronics, it allows for the creation of customizable devices with integrated functionalities, such as self-folding electronics and wearable electronics [68, 69]. Moreover, in biomedical fields, it can be used to create dynamic medical implants and devices [70, 71]. Furthermore, in the origami area, can be used to create

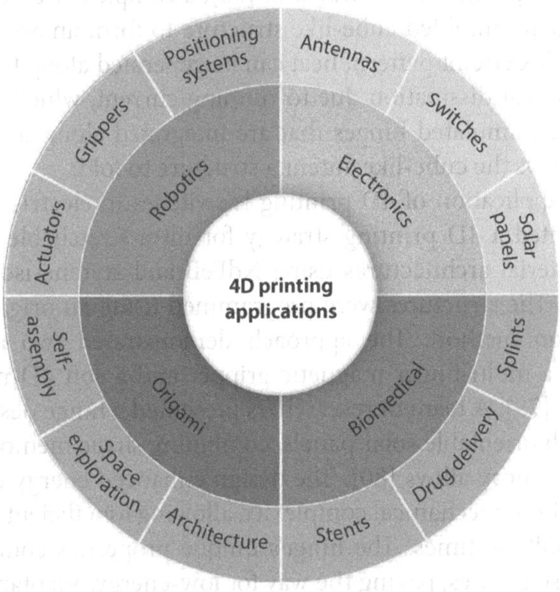

Figure 10.5 Applications of 4D printing in the areas of electronics, biomedical, origami, and robotics.

self-folding structures and deployable systems for various applications [72, 73]. In addition, in the robotics area, it can be used to create self-morphing robots and soft robotics systems, with applications in search and rescue operations and industrial applications [74, 75]. Figure 10.5 illustrates some of the main applications of 4D printing in various research areas. The following sub-sections present several examples of these applications and discuss their significance.

10.5.1 Electronics Applications

With the advantage of 3D printing, many applications could now be further improved such as the case of antennas which could be manufactured as origami shapes that transform after heating [76]. This method allows the fabrication time of antennas to be reduced as well as replacing the necessities for supporting materials as the antenna is being printed directly on the cube itself, thus minimizing manufacturing costs. For instance, a 3D structure, such as a cube, allows numerous planar antennas to be implemented onto every side of the 3D cube. This allows for radiofrequency (RF) harvesting in different directions where it is not possible with normal planar antennas with an on-demand reconfiguration of its surface via origami folding and unfolding [77]. Via the accuracy of 4D printing, detailed circuits can be implanted as intended. In [78], a complex circuit pattern was printed onto an unfolded cube-like structure to form an antenna design. In this complex circuit pattern, heat can be generated along the lines of its circuitry via heat dissipation due to running current, which in turn activates the heat-simulated hinges that are integrated along the circuit and therefore allows the cube-like antenna structure to fold.

Another application of 4D printing is switches in electrical circuits. A study presented a 4D printing strategy for ultra-stretchable magnetoactive soft material architectures using NdFeB and styrene-isoprene block copolymers. The structures were programmed using an origami-inspired magnetization method. The approach demonstrated two applications, which were a multi-finger magnetic gripper and a soft 4D-printed electrical switch [79]. A team of researchers presented a hinge design for a 6U CubeSat with orientable solar panels, combining shape memory polymers and shape memory alloys [80]. The design enhanced energy collection in satellites without mechanical complexity, allowing for efficient stowing and actuation multiple times. The hinge's unique properties could extend to other space structures, paving the way for low-energy, adaptable systems.

10.5.2 Biomedical Applications

A team of researchers presented self-assembly stents fabricated using 4D printing technology, inspired by origami principles [81]. The stents showed shrinkage ratios ranging from 40.3% to 94%, with helical angles enhancing their effectiveness. The lightweight design reduced material consumption and enhanced structural flexibility. The combination of origami principles and 4D printing presented promising opportunities for medical device biomimetic design. Another research presented a gastro-retentive drug delivery system, utilizing pharmaceutical-grade polyvinyl alcohol (PVA) with shape memory properties [82]. The system transitioned from a temporary shape suitable for oral administration to an expanded form for gastric retention, driven by body temperature and minimal fluid contact. Prototypes with various shapes were successfully manufactured using 4D printing or manual processing, showing satisfactory recovery to the original shapes in an acidic environment. While heating and mechanical deformation did not affect release performance, release timescale posed a challenge, addressed by applying a diffusive polymeric film to prolong release to over 6 hours.

The usage of 4D printing technology in the biomedical field is closely related to the advancement of 4D bioprinting technology, which aims to the application of the material transformation process for the manufacturing of biological structures [83]. Additionally, 4D bioprinting could be employed to engineer biological devices to assist in treating life-threatening diseases. Such a concept was demonstrated in [84], where researchers were successfully able to bio-print tracheobronchial splints based on the computed tomography scan of the infant patients' airways suffering from Tracheobronchomalacia. This was done to prevent external compression and collapse of the patient's airways by insertion of the bio-printed splints around the affected airways where the bio-printed splints were capable of growth in proportion to the size of the patient's airways for up to three years before degrading. Another study introduced a material programming approach for designing and 4D-printing a bio-inspired wrist-forearm splint that was based on a twining plant (*Dioscorea bulbifera*) [85]. This splint can be produced using any consumer desktop FDM 3D printer. The authors used computational design to control anisotropic extrusion, allowing for mesoscale precision. The modular nature of the proposed approach allowed for complex configurations of self-shaping structures, expanding the range of bio-inspired designs for technical systems.

10.5.3 Origami Applications

Origami, the traditional art of paper folding, finds its roots in the Japanese words *oru*, which means "fold," and *kami*, which means "paper." Origami is the art of folding a flat sheet of paper, usually squares, into complicated 3D structures, and it is also the art of geometrical transformations, which resulted in its popularity in both art and mathematics society since the 1940s. Origami has been explored to provide more technological solutions to engineering problems, and more practical applications of origami have been discovered. These applications range from the nano-scale up to the meter-scale in architecture applications [72, 86, 87]. By observing the concepts of 4D printing and origami, it can be seen that there are many similarities between them in terms of starting with a simple structure, such as a flat sheet of paper, which can be transformed into a complex and abstract design. Therefore, origami is actively being used in the 4D printing field as a method of design. The combination of 4D printing and origami also results in a field known as active origami, where an origami design is able to self-fold or self-unfold. This creates the possibility of reducing the infrastructure and construction costs of certain applications [87]. This also increases the efficiency and performance as compared to the usual industrial techniques as the 4D printed structures have the ability to self-enhance their properties [5, 88].

Yuan *et al.* [89], were able to demonstrate a self-folding table by integrating a planar square design with printed active composite hinges. When the printed design was actuated using heat, the two-dimensional sheet self-folded to form a three-dimensional table. The table exhibited structural strength by being able to support a small steel screw. A similar design was applied to an origami sailboat and crane where self-folding was also achieved. Origami tessellations are another popular structure among researchers as they enable an infinite repeatable folding pattern with variable size changes as determined by the required application [90]. Some of these tessellations include the Miura Ori [14, 91], square-twist [92], Kresling [93], Yoshimura, and waterbomb [94] designs. All of these tessellation designs differ from each other in terms of their vertex nature which depends on the number of mountain and valley folds present at one vertex of the pattern.

Another aspect where origami was proven to be useful is the James Webb space telescope [95], which was launched in December 2021. Its design was inspired by origami tessellations with 18 hexagonal mirrors joining together to form a dish structure with an aperture in the middle. Hexagonal-based tessellating structures with foldable capabilities have recently been utilized in various engineering designs for applications in the space, electronics, and soft robotics industries [96]. This is where 4D printing could be applied to further enhance the current designs as the folding mechanism could be triggered and controlled remotely once the required stimulus is encountered.

10.5.4 Robotics Applications

4D printing has been extensively used in the robotics field to create actuators, joints, grippers, and positioning systems. This is due to the high level of flexibility offered by 4D printing when it comes to allowing printed structures to change their shapes based on the shape programming process. One of the promising applications of 4D printed actuators is the area of variable stiffness actuators (VSAs). Such actuators have the ability to change their compliance within a certain spatial-temporal state. This concept was presented in [97], where the research presented three varieties of variable stiffness 4D-printed structures created using both single and multiple materials. Single-material configurations adjusted the stiffness through changes in the infill percentage and by employing various printing patterns to influence directional stiffness, as well as hinges with diverse printed patterns. Multi-material actuators integrate materials with different stiffness values within a single structure. The bending angles achieved by these actuators span from 2.05° to 88.25°. Other authors presented interesting concepts, such as self-sensing hydrogel actuators [98], and microwave activation of magnetic PLA [99].

The pattern-driven concept was then extended to develop pattern-driven bioinspired 4D printed grippers [34]. A novel concept of a 4D-printed compliance microgripper was introduced in [100]. The study presented a new method for creating a 4D-printed microgripper utilizing PLA. The microgripper featured two active beams, activated at 85 °C, while the remainder of the structure was passive. FDM was employed to fabricate both active and passive parts, with distinct printing parameters for each. The microgripper achieved a 3 mm actuation distance for each actuator's 3.925 mm deformation distance, resulting in a 0.76 ratio between the two distances. Other research teams proposed continuous fiber-embedded 4D grippers [101], and untethered biodegradable and biocompatible electro- and magneto-active hydrogel grippers [102].

4D printing is also implemented in positioning applications, such as micropositioning stages. Such devices offer the ability to manipulate the position of a small object based on the desired displacement. The first 4D-printed micropositioning stages were presented in [103]. This research presented the development of microstages designed for motion in the X-Y plane, printed using PLA. The study explored the influence of printing speed variation on the recovered shape and proposed a method to compute pre-strain induced during printing. Validation of the method's accuracy was conducted by comparing simulation results with experimental findings. Two microstage designs were introduced, with the first offering a maximum range of movement of 2.67 mm in the *x*-direction and 1.75 mm in the *y*-direction, while the second design provided a maximum range of 1.78 mm in the *x*-direction and 2.22 mm in the *y*-direction.

10.6 Conclusion and Future Outlook

4D printing has received a lot of attention during the last decade. The present status of 4D printing shows a constant development in terms of the printing technologies, printing materials, shape programming methods, as well as a wide range of applications. Despite these advances, significant difficulties need to be addressed before this technology can be implemented by industry. These challenges include materials compatibility, printing resolution, and accuracy, scalability, and cost-effectiveness. Furthermore, the integration of numerous materials and the implementation of precise shape-changing processes are technically challenging. Moreover, the standardization of printing techniques, material characterization, and design approaches is required for the widespread use of 4D printing. Looking ahead, 4D printing has great potential to be implemented in various applications. Continued research and development activities are projected to solve present issues while opening up new prospects for applications in domains, including robotics, wearable technology, soft robotics, and environmental sustainability. The latter might have a significant influence on many industries, including construction, where materials could be engineered to adapt to temperature changes or other environmental conditions, thus reducing the need for energy-intensive heating and cooling systems.

Acknowledgment

This work was supported by the Ministry of Higher Education Malaysia under the Fundamental Research Grant Scheme (FRGS/1/2023/TK09/UNIM/02/4), and the University of Nottingham Malaysia.

References

1. Fan, H., Hu, J., Wang, Y., Zhang, H., Guo, W., Li, J., Xu, S., Li, H., Liu, P., A review of laser additive manufacturing (LAM) aluminum alloys: Methods, microstructures and mechanical properties. *Opt. Laser Technol.*, 175, 110722, 2024.
2. Islam, M.A., Mobarak, M.H., Rimon, M.I.H., Al Mahmud, M.Z., Ghosh, J., Ahmed, M.M.S., Hossain, N., Additive manufacturing in polymer research: Advances, synthesis, and applications. *Polym. Test.*, 132, 108364, 2024.
3. Hu, F., Wang, W., Cheng, J., Bao, Y., Origami spring–inspired metamaterials and robots: An attempt at fully programmable robotics. *Sci. Prog.*, 103, 1, 2020.
4. Shinde, S., Mane, R., Vardikar, A., Dhumal, A., Rajput, A., 4D printing: From emergence to innovation over 3D printing. *Eur. Polym. J.*, 197, 112356, 2023.
5. Ahmed, A., Arya, S., Gupta, V., Furukawa, H., Khosla, A., 4D printing: Fundamentals, materials, applications and challenges. *Polym. (Guildf)*, 228, 123926, 2021.
6. Lee, W.H., Alshebly, Y.S., Mohamed Ali, M.S., Nafea, M., Development of a 4D Printed Variable Stiffness Gripper, in: *2022 IEEE 20th Student Conference on Research and Development (SCOReD)*, p. 151, 2022.
7. Rafiee, M., Farahani, R.D., Therriault, D., Multi-Material 3D and 4D Printing: A Survey. *Adv. Sci.*, 7, 1, 2020.
8. Vaezi, M., Chianrabutra, S., Mellor, B., Yang, S., Multiple material additive manufacturing - Part 1: A review: This review paper covers a decade of research on multiple material additive manufacturing technologies which can produce complex geometry parts with different materials. *Virtual Phys. Prototyp.*, 8, 19, 2013.
9. ISO/ASTM, Additive Manufacturing - General Principles Terminology (ASTM52900), in: *Rapid Manufacturing Association*, p. 10, 2013.
10. Gibson, I., Rosen, D., Stucker, B., 16 - Direct Digital Manufacturing, in: *Additive Manufacturing Technologies. Gibson, I., Rosen, D., Stucker, B., Direct Digital Manufacturing*, (Eds.), pp. 375–397, Springer, New York, USA, 2015. DOI: https://doi.org/10.1007/978-1-4939-2113-3_16.
11. Lee, W.H., Alshebly, Y.S., Ali, M.S.M., Nafea, M., Pattern-Driven 4D Printed PLA Actuators, in: *2022 IEEE 8th International Conference on Smart Instrumentation, Measurement and Applications (ICSIMA)*, p. 116, 2022.
12. Muehlenfeld, C. and Roberts, S.A., 1 - 3D/4D Printing in Additive Manufacturing: Process Engineering and Novel Excipients, in: *3D and 4D Printing in Biomedical Applications: Process Engineering and Additive Manufacturing*, Maniruzzaman, M., (Eds.), pp. 1–23, Wiley, Weinheim, Germany, 2019. DOI: https://doi.org/10.1002/9783527813704.ch1.
13. Jeong, H.Y., Lee, E., An, S.C., Lim, Y., Jun, Y.C., 3D and 4D printing for optics and metaphotonics. *Nanophotonics*, 9, 5, 1139, 2020.

14. Yamamura, S. and Iwase, E., Hybrid hinge structure with elastic hinge on self-folding of 4D printing using a fused deposition modeling 3D printer. *Mater. Des.*, 203, 109605, 2021.
15. Nassar, H. and Dahiya, R., Fused Deposition Modeling-Based 3D-Printed Electrical Interconnects and Circuits. *Adv. Intell. Syst.*, 3, 2100102, 2021.
16. Kim, H. and Lee, S., Characterization of Electrical Heating of Graphene/PLA Honeycomb Structure Composite Manufactured by CFDM 3D Printer. *Fashion Text.*, 7, 14, 2020.
17. Khondoker, M.A.H., Development of a Multi-material Extruder System to 3D Print Hard Thermoplastics. *Soft Elastomers Liq. Met.*, University of Alberta, Edmonton, Canada, 2019.
18. Alshebly, Y.S., Nafea, M., Mohamed Ali, M.S., Almurib, H.A.F., Review on recent advances in 4D printing of shape memory polymers. *Eur. Polym. J.*, 159, 110708, 2021.
19. González-Henríquez, C.M., Sarabia-Vallejos, M.A., Rodriguez-Hernandez, J., Polymers for additive manufacturing and 4D-printing: Materials, methodologies, and biomedical applications. *Prog. Polym. Sci.*, 94, 57, 2019.
20. Li, X., Yang, Y., Zhang, Y., Wang, T., Yang, Z., Wang, Q., Zhang, X., Dual-method molding of 4D shape memory polyimide ink. *Mater. Des.*, 191, 108606, 2020.
21. Hon, K.K.B., Li, L., Hutchings, I.M., Direct writing technology-Advances and developments. *CIRP Ann. Manuf. Technol.*, 57, 601, 2008.
22. Vaezi, M., Seitz, H., Yang, S., A review on 3D micro-additive manufacturing technologies. *Int. J. Adv. Manuf. Technol.*, 67, 1721, 2013.
23. Lee, J.Y., An, J., Chua, C.K., Fundamentals and applications of 3D printing for novel materials. *Appl. Mater. Today*, 7, 120, 2017.
24. Meyer, Y., Lachat, R., Akhras, G., A review of manufacturing techniques of smart composite structures with embedded bulk piezoelectric transducers. *Smart Mater. Struct.*, 28, 0, 2019.
25. Mohd Ghazali, F.A., Hasan, M.N., Rehman, T., Nafea, M., Mohamed Ali, M.S., Takahata, K., MEMS actuators for biomedical applications: a review. *J. Micromech. Microeng.*, 30, 073001, 2020.
26. Abavisani, I., Rezaifar, O., Kheyroddin, A., Multifunctional properties of shape memory materials in civil engineering applications: A state-of-the-art review. *J. Build. Eng.*, 44, 102657, 2021.
27. Bonetti, L., Natali, D., Pandini, S., Messori, M., Toselli, M., Scalet, G., 4D printing of semi-crystalline crosslinked polymer networks with two-way shape-memory effect. *Mater. Des.*, 238, 112725, 2024.
28. Jafarabadi, A., Ferretto, I., Mohri, M., Leinenbach, C., Ghafoori, E., 4D printing of recoverable buckling-induced architected iron-based shape memory alloys. *Mater. Des.*, 233, 112216, 2023.
29. Caputo, M.P., Berkowitz, A.E., Armstrong, A., Müllner, P., Solomon, C.V., 4D printing of net shape parts made from Ni-Mn-Ga magnetic shape-memory alloys. *Addit. Manuf.*, 21, 579, 2018.

30. Kong, D., Guo, A., Wu, H., Li, X., Wu, J., Hu, Y., Qu, P., Wang, S., Guo, S., Four-dimensional printing of polymer-derived ceramics with high-resolution, reconfigurability, and shape memory effects. *Addit. Manuf.*, 83, 104050, 2024.
31. AbuZaiter, A., Nafea, M., Mohd Faudzi, A.A., Kazi, S., Mohamed Ali, M.S., Thermomechanical behavior of bulk NiTi shape-memory-alloy microactuators based on bimorph actuation. *Microsyst. Technol.*, 22, 2125, 2016.
32. Liu, G., Zhao, Y., Wu, G., Lu, J., Origami and 4D printing of elastomer-derived ceramic structures. *Sci. Adv.*, 4, eaat0641, 2018.
33. Duerig, T.W., Stöckel, D., Melton, K.N., Wayman, C.M., *Engineering Aspects of Shape Memory Alloys*, Butterworth-Heinemann, Oxford, UK, 1990.
34. Alshebly, Y.S., Mustapha, K.B., Zolfagharian, A., Bodaghi, M., Mohamed Ali, M.S., Almurib, H.A., Nafea, M., Bioinspired pattern-driven single-material 4D printing for self-morphing actuators. *Sustainability*, 14, 10141, 2022.
35. Joharji, L., Mishra, R.B., Alam, F., Tytov, S., Al-Modaf, F., El-Atab, N., 4D printing: A detailed review of materials, techniques, and applications. *Microelectron. Eng.*, 265, 111874, 2022.
36. Zhang, Z., Demir, K.G., Gu, G.X., Developments in 4D-printing: a review on current smart materials, technologies, and applications. *Int. J. Smart Nano Mater.*, 10, 205, 2019.
37. Tan, D., Nokhodchi, A., Maniruzzaman, M., 2 - 3D and 4D Printing Technologies: Innovative Process Engineering and Smart Additive Manufacturing, in: *3D and 4D Printing in Biomedical Applications: Process Engineering and Additive Manufacturing, Maniruzzaman, M.*, (Eds.), pp. 25–52, Wiley, Weinheim, Germany, 2019. DOI: https://doi.org/10.1002/9783527813704.ch2
38. Mehrpouya, M., Vahabi, H., Janbaz, S., Darafsheh, A., Mazur, T.R., Ramakrishna, S., 4D printing of shape memory polylactic acid (PLA). *Polym. (Guildf)*, 230, 124080, 2021.
39. Lee, Y.C., Alshebly, Y.S., Nafea, M., Joule Heating Activation of 4D Printed Conductive PLA Actuators, in: *2022 IEEE International Conference on Automatic Control and Intelligent Systems (I2CACIS)*, p. 221, 2022.
40. Chan, B.Q.Y., Chong, Y.T., Wang, S., Lee, C.J.J., Owh, C., Wang, F., Wang, F., Synergistic combination of 4D printing and electroless metallic plating for the fabrication of a highly conductive electrical device. *Chem. Eng. J.*, 430, 132513, 2022.
41. Cortés, A., Cosola, A., Sangermano, M., Campo, M., González Prolongo, S., Pirri, C.F., Jiménez-Suárez, A., Chiappone, A., DLP 4D-Printing of Remotely, Modularly, and Selectively Controllable Shape Memory Polymer Nanocomposites Embedding Carbon Nanotubes. *Adv. Funct. Mater.*, 31, 2106774, 2021.
42. Cheng, Y., Li, J., Qian, X., Rudykh, S., 3D printed recoverable honeycomb composites reinforced by continuous carbon fibers. *Compos. Struct.*, 268, 113974, 2021.

43. Zhou, Y., Parker, C.B., Joshi, P., Naskar, A.K., Glass, J.T., Cao, C., 4D Printing of Stretchable Supercapacitors via Hybrid Composite Materials. *Adv. Mater. Technol.*, 6, 2001055, 2021.
44. Huang, X., Panahi-Sarmad, M., Dong, K., Li, R., Chen, T., Xiao, X., Tracing evolutions in electro-activated shape memory polymer composites with 4D printing strategies: A systematic review. *Compos. Part A Appl. Sci. Manuf.*, 147, 106444, 2021.
45. Dong, K., Ke, H., Panahi-Sarmad, M., Yang, T., Huang, X., Xiao, X., Mechanical properties and shape memory effect of 4D printed cellular structure composite with a novel continuous fiber-reinforced printing path. *Mater. Des.*, 198, 109303, 2021.
46. Zolfagharian, A., Haq, M.I.U., Nafea, M., Bodaghi, M., 4D Printing of Smart Magnetic-Based Robotic Materials, in: *3D Printing and Sustainable Product Development*, pp. 213–226, CRC Press, Boca Raton, 2023.
47. Zhang, Y., Raza, A., Xue, Y.-Q., Yang, G., Hayat, U., Yu, J., Liu, C., Wang, H.-J., Wang, J.-Y., Water-responsive 4D printing based on self-assembly of hydrophobic protein "Zein" for the control of degradation rate and drug release. *Bioact. Mater.*, 23, 343, 2023.
48. Ang, J.Y., Chan, B.Q.Y., Kai, D., Loh, X.J., Engineering Porous Water-Responsive Poly(PEG/PCL/PDMS Urethane) Shape Memory Polymers. *Macromol. Mater. Eng.*, 302, 1700174, 2017.
49. Falahati, M., Ahmadvand, P., Safaee, S., Chang, Y.-C., Lyu, Z., Chen, R., Li, L., Lin, Y., Smart polymers and nanocomposites for 3D and 4D printing. *Mater. Today*, 40, 215, 2020.
50. Sajjad, R., Chauhdary, S.T., Anwar, M.T., Zahid, A., Khosa, A.A., Imran, M., Sajjad, M.H., A review of 4D printing – Technologies, shape shifting, smart polymer based materials, and biomedical applications. *Adv. Ind. Eng. Polym. Res.*, 7, 20, 2024.
51. Ahmed, A., Arya, S., Gupta, V., Furukawa, H., Khosla, A., 4D printing: Fundamentals, materials, applications and challenges. *Polym. (Guildf)*, 228, 123926, 2021.
52. Shao, Y., Long, F., Zhao, Z., Fang, M., Jing, H., Guo, J., Shi, X., Sun, A., Xu, G., Cheng, Y., 4D printing Light-Driven soft actuators based on Liquid-Vapor phase transition composites with inherent sensing capability. *Chem. Eng. J.*, 454, 140271, 2023.
53. Patadiya, J., Naebe, M., Wang, X., Joshi, G., Kandasubramanian, B., Emerging 4D printing strategies for on-demand local actuation & micro printing of soft materials. *Eur. Polym. J.*, 184, 111778, 2023.
54. Ren, Z., Ding, K., Zhou, X., Ji, T., Sun, H., Chi, X., Xu, M., 4D printing light-driven actuator with lignin photothermal conversion module. *Int. J. Biol. Macromol.*, 253, 126562, 2023.
55. Nafea, M., Nawabjan, A., Mohamed Ali, M.S., A wirelessly-controlled piezo-electric microvalve for regulated drug delivery. *Sens. Actuators A Phys.*, 279, 191, 2018.

56. Kumar, V., Singh, R., Ahuja, I.P.S., Davim, J.P., On Nanographene-Reinforced Polyvinylidene Fluoride Composite Matrix for 4D Applications. *J. Mater. Eng. Perform.*, 30, 4860, 2021.
57. Kumar, V., Singh, R., Ahuja, I.S., PVDF Composite as the 4D Solution For Heritage Structures. *J. Inst. Eng. India Ser. C*, 104, 1193, 2023.
58. Kumar, V., Singh, R., Ahuja, I.S., 1.19 - Co-relation matrix for 4D properties of PVDF-CaCO3 composite for non-structural applications, in: *Comprehensive Materials Processing*, Hashmi, S., (Eds.), pp. 185–195, Elsevier, Oxford, UK, 2024. DOI: https://doi.org/10.1016/B978-0-323-96020-5.00139-4.
59. Samal, B.B., Jena, A., Varshney, S.K., Kumar, C.S., 4D printing of shape memory polymers: A comparative study of programming methodologies on various material properties. *Smart Mater. Struct.*, 32, 074003, 2023.
60. Alshebly, Y.S. and Nafea, M., Effects of printing parameters on 4D-printed PLA actuators. *Smart Mater. Struct.*, 32, 064008, 2023.
61. Lee, A.Y., An, J., Chua, C.K., Two-Way 4D Printing: A Review on the Reversibility of 3D-Printed Shape Memory Materials. *Engineering*, 3, 663, 2017.
62. Khalid, M.Y., Arif, Z.U., Noroozi, R., Zolfagharian, A., Bodaghi, M., 4D printing of shape memory polymer composites: A review on fabrication techniques, applications, and future perspectives. *J. Manuf. Process*, 81, 759, 2022.
63. Lee, A.Y., An, J., Chua, C.K., Zhang, Y., Preliminary Investigation of the Reversible 4D Printing of a Dual-Layer Component. *Engineering*, 5, 1159, 2019.
64. Wang, Y. and Li, X., 4D printing reversible actuator with strain self-sensing function via structural design. *Compos. B Eng.*, 211, 108644, 2021.
65. De Marco, C., Pané, S., Nelson, B.J., 4D printing and robotics. *Sci. Robot*, 3, 2, 2018.
66. Alshebly, Y.S., Nafea, M., Almurib, H.A.F., Ali, M.S.M., Faudzi, A.A.M., Tan, M.T.T., Development of 4D Printed PLA Actuators with an Induced Internal Strain Upon Printing, in: *2021 IEEE International Conference on Automatic Control & Intelligent Systems (I2CACIS)*, p. 41, 2021.
67. Alshebly, Y.S. and Nafea, M., Control of 4D Printed Actuators Twisting Behavior via Printing Direction, in: *2022 IEEE 20th Student Conference on Research and Development (SCOReD)*, p. 163, 2022.
68. Zhang, J., Yang, X., Li, W., Liu, H., Yin, Z., Chen, Y., Zhou, X., 4D printed multifunctional wearable strain sensors with programmable sensing characteristics. *Compos. B Eng.*, 275, 111346, 2024.
69. McLellan, K., Sun, Y.-C., Naguib, H.E., A review of 4D printing: Materials, structures, and designs towards the printing of biomedical wearable devices. *Bioprinting*, 27, e00217, 2022.
70. Zhang, C., Cai, D., Liao, P., Su, J.-W., Deng, H., Vardhanabhuti, B., Ulery, B.D., Chen, S.-Y., Lin, J., 4D Printing of shape-memory polymeric scaffolds for adaptive biomedical implantation. *Acta Biomater.*, 122, 101, 2021.

71. Mahjoubnia, A., Cai, D., Wu, Y., King, S.D., Torkian, P., Chen, A.C., Talaie, R., Chen, S.-Y., Lin, J., Digital light 4D printing of bioresorbable shape memory elastomers for personalized biomedical implantation. *Acta Biomater.*, 177, 165, 2024.
72. Yi, H., 4D-printed parametric façade in architecture: prototyping a self-shaping skin using programmable two-way shape memory composite (TWSMC). *Eng. Constr. Archit. Manage.*, 29, 4132, 2022.
73. Liu, Y., Zhang, W., Zhang, F., Lan, X., Leng, J., Liu, S., Jia, X., Cotton, C., Sun, B., Gu, B., Chou, T.-W., Shape memory behavior and recovery force of 4D printed laminated Miura-origami structures subjected to compressive loading. *Compos. B Eng.*, 153, 233, 2018.
74. Zolfagharian, A., Lakhi, M., Ranjbar, S., Sayah Irani, M., Nafea, M., Bodaghi, M., 4D printing parameters optimisation for bi-stable soft robotic gripper design. *J. Braz. Soc. Mech. Sci. Eng.*, 45, 224, 2023.
75. Zhai, F., Feng, Y., Li, Z., Xie, Y., Ge, J., Wang, H., Qiu, W., Feng, W., 4D-printed untethered self-propelling soft robot with tactile perception: Rolling, racing, and exploring. *Matter*, 4, 3313, 2021.
76. Kimionis, J., Isakov, M., Koh, B.S., Georgiadis, A., Tentzeris, M.M., 3D-Printed Origami Packaging with Inkjet-Printed Antennas for RF Harvesting Sensors. *IEEE Trans. Microw. Theory Tech.*, 63, 4521, 2015.
77. Wu, L., Huang, J., Zhai, M., Sun, B., Chang, H., Huang, S., Liu, H., Deformable bowtie antenna realized by 4D printing. *Electron. (Switzerland)*, 10, 1792, 2021.
78. Deng, D., Yang, Y., Chen, Y., Lan, X., Tice, J., Accurately controlled sequential self-folding structures by polystyrene film. *Smart Mater. Struct.*, 26, 085040, 2017.
79. Wajahat, M., Kim, J.H., Kim, J.H., Jung, I.D., Pyo, J., Seol, S.K., 4D Printing of Ultrastretchable Magnetoactive Soft Material Architectures for Soft Actuators. *ACS Appl. Mater. Interfaces*, 15, 59582, 2023.
80. Testoni, O., Lumpe, T., Huang, J.-L., Wagner, M., Bodkhe, S., Zhakypov, Z., Spolenak, R., Paik, J., Ermanni, P., Muñoz, L., Shea, K., A 4D printed active compliant hinge for potential space applications using shape memory alloys and polymers. *Smart Mater. Struct.*, 30, 085004, 2021.
81. Zhao, W., Li, N., Liu, L., Leng, J., Liu, Y., Origami derived self-assembly stents fabricated via 4D printing. *Compos. Struct.*, 293, 115669, 2022.
82. Melocchi, A., Uboldi, M., Inverardi, N., Briatico-Vangosa, F., Baldi, F., Pandini, S., Scalet, G., Auricchio, F., Cerea, M., Foppoli, A., Maroni, A., Zema, L., Gazzaniga, A., Expandable drug delivery system for gastric retention based on shape memory polymers: Development via 4D printing and extrusion. *Int. J. Pharm.*, 571, 118700, 2019.
83. Li, Y.C., Zhang, Y.S., Akpek, A., Shin, S.R., Khademhosseini, A., 4D bioprinting: the next-generation technology for biofabrication enabled by stimuli-responsive materials. *Biofabrication.*, 9, 012001, 2016.

84. Morrison, R.J., Hollister, S.J., Niedner, M.F., Mahani, M.G., Park, A.H., Mehta, D.K., Ohye, R.G., Green, G.E., Mitigation of tracheobronchomalacia with 3D-printed personalized medical devices in pediatric patients. *Sci. Transl. Med.*, 7, 285ra64, 2015.
85. Cheng, T., Thielen, M., Poppinga, S., Tahouni, Y., Wood, D., Steinberg, T., Menges, A., Speck, T., Bio-Inspired Motion Mechanisms: Computational Design and Material Programming of Self-Adjusting 4D-Printed Wearable Systems. *Adv. Sci.*, 8, 2100411, 2021.
86. Meloni, M., Cai, J., Zhang, Q., Sang-Hoon Lee, D., Li, M., Ma, R., Parashkevov, T.E., Feng, J., Engineering Origami: A Comprehensive Review of Recent Applications, Design Methods, and Tools. *Adv. Sci.*, 8, 2000636, 2021.
87. Ge, Q., Dunn, C.K., Qi, H.J., Dunn, M.L., Active origami by 4D printing. *Smart Mater. Struct.*, 23, p. 094007, 2014.
88. Jian, B., Demoly, F., Zhang, Y., Gomes, S., An origami-based design approach to self-reconfigurable structures using 4D printing technology. *Procedia CIRP*, 84, 159, 2019.
89. Yuan, C., Wang, T., Dunn, M.L., Qi, H.J., 3D printed active origami with complicated folding patterns. *Int. J. Precis. Eng. Manuf. Green Technol.*, 4, 281, 2017.
90. Langford, T., Mohammed, A., Essa, K., Elshaer, A., Hassanin, H., 4d printing of origami structures for minimally invasive surgeries using functional scaffold. *Appl. Sci. (Switzerland)*, 11, 1, 2021.
91. Van Manen, T., Janbaz, S., Zadpoor, A.A., Programming 2D/3D shape-shifting with hobbyist 3D printers. *Mater. Horiz.*, 4, 1064, 2017.
92. Zhao, Z., Wu, J., Mu, X., Chen, H., Qi, H.J., Fang, D., Desolvation Induced Origami of Photocurable Polymers by Digit Light Processing. *Macromol. Rapid Commun.*, 38, 1600625, 2017.
93. Tao, R., Ji, L., Li, Y., Wan, Z., Hu, W., Wu, W., Liao, B., Ma, L., Fang, D., 4D printed origami metamaterials with tunable compression twist behavior and stress-strain curves. *Compos. B Eng.*, 201, 1359, 2020.
94. Peraza Hernandez, E.A., Hartl, D.J., Lagoudas, D.C., 1 - Introduction to Active Origami Structures, in: 3D and 4D Printing, in: *Active Origami*, Peraza Hernandez, E.A., Hartl, D.J., Lagoudas, D.C., (Eds.), pp. 1–53, Springer International Publishing, Cham, Switzerland, 2019. DOI: https://doi.org/10.1007/978-3-319-91866-2_1.
95. Clampin, M., The James Webb Space Telescope (JWST). *Adv. Space Res.*, 41, 1983, 2008.
96. Wu, S., Dai, J., Leanza, S., Zhao, R.R., Hexagonal ring origami—Snap-folding with large packing ratio. *Extreme Mech. Lett.*, 53, 101713, 2022.
97. Alshebly, Y.S., Nafea, M., Mustapha, K.B., Ali, M.S.M., Mohd Faudzi, A.A., Tien, M.T.T., Almurib, H.A., 15 - Variable stiffness 4D printing, in: *Smart Materials in Additive Manufacturing*, M. Bodaghi, and A. Zolfagharian, (Eds.), pp. 407–433, Elsevier, Amsterdam, Netherlands, 2022.

98. Li, S., Yang, H., Chen, G., Zheng, J., Wang, W., Ren, J., Zhu, C., Yang, Y., Cong, Y., Fu, J., 4D printing of biomimetic anisotropic self-sensing hydrogel actuators. *Chem. Eng. J.*, 473, 145444, 2023.
99. Koh, T.Y. and Sutradhar, A., Untethered selectively actuated microwave 4D printing through ferromagnetic PLA. *Addit. Manuf.*, 56, 102866, 2022.
100. Tong, K.J., Alshebly, Y.S., Nafea, M., Development of a 4D-Printed PLA Microgripper, in: *2021 IEEE 19th Student Conference on Research and Development (SCOReD)*, p. 207, 2021.
101. Chu, G., Tian, H., Deng, X., Wang, W., Qiao, J., Li, L., Assembly-free fabrication of the deformable system composed by continuous fiber embedded 4D printing actuators. *J. Manuf. Process*, 110, 238, 2024.
102. Jang, S. and Park, S., 4D printed untethered milli-gripper fabricated using a biodegradable and biocompatible electro- and magneto-active hydrogel. *Sens. Actuators B Chem.*, 384, 133654, 2023.
103. Cheah, D.S., Alshebly, Y.S., Mohamed Ali, M.S., Nafea, M., Development of 4D-printed shape memory polymer large-stroke XY micropositioning stages. *J. Micromech. Microeng.*, 32, 065006, 2022.

Index

Printed in the USA/Agawam, MA
September 16, 2026

910661.110